PENGUIN BOOKS

REGENESIS

George Monbiot is an author, *Guardian* columnist, and environmental campaigner. His bestselling books include *Feral: Rewilding the Land, Sea and Human Life* and *Heat: How to Stop the Planet Burning*; his latest book is *Out of the Wreckage: A New Politics for an Age of Crisis*. Monbiot cowrote the concept album *Breaking the Spell of Loneliness* with musician Ewan McLennan and has made a number of viral videos. One of them, adapted from his 2013 TED talk, *How Wolves Change Rivers*, has been viewed on YouTube more than forty million times. Another, on natural climate solutions, which he copresented with Greta Thunberg, has been watched more than sixty million times.

*

Advance Praise for *Regenesis*

"George Monbiot is one of the most fearless and important voices in the global climate movement today."
—Greta Thunberg, activist and author of *No One Is Too Small to Make a Difference*

"Brilliant, mesmerizing, vital . . . a whole new way of thinking about our agriculture and our diets, our climate and our future."
—David Wallace-Wells, *New York Times* bestselling author of *The Uninhabitable Earth*

"This remarkable book, staring curiously down at the soil beneath our feet, points us convincingly in one of the directions we must travel. I learned something on every page."
—Bill McKibben, author of *Falter* and *The End of Nature*

"A world-making, world-changing book . . . It rings and sings throughout with Monbiot's extraordinary combination of passion, generosity, and justice."
—Robert Macfarlane, *New York Times* bestselling author of *Underland*

"*Regenesis* is a lively and deeply researched enquiry that confronts our dilemmas head on. . . . Transformation is urgently needed, and this book shows how it is possible."
—Merlin Sheldrake, international bestselling author of *Entangled Life*

"Monbiot shatters the shibboleths of farming and shows that the thin layer on which all terrestrial ecosystems stand is alive with organisms as diverse, fascinating, and mysterious as any found above ground."
—David Suzuki, founder of the David Suzuki Foundation

"This is an important book and a gripping read."
—Henry Dimbleby, cofounder of Leon Restaurants and the Sustainable Restaurant Association

"Monbiot writes with all the imaginative sympathy of a great storyteller as well as the overarching understanding of a moral visionary. This is a fine and necessary book."
—Philip Pullman, *New York Times* bestselling author of the His Dark Materials trilogy

"People from all walks of life should read this remarkable book. It is in my view one of the two or three most important books to appear this century."
—Professor Sir David King, former chief scientific advisor to the UK government

"*Regenesis* speaks to us like a poem. . . . It offers a magnificent political economy of global food production and concludes with a hopeful vision of a techno-ethical equilibrium between Humanity and Nature. It must be read."

—Yanis Varoufakis, author of *Another Now*

"*Regenesis* calls for nothing less than a revolution in the future of food—one that will literally transform the face of the Earth. . . . This is Monbiot's masterpiece."

—Kate Raworth, author of *Doughnut Economics*

"*Regenesis* is rigorous and restive, but also witty, original, and humane."   —Hugh Fearnley-Whittingstall, author of
The River Cottage Cookbook

"Valuable . . . Extraordinary . . . I never cease to be surprised by the unexpected perspectives Monbiot brings to bear, leading me through problems I never envisaged and solutions I never imagined."   —Brian Eno

"A fascinating and ultimately positive book . . . A harmonic vision of how changing our relationship to land use, farming, and the food that we eat could transform our lives."   —Thom Yorke

"A visionary, fearless, essential book."
—Lucy Jones, author of *The Big Ones* and *Losing Eden*

"Inspiring and compelling . . . A transformative vision of a new food future with the potential to both restore nature and feed the world."   —Caroline Lucas, MP and former leader of
the Green Party of England and Wales

"A genuinely brilliant, inspirational book."
—Sir Tim Smit, founder of the Eden Project

"Monbiot has applied his razor-sharp intellect, bountiful curiosity, and love for the land to the complex and fundamental issue of what we eat."

—Lily Cole, activist and author of *Who Cares Wins*

"Monbiot reaches for new ideas that might ignite the collective consciousness in a push to protect, rather than tragically destroy, the biosphere."
—ANOHNI

"Essential reading . . . This deeply researched book provides a blueprint for the future."
—Rosie Boycott, journalist and activist

"The writing, observation, and devotion is infectiously compelling. The learning is deep and immense."
—Mark Rylance, actor

"*Regenesis* gives us an inspiring vision of the future. . . . Monbiot has combined his gifts as an investigator, interviewer, and witty storyteller to create an exhilarating epic!"
—Robert Newman

"This passionate, extraordinary book opens up a compelling and vital new dimension: food and the way the world farms."
—Will Hutton, columnist for *The Observer* and author of *The Writing on the Wall*

GEORGE MONBIOT

Regenesis

*Feeding the World Without
Devouring the Planet*

PENGUIN BOOKS

PENGUIN BOOKS

An imprint of Penguin Random House LLC

penguinrandomhouse.com

First published in Great Britain by Allen Lane,
an imprint of Penguin Random House UK, 2022
Published in Penguin Books 2022

LIBRARY OF CONGRESS CONTROL NUMBER: 2022019003

ISBN 9780143135968 (paperback)
ISBN 9780525507567 (ebook)

Printed in the United States of America
5 7 9 10 8 6 4

Set in Sabon LT Std

To my brilliant assistant, Fi,
without whom it would all have been impossible

Contents

Acknowledgments

In researching and writing this book, I have relied to a greater extent than ever before on the generosity of other people, who have given me their time, advice, and expertise, made space for me, and put up with me during the intense and unrelenting work the project required.

Thank you above all to my wonderful family: my partner Rebecca and my daughters Hanna and Martha. Thanks too to my fantastic assistant Fiona Rowe, to Katie Kedward and Charlie Young, whose research work during the scoping phase was invaluable, and to Jo Haward, whose work on the references was so fast and efficient.

Thank you especially to my genius of an editor, Chloe Currens, whose insight, vision, and great attention and care have immeasurably improved this book. Thank you to my brilliant agent Antony Harwood, to the excellent and unfailingly accurate copy-editor Richard Mason, and to the rest of the team at Penguin.

Thank you to the farmers, practitioners, and researchers I followed, who were so patient and kind, as I took up their time and asked annoying questions: Iain Tolhurst (Tolly), Tamara and Gena, Tim Ashton, Ian Wilkinson and FarmEd, Fran Gardner, Simon Jeffrey, Paul Cawood, Pasi Vainikka and Solar Foods, FareShare and SOFEA, Stephen Marsh-Smith, Alison Caffyn and Christine Hugh-Jones, Rachel Stroer and the Land Institute, Bruce Friedrich, Maia Keerie, and Sophie Armour of the Good Food Institute. Thank you to Alexandra Elbakyan and Sci-Hub, without whom much of the research material I needed would have been unobtainable: access to knowledge is not a crime.

My great thanks to the expert reviewers, who so generously gave their time to read my manuscript, make comments and suggestions, and show me where I was going wrong: Tim Benton, Hannah

Ritchie, Tara Garnett, Mary Stockdale, Aislinn Pearson, Chloe MacLaren, Dan Blumgart, Tim Lenton, Jamie Arbib, Vicki Hird, Erik Meijaard, Tomas Linder, Simon Fairlie, Frank Ashwood, Franciska de Vries, Sarah Wakefield, and John Boardman. All the remaining mistakes are mine.

Many thanks too to my loving and inspiring sister Eleanor, to Mark Lynas, Joel Scott-Halkes, and Tong Wu, to my fellow orchardists Hugh Warwick, Zoe Broughton, Kate Raworth, Roman Krznaric, Caspar Henderson, Cristina Mateos, Phil Mann, and Amanda Smigelski-Mann, to Stewart Young, the East Ward Allotment Association, Steve Farmer, Jim Mallinson, Michael Witzel, Anna Morser, to Peter Gauvain, Clem Cheetham, and Nick Metcalfe on the *Apocalypse Cow* team, and to Franny Armstrong, Nicola Cutcher, and the rest of the scurvy crew on *Rivercide*, to my long-suffering editors at *The Guardian*, Damian Carrington, Nigel Dudley, Mike Mason, Jeremy Lent, Simon Evans, Ben Middleton, Andrew Balmford, Stuart Pengs, David Butler, Jim Thomas, the Soil Association, Piero Visconti, Duncan Cameron, Alexandra Sexton, Charles Ssekyewa, and Gunnar Rundgren.

I

What Lies Beneath

It's a wonderful place for an orchard, but a terrible place for growing fruit. In central England, far from the buffering effect of the sea, the trees are blighted by late frosts. Freezing air flows like water, but here, on this flat plot, dammed by rows of houses, it gathers and pools, drowning the orchard in cold.

Every year, as the trees come into blossom, my hopes crack open with the breaking buds. Roughly two years out of three, they wither with the flowers. Frost curls into the branches like poison gas, shrivelling and blackening the stamens.

By autumn, the orchard is a living graph of spring temperatures. Apple varieties blossom at different but regular dates. Unless a freeze is especially hard, it damages only the open flower. From the trees with and without fruit, you can tell when the frost struck, almost to the night.

Every variety belongs to the same species: *Malus domestica*, which translates literally as tamed evil. The reasons for the age-old defamation of a lovely tree are complex, but one is likely to be etymological confusion: a dialect name for fruit—*μᾶλον*, or "malon"—appears to have slipped from Greek into Latin, where it was, so to speak, corrupted: into *malum*, or evil.

This single species, too good to be true, has been bred into thousands of different forms: dessert apples, cooking apples, cider apples, drying apples, in an astonishing range of sizes, shapes, colors, scents, and flavors. We grow Miller's Seedling, which ripens in August and must be eaten from the tree, as the slightest jolt in transit bruises its translucent skin. It is sweet and soft, more juice than flesh. By contrast, the Wyken Pippin, hard as wood when picked, is scarcely edible till

January, then stays crisp until the following May. We grow St. Edmund's Pippin, which has skin like sandpaper and is dry and nutty and aromatic for two weeks in September, after which it turns to fluff, and the Golden Russet, whose taste and texture are almost identical, but only in February. The Ashmead's Kernel, crunchy, with a hint of carraway, my favorite apple, peaks in midwinter. The Reverend W. Wilks puffs up like wool when you bake it, and tastes like a smooth white wine. The Catshead, roasted at Christmas, is almost indistinguishable from mango puree. Ribston Pippin, Mannington's Pearmain, Kingston Black, Cottenham Seedling, D'Arcy Spice, Belle de Boskoop, Ellis Bitter: these fruit are capsules of time and place, culture and nature.

As every tree requires subtly different conditions to prosper, some do better here than others. Some varieties are so finely adapted to their place of origin that they perform disappointingly on the other side of the same hill. By choosing breeds that blossom at different times, we have sought in this orchard to spread the risk. Even so, in bad years, when frost strikes repeatedly, we lose almost everything.

But yes, despite the many broken dreams, it's a wonderful place for an orchard. When I arrived this morning, its beauty made me gasp. The first apple trees have come into flower: the pink buds uncurling to reveal their pale hearts. The pear and cherry trees are in full sail, carrying so much white blossom that their branches lift slightly in the breeze.

I walk the rows of trees, smelling them. Every variety has a different, faint scent: some of the blooms smell like hyacinth, some like lilac, some like *Daphne* or viburnum. I believe I can tell when a flower has been pollinated: the perfume, no longer needed to attract bees and hoverflies, is immediately cut off. The pear blossom, pure white, with twenty black stamens like tiny cloven hooves, stinks revoltingly of anchovies. The cherry petals are beginning to flake from the trees, drifting and feathering in the light wind. The new grass is streaked with shadow. Wood pigeons growl in the plum trees. To have all this within a few hundred meters of our home feels like an astonishing luxury; a luxury for which, between the five families who share it, we pay just £75 a year.

The orchard occupies three adjoining plots on an allotment site. Since 1878 in England, local governments have allotted land for people to cultivate vegetables and fruit. In principle, since 1908, we have all had the legal right to grow.*

What this legislation inadvertently spread was anarchy, in its true sense. In other words, it created thousands of self-organized, self-governing communities, otherwise known as commons. Though the local government owns the land, it is managed and run by the people who work it. Our site in Oxford is divided into 220 plots, cultivated by people who have arrived in the city from all over the world. We cross-pollinate each other's knowledge with grains of peculiar experience.

Seventeen years ago, the allotments seemed to be dying. Only one-tenth of the plots were occupied. The remnant community was desperate for people to take them on: otherwise, the local authority would reclaim the site for housing. They leased me two and a half adjoining plots, one of which was covered in monstrous brambles, snaking three meters into the air. I spent a month cutting the stems with a bush knife and hacking out the rootballs with a mattock. Beneath them sleeping beauty lay. Meadow grasses, cowslips, oxeye daisy, germander speedwell, vetch, knapweed, wood avens, scabious, yarrow, ribwort plantain, cat's ear, and hawkbit sprang from the soil. The seeds must have lain dormant for decades. I persuaded a couple of friends to join me, and we planted the plots with heritage fruit trees: mostly apples, with a few plums, cherries, and pears, a medlar and a quince.

Just as the trees became productive, I left Oxford and moved to Wales. Abandoning the orchard was among my few regrets. My friends passed it to others, who in turn passed it on. Five years later, unexpectedly, for family reasons, I returned. I didn't want to be back. But soon after I arrived, one of my best friends in the city told me that some people who had recently moved away had passed him a beautiful orchard, planted on the allotments a few years before. . . . He couldn't manage it alone, and remembered that I knew something about fruit trees.

It felt like coming home.

Now, though it covers less than one-tenth of a hectare, the orchard sometimes feels like half my world. It is the living calendar that marks my year. We have brought in three other families, creating a miniature

* In practice, in some cities, the waiting list now stretches to twenty years or more.

commons within a commons. Every couple of months, we organize a work day, with a break for lunch beneath the trees. In late winter and spring, we prune the apples and pears. In May and September, we mow the grass. In June, we thin the fruit. In October, we harvest the apples, store the sound fruit and, if the crop warrants it, spend a frantic day chopping, scratting, pressing, pasteurizing, and bottling the rest, turning some into juice and some into cider.*

In midwinter, we wassail the orchard. Wassailing is a scientific procedure deployed to ensure the trees produce a good crop the following season. The methodology consists of singing and drinking cider. According to a well-tested hypothesis, the crop the trees bear is directly proportional to the effort expended: "*For more or lesse fruits they will bring, / As you do give them Wassailing.*"[1] The hypothesis is not upheld. Then we begin the cycle again.

By mid-morning I'm six feet from the ground, with a bowsaw and long-handled pruning saw. Our lovely allotment neighbor, Stewart, has decided he is too old to manage his fruit trees, so he has passed his row, which abuts our orchard, to us, completing our three plots. His old trees are in a sorry state, the limbs congested and either sweeping the ground or rising so high that the fruit they bear is unharvestable. So I'm standing in the cherry tree, among branches so packed with blossom that you can barely see the bark, committing a desecration.

Whereas apples and pears can be pruned in the winter, stone fruit has to be pruned when the sap is rising in the spring or early summer. Otherwise, you expose the trees to infection by canker, leafcurl, or silverleaf. This means you must perform the awful sacrilege of carving

---

* Counterintuitively, apple juice is a modern product. Traditionally, the entire pressing was used to make cider (which in Britain means the alcoholic drink), though "make" misleadingly suggests an active process. Juice starts fermenting immediately. Proper cider contains nothing else. The apples provide the sugars, the flavor and, attached to their skins, the yeast. By Christmas, it's drinkable, though still sweet and fizzy. By February, it has settled into a smooth, subtle, well-balanced brew, in my dispassionate opinion the finest alcoholic drink ever to have ruined human lives. By the end of May it is a little too dry. By July it lives up to the Latin name for apples: you could use it to remove graffiti. To prevent juice from becoming cider, you need to pasteurize it. This requires energy, to bring the liquid to 70°C. Until recently, energy for heating was in short supply. The only juice people drank came straight from the press.

up a tree in flower or fruit. The snowy branches crash to the ground in a blizzard of petals.

Though this violation offends me, I love pruning. It has almost become an end in itself, as much sculpture as management. When you have completed the big, structural cuts, you trim the remaining twigs back to a bud that points in the direction you want the new growth to follow. As the tree spreads, it assumes the shape you have bidden it to take. I favor the Spanish, or goblet, style, molding the tree into a broad cup. If you get it right, this exposes every leaf to sunlight and the flow of air, eliminating woolly aphids and mildew without the need for chemical controls.

As I move through the tree, I find myself thinking about the likely history of this land. When we turned the soil, we found pieces of the white clay pipes that laborers smoked, some of them patterned with stipples, rings, and vines, bearing the mold lines and fingernail marks of those who made them. We found broken field drains, a donkey shoe, and modern oyster shells, which were sometimes hard to distinguish from the fragments of fossil *Gryphaea* we also turned up: a gnarly, hooked Jurassic oyster known in these parts as Devil's Toenails. When the seas were abundant, oysters, even in central England, were the food of the poor. One day, I found half a pearl, bored for the string on which it had hung.

Before it was surrounded by the city, then allotted equally to the townsfolk, this land was farmed, probably—to judge by the combination of field drains and dormant wildflower seeds—in rotation. Some of the surrounding place names contain the suffix -ley or -leys, which often means a temporary pasture, on which hay and forage are grown between arable crops. The oyster shells, concentrated in one part of our orchard, suggest that a tree might have stood there, beneath which the laborers sat to eat their lunch, as we do today. I picture them sprawled in their broad hats, scythes propped against the trunk, between the knuckled roots of a great oak.

We too mow the grass here only with scythes, partly to avoid using fossil fuels, partly to spare the frogs and voles. At first, we hacked at it. The harder we tried, the worse it looked. But one day, I noticed another allotment neighbor, an eighty-year-old Serbian refugee called Angela, watching us incredulously.

Despite all she has witnessed and survived, Angela manages always to find pleasure in life and goodness in people. True to her peasant roots, she presses her surplus vegetables on us, explaining that no one knows what real vegetables are these days, and we won't know how to cook them properly, but that's not her problem, as once she has given up her vegetables they are in the hands of God. We give her apples for roasting, medlars (that are better appreciated in the Balkans than they are here), and plums for brewing.

Eventually, she could bear it no longer.

"No, stop! You do it all wrong!"

She took the scythe from my hands. She felt the weight of it, lifting and lowering it slightly as if communing with the tool.

"I do this from when I little girl. I show you."

She settled the blade into the sward then appeared simply to twitch her broad hips. The grass fell flat. She trundled up the row without breaking a sweat, leaving a perfect lawn, the math laid to one side as if every stalk had been combed into place. (Math means mowing, or the cut grass produced by mowing. The stubble that remains is the aftermath.)

Now I look down from my perch in the cherry tree to the ruined limbs on the ground. I have left just four branches on the tree, more or less at the points of the compass. It looks mutilated. But it will heal. I climb down and start to process the prunings. Nothing here is wasted. We leave the heavy branches at the allotment gate, where people take them for firewood: fruitwood cuts neatly and burns sweetly. I use the sawdust in my smoker: whatever I cook in it takes on the soft dark flavor of the wood. We use some of the slimmer twigs for pea sticks, and stack the rest. After five years, the prunings break down into a rich, dry compost. We spread it around the dripline of the trees.* One spring a family of hedgehogs emerged from our stick pile. The babies were curious and unafraid. One of them waddled up to me, sniffed my outstretched hand, then tried to bite it.

---

* The dripline is the ring of ground beneath the outermost extent of a tree's branches. Because the tree acts as an umbrella, much of the water that lands on it drips to the ground along this line. As a result, the tree's feeder roots are concentrated here. If, as some people do, you stack the compost around the trunk, instead of feeding the tree, you are likely only to rot it.

Trying to grow fruit, or vegetables, as I did in prodigious quantities when I lived in Wales, reminds me every day of the constraints of biology and climate, and of the way these constraints have begun to flicker. While I have noticed no consistent change in the frosts that strike the orchard, which are all noise and no signal, other patterns have become impossible to ignore, especially the extremes of drought and rainfall that now afflict our fruit trees, the rest of the nation and much of the world. Working this tiny patch of land has helped alert me to the scale of the predicament we face, as the conditions that enable us to grow sufficient food begin to shift.

I finish stacking the pile and put my saws and loppers and helmet away. Then I take from the shed a different set of tools, to do something I can scarcely believe I have never done before. I have explored woodlands and rainforests, savannas and grasslands, rivers, ponds, and marshes, tundra and mountaintops, coastlines and shallow seas. But I have never explored, deliberately and thoroughly, the ground beneath my feet.

There are times when I struggle to understand myself, and this is one of them. Why, when I have spent over half a century immersed in the living world, seizing—or so I believed—every opportunity to discover wildlife and understand the ecologies that surround me, have I failed to explore the ecosystem that underlies so many others? Why, when I have spent thirty years growing food, have I neglected the substrate that provides, directly or indirectly, roughly 99 percent of the calories we consume?[2]

Like many people, I like to imagine that I find my own path. But we are all influenced, to a greater extent than we are usually prepared to admit, by social consensus. We think along the lines laid down by others, follow paths already trodden. We see what others see, and ignore what they ignore. We might argue passionately about the small number of issues on which the spotlight falls, but, implicitly and unconsciously, we agree to overlook other topics, often of greater importance. Few are either as important or as dark to us as soil.

A few meters from the cherry tree, I push my spade into the sward. I keep my tools sharp so, though the soil is heavy and rooty, the sod slices cleanly. I cut a small square of turf and lift out half a spit,* about a kilogram of soil. Then I settle onto my stomach in the grass and start working through it.

England is, or so I believed until I began researching this book, a dispiriting place to be a naturalist. Its visible wildlife, while once much richer than it is today, was never as varied as the wildlife in other parts of the world, especially the tropics. Now it is a threadbare remnant. This country has lost all its large land predators and most of its large herbivores. Our food webs are ragged and windowed, missing many of their strands. Uncultivated land is scarce, and even this is often mismanaged and polluted. In large parts of the country there is not much to see. Or so I thought.

I now realize that I was looking in the wrong place. While life above ground here is suppressed and depleted, below the surface lies one of the richest ecosystems on Earth. Soil at these latitudes is more diverse than the soil almost anywhere else. One scientific paper suggests there may be an inverse relationship between the diversity of plant life above ground and animal life below.[3]† The soil beneath a square meter of the orchard may contain many hundreds of thousands of animals, ranging across thousands of species. It took me a while to absorb that. Several thousand species beneath one square meter.

English soils could be as diverse as the Amazon rainforest,‡ and as little studied. Scientists estimate that only 10 percent of small soil animals have so far been identified.[4] In this orchard, there are probably thousands of undiscovered species. Many are likely to be unique to their regions: there are scarcely any microarthropods (small scuttling creatures) common to soil communities in different parts of the world.[5] We know even less about their relationships. For example, ecologists puzzle over something they call the Enigma of the Oribatids.[6] It might

---

* A spit is the length of a spade's blade.
† If this is correct, one likely explanation is that in the tropics, high temperatures and high rainfall lead to higher levels of inorganic nitrogen in the soil and higher acidity, both of which may suppress the number and range of microbes on which many soil animals feed. This does not mean that reducing biodiversity above ground enhances biodiversity below ground—far from it.
‡ Not counting the Amazon's own soils.

not sound as romantic as the Riddle of the Sphinx, but I find it as fascinating. Oribatids are one sub-group of a sub-group of the mites, which are in turn a sub-group of the Arachnids, the class that includes spiders. They are tiny and crablike and at first sight unremarkable. But in one handful of soil there might be a hundred oribatid species, all, apparently, occupying the same niche. Ecologists are accustomed to single species in single niches, as one outcompetes the others to become dominant. But here an astonishing number of related animals, in a wide range of shapes and sizes and colors, live alongside each other, apparently doing the same thing. How can this be?

Leonardo da Vinci remarked that we know more about the movement of the celestial bodies than about the soil on our own planet. This remains true today.

The first things I see are a fragment of bone, a bleached snail shell, a withered plum stone and a fragment of blue ceramic. Then I look more closely and notice a woodlouse and a little transparent millipede, its legs curling and uncurling in waves along its body, red dots along its sides like shields on a Viking longship. A chestnut centipede rushes past, carriage by carriage, into a dark siding. There are caramel beetle larvae and clusters of translucent globes, containing the faint white crescents of snail embryos. The labyrinthine stalks of seedlings work through the soil matrix, trying to find the light.

I crumble a pinch of soil into a fine sieve, then place it, in full light, over a funnel that leads into a test tube filled with gin. I prop up the test-tube rack with sticks, to prevent it from falling over, and leave it to cook in the sun.

Then I break off a lump of soil, take out my 40× magnifying loupe, and find the focal length. As soon as I do so, the earth bursts into life. The first thing I see is a springtail: a soft olive creature, rounded and slightly furry like a knitted toy, fleeing from the light. Now I've seen one, I see them everywhere: there are little gray ones less than a millimeter long; tiny white ones; a three-millimeter giant in iridescent gray and pink and blue; a humpbacked, amber species like a tiny drop of honey.

Springtails look a bit like insects, but they occupy a class of their own. Their abundance is astonishing: sometimes 100,000 or more beneath a square meter of ground. They can be male, female, hermaphodite (a bit of both), or parthenogenic, which means they can reproduce through immaculate conception. They live almost everywhere, even the Antarctic, and have survived every extinction event of the past 400 million years. In many parts of the world, they knit together the entire soil food web: in other words, they are the channel that connects much of life on land. But most people are unaware of their existence.

As I follow the springtails, a monstrous beast fills the lens. I start back. It takes me a moment to realize it's an ant. When I look around, I see I'm on the edge of the myrmecosphere, which means the soil zone influenced by ants. Close to my shoulder is one of the hummocks, about 40 centimeters high, that the yellow meadow ants started building almost as soon as I had cleared the brambles.

These anthills are like concrete. When I'm mattocking out plum suckers or resurgent brambles, I know when I've hit the edge of one, as the tool stops dead, jarring my hands. The ants bring clay from the subsoil and mix it with their saliva, making a cement strong enough to support their galleried and storied domes, the equivalent, if scaled to human inhabitants, of 100-meter towers. Into their cellars, which can extend a meter underground,[7] they carry aphids, which feed on the trailing roots of plants, and produce the honeydew on which the ants subsist.

They are ecosystem engineers, influencing all the life within their zone. In the orchard, I've noticed that germander speedwell, a little blue flower, selectively colonizes the roofs of the anthills, while the grass growing around them is thicker and darker than the grass elsewhere. The ants concentrate nutrients in and around their skyscrapers, inadvertently feeding the creatures that have adapted to live alongside them. The southeast face of every anthill is flat, and angled like a solar panel to absorb heat in the mornings.

Soon after spotting the ant, I find a white crustacean, just a millimeter long. When I look it up, I discover that it's an ant woodlouse. Unlike its relatives, it can live among these fierce creatures without being torn apart and eaten. Still more impressively, it persuades them to feed it, stroking them with its antennae and begging until they regurgitate the pellets of food they usually share among themselves.[8] Yellow ants are almost blind, and the woodlouse appears to fool them by masking itself with their scent. Its smell and the caresses of its antennae convince them that it's a hungry member of the sorority. If, however, the disguise is rumbled, and the ants attack, it lifts the two horns on its bottom and squirts glue into their faces, jamming their jaws.

I expose a long pale centipede, terrifying under magnification, like a medieval Great Worm. It snaps its fangs,* through which venom flows, then slithers away with a horrid combination of sinuosity and scuttling. By comparison, a docile, flat-bodied millipede, pinkish brown, armored with wide, overlapping mail, guarding her clutch of eggs, is as comfortably rustic as a farmyard hen. Little white potworms squirm out of the light.

Mites are everywhere, round and crabby. In soils like this, they are even more numerous than springtails: in some places, there are an astonishing half million per square meter.[9] Some, like hermit crabs, have tiny feet that barely emerge from their carapaces, others, long groping forelegs. They are brown, pink, mauve, yellow, orange, or white. In the soil, there seems to be a white version of everything. The white animals generally live at greater depths, where everything is blind (barring a crude ability to distinguish light from darkness), so there is no need for disguise. All that an animal creates incurs a cost in energy and resources, including color and eyes. If they can do without, natural selection ensures that they will.

I take the test tube from its rack and hold it against a sheet of black paper. With my lens, I can just detect tiny white filaments: nematode worms, driven from the soil by the light and heat of the sun, down the funnel and into the gin. These too are fantastically abundant, and critical to soil food webs. When conditions are right, they can multiply twelvefold in one day.[10]

I feel huge and violent and slow as I break into the soil's hidden chambers. All of its animals hate the light, and they move with surprising speed when it falls on them. Otherwise, in this voracious jungle, they would be immediately snapped up. I see the carnage that soil predators have left: the hollow scutes of millipedes, the wing shields of beetles, empty snail shells, the armor scattered after battle.

Then I notice something that looks like a creature from a Japanese anime: long and low, white, with two fine antennae at the front and

---

* Technically these are modified forelegs, called forcipules.

two at the back, poised and sprung like a virile dragon or a flying horse. I half-expect to find a miniature Studio Ghibli heroine riding on its back: by now, nothing would surprise me. It has six legs, but it isn't a springtail, and resembles no insect I've ever seen. When I look it up, I discover that it's a bristletail, or dipluran. It belongs to an entire class of life—a group with the same rank as insects or mammals—of which I knew nothing.* How could it be, after a lifetime immersed in natural history and a degree in zoology, that I have never heard of such a thing? But this is not the most spectacular exposure of my ignorance.

Soon afterward, I spot an animal that at first I take for a tiny white centipede. Now that I'm looking, I see loads of them. When I peer at one closely through the lens, I notice that instead of the fifteen pairs of legs or more a centipede possesses, it has twelve, and rather than an armored head bearing wicked curved jaws, it has the soft round face of a herbivore or detritus eater. Leafing through a textbook on soil ecology, I find a photo, and the answer astonishes me. It's a creature called a symphylid, a member not just of a class I had never encountered before, but, according to some authorities, an entire phylum.†

A phylum is a big deal. Human beings belong to the family Hominidae, the great apes. This family, in turn, is part of the Primate order: apes, monkeys, lorises, tarsiers, bushbabies, and lemurs. This order is a subset of the class we call Mammals, encompassing everything from shrews to whales. The mammals are one component of the phylum Chordata, which brings us together with birds, reptiles, amphibians, fish, lancelets, and sea squirts. Now I find myself looking (some sources say) at a phylum, a grouping comparable to the Chordata, and probably much more numerous, which until today had been unknown to me.

I'm struck by an astonishing thought. I can see, in this half-spit of soil, more of the major branches of life than I've seen in the Serengeti, or in any other ecosystem. Here are insects and crustaceans, mites and spiders, chilopods (centipedes) and diplopods (millipedes), springtails

---

* The taxonomy of soil animals keeps changing, so by the time this book is published, it might be out of date. At various points over the past few years, diplurans have been treated as a class, a sub-class, and an order.
† Again, this position keeps shifting. Sometimes the symphylids are considered a class, sometimes a phylum.

and earthworms, nematodes, mollusks, and creatures I never knew existed.

The soil can support such abundance because of its gigantic surface area. In the extreme case—the finest clays—a single gram (half a teaspoon when dry), would, if all its surfaces were laid out flat, cover 800 square meters, an area slightly larger than our orchard. Just as importantly, far from being the undifferentiated mass I once perceived, soil is a cosmopolitan city of zones and structures, in which distinct cultures inhabit adjacent parishes. One of these zones is the myrmecosphere, the ant borough, itself divided into subordinate precincts. But even more ecologically important are the narrow wards surrounding the roots of plants, known as the rhizosphere. It is upon this zone that humanity depends. As I pull the clod apart, the roots are so dense that it feels like fabric tearing.

I turn my attention to a tiny root hair. To the naked eye, it's a single strand, as thin as cotton thread. But under the lens, I see it is caged and frosted by much finer hairs, glittering like crystals in the sunlight. Every rootlet has them, even around the growing tips, which at this time of year cannot be more than a day or two old. Some look like whiskers, some are woven so tightly that they remind me of frayed nylon mesh around an ironing cable. They are filaments—hyphae—of the fungi whose lives are knitted into the lives of plants.

These are not, in most cases, fungi whose fruits we will see, though mushrooms and toadstools also form relationships with plants. The great majority—perhaps millions of species—live only within the soil, and many of them lace through and proliferate from the plant roots on which they depend. Most plants rely on these fungi to gather minerals and moisture from the soil.[11] The plant feeds the fungi with carbohydrates and lipids* that it makes through photosynthesis; the fungi feed the plant with nitrogen, phosphorus, and other elements they scour from the ground and transport with far greater efficiency than plants can manage. Their tiny filaments creep into pores and

---

* The chemicals that are the building blocks of fats and many other crucial compounds.

crevices too tight for even the finest root hairs to explore, and the enzymes and acids they release break mineral bonds that plants cannot split.

This mutually beneficial, symbiotic relationship is as old as the first land plants, some 460 million years.[12] When algae emerged from the water, they had no roots: in the ocean, they could absorb nutrients directly from the water. To survive, they needed to form relationships with the fungi that had long colonized the land and were, in effect, nothing but root. Just as we now know that we are not the singular beings we assumed ourselves to be, but a community composed of billions of microbes and the multicellular system that houses them, so we must now see plants not as rugged individuals but associations of unrelated creatures, combining forces to create life forms so complex that we are only beginning to understand how little we know.

In every gram of soil in places like our orchard, where plants are well established, there is around a kilometer of fungal filaments:[13] one kilometer in less than a teaspoonful. The filaments of each fungus form a dense net called the mycelium. In some forests, the mycelium of a single fungus can extend through several square kilometers of soil, though most are much smaller. They are constantly growing and retreating, forming new relationships, changing the terms of established ones, meshing with each other, shifting nutrients from one place to another, securing their own survival while serving the plants that host them. Some of them stitch together the roots of hundreds of plants.

The discovery that sugars sometimes move from the roots of strong, healthy trees into the roots of weak or sick ones generated great excitement among people who saw it as evidence of altruism in plants. But, as Merlin Sheldrake suggests in his wonderful fungus book *Entangled Life*, a more likely explanation is that fungi are, in effect, farming their hosts, shifting food from one plant to another, to ensure that all those on which they depend remain alive.[14]

Sheldrake also explores the possibility that the fungal mycelium is a form of intelligent life. It possesses directional memory. It can navigate labyrinths. It can send messages from one end of the network to another, changing its responses far from where it receives a stimulus. After discovering that fungal hyphae can conduct electrical pulses at intervals similar to those moving through an animal's sensory nerve cells,* some researchers see the millions of junctions within a mycelium as decision gates or processors, and the network as something resembling a computer.

Fungi are crucial to the health of the plants with which they grow. Perhaps to an even greater extent than their green partners, they mesh the soil together,[15] defending it from erosion, absorbing the rain that falls on it, locking up the carbon it contains.

All this, you might think, is remarkable enough. But what I cannot see, even with my loupe, is still more extraordinary.

Here is a fact that changes everything we once thought we knew about the living systems that sustain our lives. Of all the sugars that plants make through photosynthesis, they release between 11 percent and 40 percent into the soil.[16] They don't leak them accidentally. They deliberately pump them into the ground. Stranger still, before releasing them, they turn some of these sugars into compounds of tremendous complexity, with impossible names such as

2,4-dihydroxy-7-methoxy-2H-1,4-benzoxazin-3(4H)-one

Making chemicals like this requires energy and resources. At first sight, tipping this expensive brew into the ground looks crazy: in human terms, like pouring money down the drain. Why would they do this? The answer unlocks the gate to a secret garden.

These complex chemicals are not dumped randomly in the soil, but into the zone immediately surrounding the roots,[17] the rhizosphere. They are released to create and manage a series of marvelously intricate relationships with the creatures on which all life stands: microbes.

Soil is crammed with bacteria. Its earthy scent is the smell of the chemicals they produce. Petrichor, the smell released by dry ground when it is first touched by rain, is caused in large part by an order of bacteria called the Actinomycetes. The reason that no two soils smell the same is that no two soils have the same bacterial community. Each, so to speak, has its own terroir. Biologists call soil microbes "the eye of

---

* Roughly four action potentials per second.

the needle," through which the nutrients in decomposing materials must pass before they can be recycled by the rest of the food web.[18]

Microbes live throughout the soil, but in most corners, most of the time, they exist in limbo, waiting, in a state of suspended animation, for the messages that will wake them up. When a plant root pushes into a lump of soil and starts pumping out signaling chemicals and sugars, it triggers an explosion of activity. The bacteria responding to its call consume the rich soup the plant feeds them and proliferate at astonishing speed, to form some of the densest microbial communities on Earth. There can be a billion bacteria in a single gram of soil in the rhizosphere.[19]

These bacteria gather and unlock many of the nutrients on which plants survive. Bacteria in the rhizosphere, alongside the fungi with which the roots are meshed,* and other microbes, capture iron, phosphorus, and other elements in the soil and make them available to plants. They break up complex organic compounds, allowing them to be absorbed by the roots.[20] Uniquely, bacteria can turn the inert nitrogen in the air into the minerals (nitrate and ammonium) that are essential for making proteins. No part of the food web can survive without bacteria.

Soil bacteria also produce growth hormones and other specific chemicals that help plants grow. The complexity of some of the compounds the plant releases into the soil is explained by the fact that it seeks not to awaken bacteria in general, but the particular bacteria that are most effective in promoting its growth.[21] Plants speak in chemical languages that only the microbes to whom they wish to talk can understand.

The language changes from place to place and time to time, depending on what the plant needs.[22] When plants are starved of certain nutrients, or the soil is too dry or too salty,[23] they will call out to the bacteria that can overcome these constraints. Some biologists describe this as their "cry for help." In response to these chemical cries, a specific community of bacteria proliferates around their roots.

When you take a step back from these facts, you see something that

---

* Bacteria also appear to stimulate the relationships between plants and fungi, and, in some cases, to destroy the toxins that inhibit fungal growth.

transforms our understanding of life on Earth. The rhizosphere lies outside the plant, but it is as essential to its health and survival as the plant's own tissues. It is, in effect, the plant's external gut.[24]

Some of the similarities between the rhizosphere and the human gut, where bacteria also live in astonishing numbers, are uncanny. In both systems, the microbes break down organic material into the simpler compounds the plant or the person can absorb. Though there are over 1,000 phyla (major groups) of bacteria, the same four phyla* dominate the rhizosphere and the guts of mammals.[25] Perhaps these four bacterial groups have characteristics that make them more prepared than others to cooperate.

In humans, the infant immune system is less active than that of adults, enabling a wide range of bacteria to establish in our guts. Similarly, young plants release fewer defensive compounds into the soil than older ones, allowing a broad variety of microbes to colonize their rhizospheres.[26] Human breast milk contains sugars called oligosaccharides. At first, scientists struggled to understand why mothers express these compounds, as babies can't digest them. It now seems that their sole purpose is to feed the bacteria with which the child will grow. They selectively cultivate a particular bacterial species† with a crucial role in helping the gut to develop and calibrating the immune system.[27] Similarly, young plants release large quantities of sucrose into the soil, to feed and develop their new microbiomes.

Like the human gut, the rhizosphere not only digests food, but also helps to protect plants from disease. Just as the bacteria that live in our guts outcompete and attack invading pathogens, the microbes in the rhizosphere create a defensive ring around the root. Plants feed beneficial bacteria species, so that they crowd out pathogenic microbes and fungi.[28]‡

Sometimes plants deploy chemical warfare, releasing compounds that poison or suppress harmful microbes, but encourage helpful ones.[29] So precise are some of these chemical attacks that they can knock out a pathogenic variety of a bacterium species, but not a beneficial genetic

---

* The Firmicutes, Bacteroidetes, Proteobacteria, and Actinobacteria.
† *Bifidobacterium longum infantis.*
‡ The effect is called colonization resistance.

variant of the *same species*.[30] Sometimes plant and bacterium work together against a common enemy, both producing the same defensive chemical.[31] Sometimes the distress flares fired by plants provoke friendly microbes to attack their rivals with antibiotics.[32] Sometimes, if a harmful fungus has managed to invade the roots, the plant will stand down its usual defenses and allow certain bacteria species to invade as well, which then fight and suppress the fungus inside the root tissues.[33]

The pathogens fight back, hitting the plant's auxiliary microbes with lethal "effector proteins."[34] Some pathogenic species have evolved to thrive on the compounds that are meant to suppress them. Some fungi and insect pests use the plant's distress signals to locate and attack it.[35]

Plants also cry out for help from larger creatures. When insects attack their roots, they release volatile chemicals into the soil that attract certain species of nematode:[36] the tiny white worms I found in my test tube. These nematodes use their sharp beaks to pierce the skin of underground caterpillars. Then they wriggle into the body cavity and regurgitate the luminous, symbiotic bacteria that live in their guts. The bacteria produce an insecticide that kills the larva and antibiotics, which wipe out the microbes already living inside the insect. Then they digest the caterpillar from the inside, and the nematodes eat the proliferating bacteria.

The nematodes' population explodes, sometimes producing 400,000 young within the rotting hulk of a single caterpillar.[37] They burst from its sagging skin into the soil, seeking new prey. These prey might be easy to find, because the luminous bacteria make the caterpillars they infect glow blue. The glow seems to attract other caterpillars, which can then be attacked in turn.

After the Civil War battle at Shiloh, Tennessee, in 1862, thousands of injured soldiers were left lying in the mud, in some cases for two days and two nights, as the number of casualties on both sides was so great that it overwhelmed their armies' capacity to retrieve and treat them. Many died from their injuries and the consequent infections. But at night, some of the injured men noticed a strange blue glow emanating from their wounds. Their ghostly penumbra could be seen from a distance. Field surgeons observed that the soldiers who luminesced healed more quickly and had a higher survival rate than those who didn't.[38] They called it the Angel's Glow.

An explanation for the Angel's Glow was proposed 139 years later, when a seventeen-year-old high-school student, William Martin, acting on a hunch, persuaded his friend Jonathan Curtis to help him investigate.[39] Their paper, which won a national science prize, argued that the soldiers appear to have been attacked by insect-eating nematodes in the soil contaminating their wounds. The nematodes regurgitated their bacteria, and the antibiotics these microbes produce are likely to have destroyed the other pathogens infecting the wounds. Because the luminous bacteria have evolved to infect insects, whose body temperature is lower than that of humans, the students speculated that only hypothermic soldiers were inoculated. When they were brought in for treatment, and warmed up, the bacteria that had saved them died, preventing complications. (A related species, adapted to mammalian temperatures, causes severe infections.)[40]

Many of the antibiotics used in medicine were developed by soil bacteria[41] for use in their brutal underground battles, most of which are fought in the rhizosphere. As some of these crucial drugs begin to lose their efficacy—because the germs we seek to kill with them become resistant—we urgently need to discover new ones. The rhizosphere is likely to be a rich source. Using genome mining—prospecting a creature's genetic code for clusters of genes that make complex chemicals—researchers have already started to discover new antibiotics in the bacteria that live with plants.[42] As only half the major groups of soil bacteria have so far been grown in laboratories,[43] we have little idea of what the rhizosphere might offer.

Another way in which microbes in the rhizosphere—their "external gut"—protect plants from attack is to stimulate the plant's immune system. If its leaves are attacked by fungi or insects, one of the plant's first responses might be to release hormones into the soil, crying out for help to the bacteria living there. This looks like a strange way to react: the bacteria cannot move out of the soil to attack the pathogens on the leaves above. But they bounce the plant's signal back with a chemical message of their own, which fires up its immune response.[44]* This allows the plant to produce defensive chemicals in its leaves, and to shut the pores (the stomata) through which fungi might invade.[45]

* This process is called Induced Systemic Resistance.

It seems like a cumbersome way of fighting off a pest. But because the plant's immune system co-evolved with bacteria, and is trained and primed by them throughout its life, it can't work any other way. This process, too, is similar to relationships in the human gut. Bacteria in the colon, some of which are friendly, some pathogenic, and some of which switch between roles, educate our immune cells, and send chemical messages that alert them when pathogens attempt to break through the colon's protective mucus layer and attack the gut walls.[46]

We now know that a combination of excessive hygiene, the overuse of antibiotics, and a shift from varied diets containing plenty of fiber to less diverse, low-fiber diets damages our gut biomes, reducing the number of species they contain. This harms our dietary health and immune systems. Similarly, in the last few years agricultural scientists have discovered that plants seem to be less capable of fighting off attacks by certain pathogens when they grow in damaged soils with a low diversity of microbes.[47] Where the soil has been harmed by too much fertilizer, by pesticides or fungicides, excessive plowing or crushing by heavy machinery, their cry for help is more likely to be exploited by parasites and pests. In both cases a dysbiosis is caused.[48] This is a medical term, meaning the collapse of our gut communities. But it could be applied to the unraveling of any ecosystem.[49]

An interesting line of research suggests that soils with a rich and well-balanced microbiome suppress pathogenic bacteria that cause disease in people,[50] making the transmission of human diseases through food less likely.[51] Our health depends, in ways that are obvious and ways that are not, on the health of the soil.

Researchers have discovered that, like healthy and unhealthy gut biomes, soils can be either "suppressive" of disease or "conducive" to disease. When plants die, they can bequeath a legacy of the bacteria they have cultivated in the soil, protecting those that grow in their stead. Some researchers are now experimenting with the agricultural equivalent of fecal implants. Just as doctors take stool samples from healthy people and transplant them into the guts of unhealthy patients, some agricultural scientists speculate that implanting suppressive soil into unhealthy, "conducive" ground could suppress pathogenic bacteria and fungi.[52]

*

Something catches my eye in the hole I dug. It's a huge lobworm, dangling into the void, doubtless wondering where its burrow has gone. I feel suddenly guilty. I have learned that earthworm burrows can last for many years, sometimes decades, and are used, like our homes, by successive generations.[53] They form part of another, crucial soil structure: the earthworm zone or drilosphere.

Every hectare of stable, grassy land like this might be reamed by 8,000 kilometers of earthworm burrows.[54] The burrows tend to aerate the soil and help water to trickle through it. One experiment showed that after worms were introduced to soil from which they had been missing, within ten years the infiltration rate of the water landing on the ground almost doubled.[55] This means that less water flashes off the surface, so less soil is carried away, and more water reaches the roots of plants. One estimate suggests that worm burrows halve the rate of soil erosion. But their effects vary from place to place and season to season. In other cases, earthworms can make the soil less porous, or raise erosion rates, by bringing loose soil to the surface.

Earthworms can pull down into their burrows almost all the leaves and stems and twigs that fall on the ground.[56] Like birds, they swallow small stones and pieces of grit, and use them to grind up these pieces of dead plant in their gizzards. The bacteria that live in their guts help digest them, and some species then excrete everything they can't absorb onto the surface of the soil, in the form of casts.

The combined effect of this activity is extraordinary. In places like this orchard, earthworms can bring to the surface 40 tons of soil in every hectare, every year.[57] In tropical savannas, the turnover can reach 1,000 tons.[58] Dilapidated buildings slowly disappear into the ground not because they sink, but because the soil, continually squirted from the surface by worms, rises around them.* Because of the organic material the earthworms eat, their casts are much richer in minerals than the rest of the soil. By grinding up dead plants, they make their nutrients available to bacteria and fungi, which make them available, in turn, to living plants. Where earthworms exist, the

---

* This effect was noted and measured by Charles Darwin in his wonderful book *The Formation of Vegetable Mold through the Action of Worms, with Observations on their Habits* (1881).

weight of plants and animals above ground, on average, is 20 percent greater than where they don't.[59]

Earthworms also release plant growth hormones,[60] though it is not yet clear whether they do so directly or provoke bacteria to make them. Sometimes worms make plants more resistant to parasitic nematodes[61] and sucking insects, either by unlocking nutrients or by triggering their immune systems with chemical signals.[62]* In turn, plants might use their chemicals to control the behavior of worms.[63] The harder we look at any ecosystem, the greater the complexity we discern.

In my lump of soil, I find a leathery ochre case shaped like a lemon, about seven millimeters long. It reminds me of the dried, inflated pigs' bladders once used as footballs. Using my loupe, inside it I can see a pulsing red streak, alternately weak and strong, like blood pumping through a vessel. It's a baby worm, developing inside its cocoon. Earthworm reproduction is as weird as everything else in the soil. After earthworms mate (any worm within a species can mate with any other, as all are both male and female), the saddles around the middle of their bodies thicken and harden. Then a casing containing the eggs and sperm slides off the saddle and over the worm's head, pinches together at both ends when it slips off, and forms the cocoon.

When I started working through this lump of soil, I was reminded of something that I couldn't quite place. Now it comes to me: it feels like the first time I snorkeled. Then, as now, when I broke the surface I found myself in a new world, imperceptible from above. As soon as I remember this, the soil begins to look like a coral sea. Like the sea, with its reefs and open water, it has more structured and less structured zones: places of intense biological activity (such as the rhizosphere, the drilosphere, and the myrmecosphere) and the bulk soil through which large predators roam: centipedes and beetles instead of sharks and dolphins.

Like coral reefs, the most structured regions are rich in symbiotic relations. Just as coral is a combination of minerals derived from rocks, and animals, plants and microbes cooperating and competing to form structures from these minerals, soil is an ecosystem built by

---

* At other times they seem to make plants more susceptible to pests.

living beings from dead materials.[64] On its biological relationships, the soil's health and fertility—and thereby the survival of most of the world's terrestrial life—depends. It may not be as beautiful to the eye as coral, but once you begin to understand it, it is as beautiful to the mind.

In truth, we scarcely know it. So neglected has this ecosystem been, so little money and effort has been invested in comprehending it, that we are only beginning to unearth its complexities. The small funds available for studying soil life have mostly been spent on finding new ways to kill it: in other words, to destroy agricultural pests. As one of my university lecturers told me, "I study insects because I love them. But the only funding I can get is to kill them." By contrast to the many professional groups investigating other living systems, there is no soil ecology institute anywhere on Earth.

Soil, which we once saw as a homogeneous mass, is composed of structures within structures within structures. Earthworms, roots, and fungi create clumps of soil, glued together with the fibers and sticky chemicals they make, called aggregates.[65] Within these aggregates, tiny animals like mites and springtails create smaller clumps. Within them, bacteria and their microscopic predators—creatures I cannot see even with my loupe, such as tardigrades, ciliates, and amoebas— form still smaller aggregates.

Between these clusters are holes of different shapes and sizes. Around them are films of water and the complex chemicals released by plants and animals. Each of these clusters and voids and films has its own properties, creating millions of tiny niches that different species can exploit.

In 2020, scientists proposed what could be seen as the first steps toward a Theory of Soil.[66] This means that they began to understand what soil is. That might sound like a strange statement. But it has taken us until now properly to grasp that the substrate on which our lives depend is a biological structure.

Microbes create aggregates by sticking tiny particles together with the carbon-based polymers, or cements, they excrete. In doing so, they stabilize the soil and assemble habitats for themselves. Over time, this process builds an ever more complex architecture: pores and passages through which water, oxygen, and nutrients can pass. In other words, soil is like a wasps' nest or a beaver dam: a system built by living creatures to secure their survival. But unlike those simpler structures, it becomes an immeasurably intricate, endlessly ramifying catacomb, created by bacteria, plants, and soil animals, working unconsciously together. In other words, soil behaves like Dust in a Philip Pullman novel: it organizes itself spontaneously into coherent worlds. These are built on the principle of fractal scaling. This means that the structure is consistent, regardless of the magnification used to observe it.

The self-organized, adaptive world that microbes, plants, and animals build to suit themselves helps to explain soil's astonishing structural resilience in the face of droughts and floods: it survives crises that would otherwise reduce it to amorphous powder. But these findings could also explain why soil can start to break down when it's farmed. When farmers or gardeners apply nitrogen fertilizer under certain conditions, the microbes respond by burning through the carbon in the soil, much of which is stored in the polymers that build the catacombs.[67] Without cement, the structure—and the system—begins to disintegrate. The pores cave in. The passages collapse. Oxygen and water can no longer permeate. Because soil is fractally scaled, as the micro-structure breaks down so does the meta-structure: it becomes sodden, compacted, airless. Paradoxically, plant roots in over-fertilized soils can struggle to reach the nutrients they need.

Multiplying soil's spatial complexity is its complexity through time. The opportunities in a speck of soil can change dramatically from hour to hour, as it dries out or becomes saturated, as bacteria consume the organic matter it contains, as a root hair breaks into it and releases sugars and complex chemicals, as a worm engulfs and excretes it, as an ant colony sticks it together with saliva, or as a larger soil animal, like a mole, a rabbit, or a badger, digs it out and turns it over.

These fluctuations in space and time create what some ecologists call "hot spots" and "hot moments":[68] places and instances of intense biological activity. These endless variations contribute to a marvelous biological concept: the Hutchinsonian hypervolume.[69, 70] This describes the multi-dimensional opportunities that permit the survival of different creatures.[71] Broadly speaking, the more complex a system is across space and time, the greater the diversity it can support.

The massive Hutchinsonian hypervolume of healthy soil might

explain the mystery I mentioned earlier: the Enigma of the Oribatids. How can hundreds of species of one group of mites live together in the same place at the same time, apparently doing the same thing, without one or a few dominating the rest and driving them to extinction? A possible answer is that—because our understanding of the soil is so weak—it might look as if they are living together in the same place and the same time, but they aren't. They might each be exploiting minuscule hotspots and hot moments we have failed to detect.[72]

I have to keep reminding myself that soil does not exist for our benefit. It is not trying to help us grow food. Like all complex, self-organized systems, it seeks its own equilibrium. When my allotment neighbors, ecologically sensitive as many of them are, talk of creating a "well-tilled soil" or a "fine tilth," they are, unwittingly, talking about demolishing its complex structures, destroying many of the niches occupied by this astonishing profusion of life. When we talk of "breaking the soil" to grow crops, the phrase is appropriate in ways that were never intended.

After two hours of exploration, I discover that I have not even removed the kilogram of soil from my spade, so quickly and so deeply was I sucked into this new world. And I haven't worked through half of it. But my back has stiffened, and it's time for lunch. A faint net of cloud now catches the sun. I push the soil back into the hole and replace the turf. When I stand up, I realize I am covered in cherry petals, which have snowed on me from the remaining branches of the tree.

Learning about the soil has taught me, to a greater extent than ever before, that we establish our truths from information that's patchy and shallow, beneath which lie realities we scarcely imagine. Widely accepted claims are based on hearsay and myth, while scientific findings, however dramatic and intriguing, are scarcely known beyond a small circle of specialists.

In researching this book, I've discovered that this gulf between perception and reality applies to almost every aspect of our food systems. Our beliefs about food and farming are dominated by fables and metaphors that describe not the world as it is, but an idealized, simplified planet, prompting us to make catastrophic mistakes. What follows is my attempt to tell a new story, a Regenesis, about what we eat and how we grow it, a story that bridges the gap between scientific findings and popular beliefs, and embraces the fascinating complexities of the living world.

Every question about the soil ramifies into further questions. Every answer uncovers a riot of transgressive relationships, and opens a new field of study. The more we understand about life on Earth, the more intricate and connected it turns out to be, and the greater its role in creating the physical environment. As the conservationist John Muir famously remarked, "When we try to pick out anything by itself, we find it hitched to everything else in the Universe." The soil might be the most complex of all living systems. Yet we treat it like dirt.

Most of us perceive soil as a dead and passive substrate: a tabula rasa that achieves its purpose and potential only when crops are standing in it. We imagine that its role in producing food is confined to anchoring the roots of plants and absorbing the synthetic chemicals we apply. If we encounter its life forms, we tend to react with horror, and call them disgusting. If we want to insult someone gravely, we might describe them as a worm: an animal on which, perhaps above all others, our lives depend. But understanding it is crucial to addressing some of the greatest questions that confront us: how we might feed ourselves in a world whose natural and human systems are changing at astonishing speed, how we might do so without destroying the basis of our subsistence, and how we might, while securing our own survival, protect the rest of life on Earth. The future lies underground.

2

What Lies Ahead

Every complex system is more elaborate and more fascinating than we could possibly imagine.* All, by definition, have features that cannot be predicted by studying their parts in isolation, as soil ecology shows us. One complex system, which no one sought to create in its entirety, but which, like all such networks, has organized itself, is the global food system: the means by which we grow, trade, process, pack, distribute, buy, and eat our food. As I began to explore it, I kept stumbling across astonishing, unexpected properties. Let me give you an example.

Scientists at Germany's Max Planck Institute have found that as much as 40 percent of the rainfall in parts of East Africa appears to be caused by farmers watering their fields in India, Pakistan, and Bangladesh, between 4,000 and 6,000 kilometers away.[1]

When water is confined to river channels, little evaporates, as rivers tend to possess a small surface area. When it is locked in aquifers beneath the ground, scarcely any turns to vapor. But when farmers pump it out of a river or out of the ground, then spread it across their fields, they greatly increase the water's surface area. Under the sun and wind, much of it then steams into the air. The crops they grow in places that would otherwise be arid draw water through their roots and release it from their leaves, a process called transpiration.

---

* A complex system is not the same as a merely complicated system, such as an engine, which is designed to behave in a certain way, and responds to a stimulus (like your foot on the accelerator) in a linear and predictable fashion. The properties of a complex system cannot be determined by looking only at its parts. Its components form a self-organized network, which responds to pressure in spontaneous and non-linear ways.

From February to April, the vapor released from the irrigated fields in India, Pakistan, and Bangladesh is picked up by the prevailing winds, which blow southwest across the Arabian Sea. After traveling thousands of kilometers, this air hits the coast of East Africa. It rises and cools. The vapor it carries condenses, and some of it falls as rain.

The water inadvertently dispatched by farmers in South Asia increases the rainfall in East Africa by up to a millimeter a day. This might not sound like much, but in the arid parts of the region it appears to be crucial to the survival of farmers and herders, and the people they feed. The growing local population may have become dependent on this extra rainfall. The water vapor also reduces temperatures by about half a degree,* through evaporation and cloud cover, which could be vital in a place where both people and livestock face severe heat stress. If irrigation in South Asia failed, the consequences could be grave not only locally but also far across the ocean.

Between April and May, the winds turn and the gift from South Asia is rescinded. The extra millimeter or so of water that fell on East Africa now falls instead on China. Irrigation in India, Pakistan, and Bangladesh also affects their own monsoons, as well as rainfall across South East Asia. "When we try to pick out anything by itself . . ."

All complex systems are surprising, all are, in some ways, alike. Whether they are ecosystems, atmospheric systems, ocean currents, financial systems or human societies, the same principles govern their behavior.

The first of these is that they possess emergent properties. This means that their components, however simple they each might be, behave in complex ways when they come together. They also organize themselves, spontaneously creating order, without central control. Even if the system were catalyzed by human action, when it reaches a certain level of complexity, it can no longer be controlled from the top. Through the networks created by billions of decisions, the system takes over.

One aspect of the human tragedy is that our attempts to solve our problems accidentally create systems whose complexity escalates faster than our understanding of them. They behave in strange,

---

* All temperatures in this book are expressed in Centigrade.

counterintuitive ways, sometimes producing outcomes that are the opposite of what anyone intended.[2]

In 1939, just as complex systems first began to be described by scientists,[3] John Steinbeck, in *The Grapes of Wrath*, instinctively discerned how they had begun to dominate our lives:

> It happens that every man in a bank hates what the bank does, and yet the bank does it. The bank is something more than men, I tell you. It's the monster. Men made it, but they can't control it.

Complex systems have thresholds. They can absorb a certain amount of change without altering the way they behave, then suddenly their self-organization breaks down, and they collapse. These thresholds can be hard to identify, until they have been crossed. They are often described as tipping points. A system might be secure under some conditions: its self-organizing properties stabilize it. But when conditions change, and it is pushed toward a threshold, the self-organizing properties have the opposite effect, amplifying chaos.[4] They start transmitting shocks through the network, which compound and aggravate each other. At this point, a small disturbance can then tip the entire system over the threshold, and it collapses, suddenly and unstoppably.

One result of collapse is that a system can flip into an entirely new, stable state. Because this new state, in many cases, has its own self-reinforcing properties, which stabilize and secure it, a flipped system can be difficult, sometimes impossible, to flip back. In general, far more energy is needed to reverse the flip than was needed to cause it.[5] This situation is called hysteresis.

A classic example in ecological science is the pollution of a lake by fertilizer leaching off the land. When pollution levels reach a critical point, the lake ecology suddenly tips, as microbes feeding on the fertilizer turn the water from clear to cloudy. But you can't tip it back simply by reducing the level of fertilizer in the water to below that critical point. To clear the water, you must cut the extra nutrients almost to zero.[6] Tipping a system into a new stable state is like knocking away the pebble that wedged a boulder in place at the top of a hill. Reversing hysteresis is like pushing the boulder back up the hill.

The collapse of one system can trigger the collapse of the other complex systems with which it interacts.[7, 8] The U.S. Dust Bowl, for example, was caused by years of incremental damage to the soil from intense cultivation. This damage pushed the soil system close to a tipping point. When droughts struck in the 1930s, the soil structure collapsed, swiftly and catastrophically.[9] Within a few years, the wind whipped, on average, an astonishing 1,000 tons of soil off every hectare of the southern plains.[10] Four million hectares lost almost all their fertile topsoil. Food production collapsed, followed by the farming communities that depended on it. In 1937 a U.S. government report pithily described the system's hysteresis: "One man cannot stop the dust from blowing, but one man can start it."[11]

Had the global financial system been allowed to cross a critical threshold in 2008, its collapse would have triggered cascading failure across human society. Only a global bailout amounting to trillions of dollars pushed it back into a safer state. Even before hysteresis occurred, far more energy (or money) was needed to stop the collapse than was needed (via the U.S. subprime crisis) to cause it. Mass extinction events appear to be another example of contagious collapse. When one ecosystem or Earth system topples, it can pull down the systems that interact with it.[12]

While the precise thresholds are difficult or impossible to locate, we now know enough about the behavior of systems to predict whether they are resilient or fragile. Fascinatingly, and chillingly, we no longer need to know what the system is (whether it is a polar ice shelf or an insurance network) to decide whether or not it is likely to collapse.[13] We just need to know the mathematical values of its components.[14]

Scientists represent complex systems as a mesh of nodes and links. The nodes are like the knots in an old-fashioned fishing net, while the links are the strands of twine that connect them. In a food web, for example, the nodes might be species of plants or animals, and the links are the feeding relationships between them: who eats whom. In the financial system, the nodes are the banks and other major players, and the links are their commercial and institutional relationships (often also a matter of who eats whom).

If the nodes behave in a variety of ways, and their links to each other are weak, the system is likely to be resilient. If the nodes behave in similar ways and are strongly connected, it is likely to be fragile.[15]

This is because the behavior of similar nodes is likely to synchronize, as they are shaken by the same disruption, while strong links ensure the disruption resonates through the network. For example, in the approach to the 2008 crisis, the big banks developed similar strategies and similar ways of managing risk, as they pursued the same sources of profit.[16] They became strongly linked to each other (partly through securitization and derivatives trading) in ways that regulators scarcely understood.[17] When Lehman Brothers failed, it threatened to pull everyone down.

Some nodes—those with the most connections to others (the biggest banks, for example)—turn out to be much more important than others.[18] If they implode, their failure tips the whole system into collapse.

Another important issue is "modularity": to what extent is the system divided into compartments?[19] If different parts of a system have a degree of isolation from each other, the network as a whole is more likely to be resilient, as shocks are less likely to spread.[20]

Ideally, the network will contain circuit breakers, like fuses in an electrical system, that prevent the spread of contagious collapse. There should be a backup system, working within or alongside the main network, that operates on entirely different principles.[21] There should be plenty of redundancy (spare capacity) in the system: this acts as a kind of shock absorber.

When you take a step back from these rules of resilience, you notice something alarming. Our efforts to improve the performance of our own small corner of a system often weaken the system as a whole. Enhancing the efficiency of a business or a process is another way of saying that we are reducing its redundancy. Connecting a business more strongly to others might be essential to its economic survival, but it can make the network more fragile. Setting common standards makes trade quicker and easier, but it ensures that all the nodes (in this case nations) start to behave in the same way. Globalization grants people and nations greater reach, but it destroys modularity and sweeps away circuit breakers. Creating a "network of networks," an oft-proclaimed aim of governments, banks, tech firms, and corporations seeking mergers and acquisitions, could scarcely be better designed to spread contagious collapse.[22]

It seems to be an inherent property of the systems we create that over time they become more complex, more connected, and less comprehensible to the human mind. Even before we consider the effects of stupidity or malfeasance, simply through the apparently rational pursuit of our own interests we threaten the networks on which we depend. In seeking security, we expose ourselves to catastrophe.

So here are two questions fundamental to our survival, which are seldom voiced in public life. Is the global food system resilient? In other words, can it withstand major shocks? And is it becoming more robust or less robust?

One of the fastest cultural shifts in human history is the convergence toward a "Global Standard Diet."[23] A few decades ago, people in different countries—or different parts of the same country—ate radically different diets. They weren't always good diets. Many were insufficient and monotonous. But they were distinct. From nation to nation, even, in some places, valley to valley, the food people ate both shaped and was shaped by their discrete farming systems, histories, and traditions. Originating in rich nations in the 1960s, then spreading rapidly across the rest of the world, one diet has body-slammed the peculiarities of place and culture out of its path.

While many of us now have access to a much wider range of foods than our grandparents knew, globally our diets have become more alike.[24] In other words, our food is locally more diverse, but globally less diverse.[25] In almost every nation, we have started eating food that's denser in energy, with more accessible calories. We now eat far more vegetable oil, fat, and protein than we did sixty years ago (though, surprisingly, not much more sugar), fewer roots and tubers, but a few more vegetables of other kinds, and a little more fruit.[26] Most of our food comes from a tiny number of species. Just four plants—wheat, rice, maize,* and soybeans—account for almost 60 percent of the calories grown by farmers.[27]

These crops have concentrated in the regions where their production is most efficient. Just four countries (the U.S., Argentina, Brazil,

---

* In the U.S., maize is known as corn. Or, given the relative volumes grown, perhaps I should say that in Europe, corn is known as maize.

and France) harvest 76 percent of the corn exported to other nations.*
Five countries (Thailand, Vietnam, India, the U.S., and Pakistan) sell
77 percent of the world's rice, and five (the U.S., France, Canada, Russia, and Australia) supply 65 percent of the wheat.[28] Only three nations,
Brazil, the U.S., and Argentina, grow 86 percent of the world's soybeans
(which in turn supply three-quarters of its feed for farm animals).[29] We
are, therefore, dependent to an astonishing degree on the United States
for our global standard food supply.

In just eighteen years, the number of trade connections between the
exporters and importers of wheat and rice has doubled.[30] Roughly 40
percent of the world's people now rely on food from other nations,[31]
and global imports of cereals are likely to double again by 2050.[32]
Countries that once produced a little less or a little more than the
food they needed are now polarizing into super-importers and super-
exporters.[33] Some nations, especially in North Africa, the Middle East,
and Central America, rely on imports because they no longer have
enough fertile land or water to grow their own crops.[34] Others, especially in sub-Saharan Africa, might have enough land and water, but
their yields are low, or they are undercut by cheap imports from countries with bigger farms and government subsidies. For rich nations,
like my own—the United Kingdom—food imports cost little by comparison to exports of valuable goods and services, so it makes sense
to buy.

Every individual decision, in other words, may be economically
rational, but the result is a system that has become less resilient. There
are plenty of warnings in the scientific literature. They show that some
nodes (the major exporters) have become bigger and more important,
while their links to other nodes (the importers) have become much
stronger:[35, 36] these are classic causes of declining resilience.[37] They
show that rising trade is knocking down the compartments that used to
exist between national food production systems: in other words, the
system is becoming less modular.[38] They show a growing vulnerability
to external shocks.[39, 40] They show that these shocks could now proliferate across the entire network, becoming "globally contagious."[41]

* A different data set places Ukraine ahead of France (http://www.fao.org/faostat
/en/#data/FBS).

But these warnings have scarcely ruffled the surface of public consciousness. A general failure to understand the nature of complex systems allows the belief to persist that what is good for one is good for all. The invisible hand of private interest, which Adam Smith believed will "advance the interest of the society, and afford means to the multiplication of the species,"[42] could, when extended far enough within a complex system, have the opposite effect.

The Global Standard Diet creates the Global Standard Farm, and the Global Standard Farm promotes the Global Standard Diet. Farmers worldwide are converging on identical techniques, using the same machinery, the same chemicals, and the same varieties of the major crop plants. Since 1900, the world's crops, according to the UN, have lost 75 percent of their genetic diversity.[43] This genetic narrowing can make crops more susceptible to diseases, such as the Ug99 stem fungus, a virulent pathogen afflicting wheat, which, originating in Uganda, has now swept across Africa and parts of Asia, assisted by global trading networks that sometimes distribute diseases almost as quickly as they distribute food.[44] As the same herbicides are used to treat the same crops, the same herbicide-resistant superweeds spring up around the world, and now threaten in some places to overwhelm farmers' efforts to control them.[45] Because of the convergence toward identical crops and identical growing techniques, farming's backup systems—different ways of growing food, different ways of selling it—are shutting down.

Farmers everywhere are advised to "close the yield gap," which means maximizing the amount of food their crops can produce. This appears to make sense, but these gains in efficiency ensure that the redundancy (the spare capacity) within the system declines. Already there are signs that—despite massive investments in research and development—major crops in some places are approaching a "yield plateau": a level beyond which production can no longer rise.[46] Yield plateaus, one study found, have already been reached in around one-third of the world's rice and wheat farms.[47]

As crops approach their plateaus, returns on effort diminish.[48] Fertilizers have a massive impact on production when yields are low, but every new increment is less effective.[49] Beyond a certain point, the money that farmers must invest in improving their yields outweighs any gains they make.[50]

These trends have also triggered a classic self-accelerating feedback loop, typical of complex systems. As diets converge, and the farming methods that supply them converge, the biggest players globalize their businesses and destroy their smaller competitors. Corporations that supply the universal seeds, machinery, and chemicals enjoy ever greater economies of scale. So do the companies that trade and process the universal farm products. Market power translates into political power: the companies use their wealth to lobby governments and shape trade treaties. They secure wide intellectual property rights (patenting seeds and breeds as well as chemicals and machinery). They gain permission to merge and swallow each other. Their products then achieve even greater dominance.[51]

In other words, their growth relies on ripping down circuit breakers, back-up systems, and modularity, and streamlining a system whose major nodes are already too big and whose links are already too strong.[52] It's an accelerating cycle that inexorably destabilizes the system.

The result is a corporate sector even more concentrated and connected than the financial sector was before the 2008 crash. Four companies—Cargill, Archer Daniels Midland, Bunge, and Louis Dreyfus—control, on one estimate,[53] 90 percent of the global grain trade. They are consolidating vertically as well as horizontally, buying into seed, fertilizer, processing, packing, distribution, and retail businesses. But even this portion of the world is not enough—they continue to snap up their smaller competitors.[54]

Another four companies—ChemChina, Corteva, Bayer, and BASF—control 66 percent of the world's agricultural chemicals market,[55, 56] while a similar cluster (with BASF replaced by LimaGrain) owns 53 percent of the global seed market. Some of the mergers that created these giants were designed to integrate seed and chemical businesses, so that the products could be sold as a single package.[57] Some of their seed varieties are genetically engineered to be grown with the herbicides they manufacture. When farmers purchase a seed and chemical package from these conglomerates, they buy, in effect, a set of decisions about how they will farm. Global standardization advances every year.

Three corporations—Deere, CNH, and Kubota—sell almost half the world's farm machinery.[58] Another four companies control 99 percent of the global chicken-breeding market, and two supply almost all the ducks.[59] Four firms run 75 percent of the world's corporate abbatoirs and packing plants for beef; four others control 70 percent of corporate pork slaughter.[60] These firms too are integrating vertically, either buying up farms or contracting farmers to supply their meat, under strict and unvarying conditions, often using the standardized feed and other products they supply. Traders take over the feed mills and refineries with which they once did business. Supermarkets dominate and control the growers who sell to them. Fast-food chains elbow out independent restaurants.

While mergers and acquisitions have been accelerating in many industries,[61] the food sector has consolidated further and faster than most.[62] One reason is that sectors with a high rate of technological change can use their intellectual property—such as patents on genetically engineered seeds—to lock competitors out of the market.[63] In doing so, they accidentally create complex systems of their own, which shift quickly and interact with other systems in ways that are often opaque and unpredictable.

The fragility of the food system is exacerbated by a global shift to just-in-time delivery. Around the world, borders have opened and roads and ports have been upgraded, streamlining the global trade network. Thanks to this neatly integrated system, companies have been able to shed the costs of warehousing and inventories, switching, in effect, from stocks to flows. In good times, this works. But if deliveries are interrupted or there's a sudden surge in demand, shelves can empty suddenly, sometimes with disastrous consequences.[64]*

---

* For example, when the COVID-19 pandemic hit the United Kingdom, it exposed an unforeseen consequence of privatizing the supply of medical masks, gowns, gloves, and other protective equipment. Instead of stockpiling supplies, the companies charged with providing this equipment had cut their costs by minimizing storage. When demand was steady and predictable, their just-in-time systems looked rational and efficient. But when emergency struck, their minimized systems could not be scaled up fast enough to meet the demand, not least because health systems all over the world

As if the system were not vulnerable enough, over the past thirty years governments have gradually transferred responsibility for maintaining stocks of food to the private sector.[65] It's true that there were problems with the way some states ran their strategic food reserves, including corruption, perverse incentives, and storage costs. But handing the job to private companies replaces an imperfect system with a dysfunctional one. In Malawi, for example, the International Monetary Fund encouraged the government to reduce its stocks. When grain was needed, the Fund argued, it could be bought from private traders.[66] But in 2001, when harvests failed and government silos were empty, food prices soared (as they do when supplies are short). By early 2002, as a result, people began to die of hunger.

There are three main problems involved in handing responsibility to the private sector. The first is that when companies store grain, they do so to make a profit, rather than to save human lives. Companies have an interest in keeping food prices high, and they often do it by limiting the amount they release onto the market. As a result, private stocks of grain appear to be less effective than public ones at dampening volatile prices.[67]

The second problem is that the size and nature of private food reserves tend to be proprietary secrets: companies speculating in commodity markets have an interest in preventing other people from discovering how big their stocks might be.[68] So food storage becomes invisible, even to governments, leaving us with little idea of how well we would cope in a major crisis.

The third problem is that the companies on which governments rely to hold stocks are, in the main, the same corporations that have monopolized the global food trade: the nodes become even bigger.

The food industry is becoming more tightly coupled to the financial sector.[69] The same institutional investors crop up throughout the global food system, buying interests in farming, trading, processing, and retailing. They too seek integration, ensuring that their market power in one sector reinforces their market power in another.[70] When

---

were trying to source the same equipment. This failure was lethal: hundreds of health and care workers died in this country, partly as a result of inadequate protection.

sectors become mutually dependent, they increase what scientists call the "network density" of a system, making it especially vulnerable to cascading failure. "Hyper-connections" in a network of networks generate "hyper-risks."[71]

Speculation in commodity futures exchanges might once have buffered global markets against risk. By fixing prices for crops before they are harvested, it helped protect both farmers and traders from volatility. But over the years, price speculation has become an end in itself, and is likely now to be a destabilizing force. It's hard to find exact figures, but the limited public information suggests that, on the biggest futures exchange in Chicago, every year between 65 and 215 times as much wheat is traded in the U.S. as harvested: in other words, the same crop is exchanged many times.[72]

Some financiers have also been buying land. Global farm seizures in the twenty-first century, driven to a large extent by ultra-rich people and institutions adding land to their portfolios,[73] often involving corruption and coercion, have pushed millions of cultivators off their land.[74] Since 2000, 9 percent of Africa's land—often the best land, often in countries where many people go hungry—has been bought or seized.[75] Today, over 70 percent of the world's farmland is owned or controlled by just 1 percent of its "farmers."[76] I put farmers in scare quotes because some of the biggest owners have no experience or even direct connection with growing crops or raising animals. They include investment banks, pension funds, hedge funds, and private-equity vehicles.

In fact, it's often hard, amid the complex corporate structures and cross-shareholdings, to discern who the owners are. Even in rich nations with a strong administrative state, large tracts of land are now held by opaque companies registered in tax havens.[77] These speculators contract other people or companies to farm the land. Neither the new owners nor their contractors are likely to take much interest in protecting the soil and sustaining its fertility. The land is just one asset in a portfolio of investments. If it fails to deliver the same return as their other speculations, they either work it harder or sell it to another faceless asset manager.

Small farmers, on average, grow a wider range of crops, using a wider variety of techniques, than big ones.[78] The land grabbers who replace them tend to focus on what traders call "flex crops": commodities that can be switched between different markets. For example, soybeans and corn can be used to produce food, animal feed, or biofuel.[79] The system becomes more generalized, more homogeneous, less buffered. The biggest problem the global food system faces appears to be the global food system.

How can you detect whether a complex system might be approaching a tipping point? It begins to flicker.[80, 81] In other words, its behavior becomes more volatile:[82] the small, random changes that a system would previously have absorbed are amplified into bigger and bigger shocks. Flickering is what the global food system is now doing. One paper reports that "the frequency of shocks has increased across all sectors at a global scale" since the 1970s.[83] The system seems to be losing its resilience.

Until 2014, something wonderful seemed to be happening. Malnutrition looked as if it were heading for extinction. The number of chronically hungry people was falling steadily, and governments celebrated what looked like inexorable progress toward the UN's target of sufficient food for everyone by 2030.* But then the trend began to turn.[84] Within five years, the number of chronically hungry people had risen by 60 million, to 690 million.

Did food start running out? Far from it. Global food production has been rising steadily for more than half a century, comfortably beating population growth. In 1961, there were 2,200 kilocalories a day available for every person on Earth. By 2011, this had risen to almost 2,900 kcal.[85] Crop production as a whole has risen much higher: to an astonishing total of 5,400 kcal per person per day. But almost half this bounty is lost through feeding the food to farm animals, using it for other purposes (such as biofuels), and through waste. Even so, in principle, there is more than enough for everyone, if it were affordable and well distributed. The new hunger, it seems, is caused by systemic instability.

---

* Sustainable Development Goal 2.1: "By 2030 end hunger and ensure access by all people, in particular the poor and people in vulnerable situations including infants, to safe, nutritious and sufficient food all year round."

Take the years 2008 and 2011, which saw the biggest spikes this century in the price of food and, as a result, in hunger. The global wheat price rose by 33 percent in 2008 and 38 percent in 2011. But these major shocks had two extraordinary features. Ostensibly, they were triggered by heatwaves and droughts in some food-growing regions in 2007 and 2010. But these disruptions were by no means the most extreme in recent years.[86] Even more remarkably, in both cases the total volume of wheat available in international markets rose: by 5.5 percent and 3.2 percent respectively.[87] What appears to have happened is that the impact of small shocks in some growing regions was magnified across the global food system by commodity traders.[88, 89] Major producers then panicked and restricted their exports.[90] The shock wave rolled through the network, becoming bigger as it traveled. It landed almost entirely on the poorest nations, whose imports fell sharply, even as richer countries sustained or raised the amount they bought.

In a similar way, the global food system might have amplified the chaos caused by the COVID-19 pandemic. It's true, of course, that the pandemic was challenging for many economic sectors. But a combination of empty shelves in some nations and a sudden loss of income was devastating for many households. While world harvests were good in 2019 and 2020, logistics chains broke down in places, some countries stopped their exports,[91] and millions went hungry, even as crops rotted in the fields.[92] The global impacts on the food system were less severe than some of us feared, thanks in part to an element of luck. Had the systemic chaos caused by the first waves of the pandemic coincided with the heat domes,[93] droughts,[94, 95, 96] and floods[97] that afflicted several important growing areas in 2021, causing major spikes in prices,[98] the number of people suffering from malnutrition might have been much higher.

Here are the pressures that this fragile system, which could be approaching a tipping point, must now withstand. Among them are issues scarcely known and understood beyond a small group of specialists. The survival of many of the world's people could depend on bringing them to light.

By 2050, the human population of the planet will rise to between

9 and 10 billion. In principle, the world already produces enough food for between 10 and 14 billion.[99] The problem is that an ever smaller proportion of this embarrassment of riches is feeding people directly. Why? Because while the human population growth rate has fallen to 1.05 percent a year,[100] the growth rate of the livestock population has risen to 2.4 percent a year.[101] By 2050, to put it in brutal terms, the extra humans on the planet will weigh a little over 100 million tons, whereas, unless the current trend is disrupted, the extra farm animals will weigh 400 million tons.[102] The biggest population crisis is not the growth in human numbers, but the growth in livestock numbers.

This escalating pressure is caused by Bennett's Law, which states that the consumption of fat and protein rises with people's incomes.[103] On average, the world's people eat 43 kilograms of meat per year.[104] In the UK, we eat a little more than our average adult bodyweight: 82 kg. U.S. citizens eat on average 118 kg per year.

Whereas in the richest countries, meat consumption has stabilized and in some places declined a little, the rest of the world is catching up. In fifty years, the number of cattle on Earth has risen by around 15 percent,[105] while the number of pigs has doubled, and the number of chickens has increased fivefold.[106] By 2050, according to the UN, world meat consumption is likely to be 120 percent greater than it was in 2000.[107]

These animals must be fed. Already, roughly half the calories farmers grow are used for raising livestock.[108] Rich nations like the UK claim to produce a large part of the meat and eggs and dairy they eat, but they can do so only by importing feed. Much of this takes the form of soybeans from South America, and soybeans' expansion has been devastating to rainforests, wetlands, and savannas. Because we eat so much meat, the UK's diet requires nearly 24 million hectares of land.[109] But we farm only 17.5 million hectares here.[110] In other words, our farmland footprint is 1.4 times the size of our agricultural area. If every nation had the same ratio of consumption to production, feeding the world would require another planet the size of Mercury.

I believe—and hope—this profligacy will horrify our descendants. But they might be even more shocked to discover that a further 41 million hectares, four and a half times the UK's agricultural area, is devoted worldwide to crops that are grown to be burned: they are used to produce the world's biofuels.[111] To me, burning food is the definition of decadence. The crops used for fuel could feed almost half the people who are chronically hungry.[112]

Unless there is a radical change in the way we produce our food, of the kind I propose in this book, by 2050 the world will need to grow around 50 percent more grain.[113] In principle, and assuming nothing else changes, this is just about possible on the land used for farming today. While the rise in yields has slowed, at current rates the harvest of the four major crops (corn, wheat, rice, and soybeans) will, on average, be 50 percent bigger in 2050 than it is today,[114] which happens to match expected demand. But it is unsafe to assume that nothing else will change.

An immediate impact of global heating on the food system is that some parts of the planet are likely to become too hot for outdoor work. As our normal body temperature is 36.8°C, when the air rises above 35°C, the temperature gradient is too shallow to draw heat away from our bodies through radiation. At this point, we can lose heat only through evaporation—in other words, sweating. But because evaporation also requires a gradient—in this case a moisture gradient—sweating no longer cools us when humidity reaches 100 percent. The measurement of heat at 100 percent relative humidity is called the wet-bulb temperature. Human beings die of heat stress beyond a wet-bulb temperature of roughly 35°C.

In reality, most of us would die before that point. A 35°C wet-bulb temperature can kill a fit and healthy person lying perfectly still. During the heatwaves in southern Europe in 2003 and Russia in 2010, many people died, though the wet-bulb temperature never rose above 28°C.[115]

Already, at two weather stations in the Persian Gulf, wet-bulb temperatures have exceeded 35°C several times. Many other stations beside shallow seas in the subtropics—the Red Sea, the Gulf of Oman, eastern India and Pakistan, the Gulf of Mexico, and the Gulf of California—have recorded wet-bulb temperatures above 31°C.[116] So has a region that doesn't at first sight meet the necessary conditions: the western side of South Asia. In this case, the vast scale of irrigation seems to have intensified the humidity caused by the summer mon-soon: another unanticipated (though in this case harmful) effect of watering the land.[117]

Since 1979, the number of extreme wet-bulb heat shocks has more than doubled.[118] In large parts of Africa there is almost no monitoring of severe heat events. People are likely to have been dying of extreme heat in high numbers already, but their cause of death has not been registered.[119]

One of the reasons why we have underestimated the likely impact of heat shocks is that we tend to generalize temperature rises: 2° or 3°C of global heating doesn't sound like a major shift. But this 2° or 3°C is aver-aged across the planet. Because the land heats faster than the oceans, and because much of the world's population growth is taking place in some of the Earth's hottest countries, one study estimates that 3°C of global heating above pre-industrial levels translates into an average extra temperature experienced by human beings in 2070 of 7.5°C.[120]

Since farming began, humans have concentrated in places with an average annual temperature of around 13°C, which tends to create the best natural conditions for growing crops and raising livestock. Vast numbers have made their homes in this temperature band. But it is about to shift, swiftly and catastrophically. According to this study, the band will move further toward the poles in the next fifty years than it has done in the past 6,000 years. If people are unable to migrate, one-third of the world's population could be confined to places with an average annual temperature of 29°C: in other words, as hot as the hottest parts of the Sahara are today. Among them will be 1.2 billion people in India, nearly half a billion in Nigeria, 185 mil-lion in Pakistan, and 150 million in Indonesia.

How will people farm in these conditions? It might be possible inside an air-conditioned tractor, but the great majority of the world's farmers cannot afford such a thing. In Nigeria, for example, where 72 percent of small farmers earn less than two U.S. dollars a day,[121] a used modern mini-tractor costs around 3 million naira, or roughly $7,000.[122] For much of the year, a high proportion of the world's smallholders—who are concentrated in the hot parts of the world—will not be able to work. Smallholders (people who farm less than two hectares) produce around one-third of humanity's food.[123] Shifts from peasant to corporate agriculture are often accompanied by hunger, as many of the rural poor work outside the cash economy, producing their own food. If they lose their footing on the land, they can lose everything.

But even if corporate farmers with air-conditioned tractors took over, could anything be grown? Here, too, our perception of the problem has been distorted by the averaging of global temperatures. Most of the published studies look at the impacts of 1.5° or 2° or 4° of global heating on crop production. But if 2° or 4° of average heating means greater temperature increases in some of the major growing regions, the possible impacts might not have been fully captured. Temperature rises so far have probably caused a small reduction in the yields of major crops.[124] The damage has been lower than some people predicted, partly because crops have been shifted to places that suit them better. For example, even as the planet has heated, wheat fields on average are colder than they were before, because wheat growing has migrated to cooler places.[125]

In some cases, yields will improve as temperatures rise by 2°C or even 4°C; in other cases, they'll fall. Some scientific papers give the balance as positive,[126] others as negative.[127, 128, 129] But there's an urgent need for studies that are better matched to the likely temperature rises in the world's major bread baskets, which could in some places be higher than 4°C, even if overall heating doesn't exceed 2°C.[130] Above all, we need to know more about the impacts of rising temperatures on crop yields in the hot places where most of the world's hungry people now live.

And there's another problem: the weight of grain we produce tells us little about how well it might feed us. Even if rising yields can be sustained, experiments and modeling studies show that a combination of higher temperatures and higher concentrations of carbon dioxide in the air will greatly reduce the amount of minerals (such as iron, zinc, calcium, and magnesium), protein, and B vitamins that crops contain.[131, 132] The reason seems to be that plants grow faster in these conditions, and have less time to absorb nutrients.[133]

People in rich nations, where we fortify some of our food, and generally eat more protein than we need,[134, 135] will be less affected than people in poor nations, many of whom already suffer from vitamin, mineral, and protein deficiencies.[136] This contributes to the injustice of a crisis mostly caused by the rich and mostly felt by the poor. One study estimates that an extra 122 million people could suffer from protein deficiency by 2050 as a result of rising greenhouse gases;[137] another suggests 148 million people.[138] Anemia in poor countries is a major health problem, especially for girls and women; if crops contain less iron, many more will suffer.[139] Zinc deficiency already affects over a billion people: it can cause premature birth, stunting, and weakening of the immune system.[140] A further 130 million people could become deficient in folate, one of the B vitamins.[141] Folate deficiencies can be devastating to pregnant women and their fetuses. One paper describes the falling concentrations of protein and minerals in crop plants as "existential threats."[142]

Yet another issue is concealed by the fairly reassuring predictions about the impact of rising temperature on crop yields. As the planet warms, the number of extreme weather events increases. A study of insurance payouts for corn and soybean losses in the U.S. shows that the 1°C rise in global heating we've already experienced has almost doubled the crop losses caused by droughts and heatwaves.[143] While there were fewer payouts for frost and deluge, the overall effect of extra heating appears to be highly damaging, reducing the food supply and, crucially, causing it to fluctuate more violently.

Worldwide, there are likely to be more cyclones, worse hurricanes, more droughts, more floods.[144] In 2021, two years after Cyclone Idai tore through Mozambique, over 100,000 people, many of them farmers, were still stranded in resettlement camps.[145] Since then, three more devastating cyclones have struck Mozambique, each of them leveling crops and homes and driving hundreds of thousands from their land. In some regions, moderate weather has given way to a violent cycle of flood and drought:[146] instead of rain bringing farmers relief from droughts, it now drowns their crops. When the waters recede, the drought resumes. Drought, in some regions, brings fire, which destroys homes and crops and kills farm animals. Scientists have discovered an unexpected impact of wildfires: even hundreds of kilometers downwind of a major conflagration, the ozone pollution and aerosols released can affect the health of plants and reduce crop yields.[147]

Extreme weather threatens not only the production of grain, but also its transport. Around 55 percent of the cereals and soybeans that are traded internationally are shipped through at least one "chokepoint": the Panama Canal, the Suez Canal, the Turkish Straits, the Strait of Gibraltar, the Bab-el-Mandeb, the Strait of Hormuz, or the Malacca Straits.[148] Some of these chokepoints have already been severely affected by weather. The Turkish Straits have been restricted by high winds and the Panama Canal by drought.[149] In 2021, a sudden gust of wind during a sandstorm helped drive a container ship—the *Ever Given*—across the Suez Canal, wedging its bow into the bank.[150] It was stuck for six days, during which time several billion dollars' worth of cargo was delayed, as ships at either end of the canal were unable to pass. If the *Ever Given* had had to be unloaded to lighten it in order to float it off the bank, which would have been necessary if the diggers and tugs had failed to release it, the operation would have taken weeks, and might have caused a serious disruption to food supplies. The impacts of fighting, political conflict, and piracy, all of which have in some cases been exacerbated by climate breakdown, multiply when they encroach on chokepoints.

One-fifth of the world's wheat exports and one-sixth of its corn exports pass out of the Turkish Straits, which are just a kilometer wide at their narrowest point. A quarter of the soybeans and a quarter of the rice traded worldwide pass through the Straits of Malacca, whose bottleneck is a slightly more generous 2.5 kilometers.[151] The Panama Canal, just one-third of a kilometer across, carries 40 percent of U.S. corn exports and half its soybeans. If the canal had to close entirely, many of the ships that would have passed through it would have to be rerouted through the Straits of Malacca (crowded at the best of times), causing, in some months, an 80 percent increase in traffic. As ships stack up in the lanes, waiting to move through, the just-in-time food chain could snap.[152]

Some countries, especially in the Middle East, North Africa, and the Horn of Africa, are largely dependent on the food passing through the world's dire straits, and hunger there would quickly rise without it. Even China, though partly surrounded by open ocean, would struggle: almost half the soybeans it imports pass through the Straits of Malacca; much of the rest are carried through the Panama Canal.

*

But I have yet to mention the least safe of our assumptions. Farmers all over the world are blithely told to raise their yields with the help of irrigation. A global study discovered that closing the yield gap worldwide would require 146 percent more freshwater than is used today.[153] Does this water exist?

Over the past 100 years, our use of water has increased sixfold.[154] Irrigation already consumes around 70 percent of the water people withdraw from rivers, lakes, and aquifers (the natural reservoirs under the ground).[155] Because so much water is used for farming, rivers such as the Colorado and the Rio Grande fail to reach the ocean, while lakes like the Aral Sea are shrinking. Irrigation demand is one of the reasons why species living in freshwater are becoming extinct at roughly five times the rate of species that live on land.[156]

In California's Central Valley, which produces, by value, 8 percent of the food in the U.S., the water table is falling by 3 centimeters a year.[157] It's worth noting that while people are rightly concerned about the thirsty habits of almond and pistachio trees, more than twice as much irrigation water is used in California to grow forage crops, especially alfalfa, to feed livestock.[158]

Already, 4 billion people suffer from water scarcity for at least one month every year.[159] Thirty-three major cities, including São Paulo, Cape Town, Los Angeles, and Chennai, are threatened by extreme water stress: during droughts, some of them could lose their supplies altogether.[160]

At the same time, crucial water sources are disappearing as a result of global heating. Around one-third of the world's irrigated farmland depends on the water running off mountains. As groundwater is over-used and demand increases, the importance of mountain water will rise: another blithe assumption is that it will provide around half the world's needs by the middle of the century.[161] But mountains, on average, are heating faster than the rest of the planet's surface,[162] and the glaciers and snowpack that supply much of this water are shrinking.

In the world's largest irrigated farming system, along the Indus River, the threat of water wars is as real as the threat of oil wars in the Middle East. Already, 95 percent of the river's flow is extracted to feed and clothe* people in Pakistan, India, China, Afghanistan, and,

*Cotton is a major crop.

through exports, several other nations.[163] Water stress in this catchment is already intense, especially in Pakistan. As the economy and the population grow, by 2025 the demand for water here is expected to be 44 percent greater than the supply.[164]

An essential fraction of the water that keeps the people of this region alive flows from melting glaciers in the Himalayas, the Hindu Kush and the other high mountains of the Indus catchment. The meltwater arrives at crucial moments: in the weeks before the monsoon, up to 60 percent of the water withdrawn from the river for farming arises from melting glaciers and snow.[165] During the region's long droughts, around half the river's flow is supplied by meltwater.[166]

But one of the reasons why farming has been able to intensify and cities have been able to grow in the Indus Valley is that the glaciers have been melting roughly one and a half times faster than they have been accumulating. Thanks to global heating, the river's flow has been higher than it would otherwise have been. But this bounty, on which the population now depends, cannot last. By the end of the century, between one-third and two-thirds of the ice mass in the Hindu Kush and the Himalayas will likely have disappeared.[167] The volume of water running off them will probably peak around the middle of the century, and then decline.[168]

In 1960, India and Pakistan struck the Indus Water Treaty, dividing the river's headwaters and tributaries between them, and, it seems, averting major conflict. But the treaty has a flaw: it says nothing about what happens in the event of a major decline in the flow of these rivers, caused by climate breakdown. Now, as both nations seek to expand their use of water, the Indus catchment is again becoming a dangerous flashpoint. Some analysts believe that water is a primary motivation for repeated conflicts in Kashmir.[169] There are three nuclear powers in the catchment (India, Pakistan, and China) and several highly unstable regions, divided by economic, tribal, and religious tensions and afflicted by hunger and extreme poverty. New agricultural, industrial, and urban developments are being built on the expectation of sufficient water. It is hard to see this ending well.

In the western United States and Canada, Central Asia, Chile, Argentina, Turkey, northern Italy, and southern Spain, vanishing mountain ice and snow could have devastating impacts on crop production. Snowmelt in temperate regions often arrives just as crops need it most: during the early stages of growth. Farmlands that depend heavily on melting snow produce 10 percent of the world's irrigated rice, a quarter of the irrigated corn, and a third of our irrigated wheat.[170]

The answer to declining water supplies, farmers are repeatedly told, is to improve the efficiency of irrigation: for example, to dripfeed water rather than flooding the fields and to line irrigation canals with concrete to stop leaks. But just as energy efficiency tends to increase energy use,* so water efficiency tends to increase water use, an effect known as the Irrigation Paradox.[171] The reason is that, as efficiency improves, and the same amount of water can be used to grow more crops, irrigation becomes cheaper, attracts more investment, encourages farmers to grow thirstier plants, and expands across a wider area. In the academic jargon, "demand hardens." The redundancy of the system declines as farmers' dependency on every drop rises. This is what happened, for example, in the Guadiana Basin in Spain, where a €600 million investment to reduce the use of water by improving irrigation efficiency has instead increased it.[172]

This is not the only perverse outcome. As efficiency improves and leaks are stopped—especially when irrigation canals are lined or replaced with pipes—less water soaks away to recharge the aquifers. Rivers are less likely to reach the sea. Cities downstream become thirstier.

These issues are sharpened by the prediction that climate breakdown is likely, on the whole, to make wet places wetter and dry places drier.[173] Regions that already suffer from water stress, such as the lands surrounding the Mediterranean, southern Africa, eastern Australia, and the drier parts of Mexico and Brazil, are drying out, after just 1°C of heating. Another degree, one estimate suggests, would parch 32 percent of the world's land surface.[174] One paper forecasts that, in the worst case, the proportion of the land used to grow wheat today that suffers from severe droughts will rise from 15 percent to 60 percent by the end of the century.[175] Even in the best case—in other words, if countries keep their promises under the 2015 Paris Agreement

---

* An effect known as the Jevons Paradox or the Khazzoom-Brookes Postulate.

on climate change—the frequency and intensity of droughts in these regions would double by 2070. By the middle of this century, severe droughts could simultaneously afflict an almost continuous belt of land from Portugal to Pakistan.

Altogether, according to a paper in the journal *One Earth*,[176] the impacts of climate breakdown could push one-third of the world's food production out of its "safe climatic space" by the final two decades of this century. "Safe climatic space" means the conditions that allow humans and their activities to persist.

There's a widespread belief that these great threats could be overcome by growing new varieties of crops, resistant to drought and heat stress. But this could be an example of the Adaptation Illusion.[177] Much of the yield gain in major crops that researchers expect under normal conditions will come from new drought- and stress-resistant traits. These, for example, allow crop plants to withstand competition with each other when they are grown closer together, or to be grown at hotter times of the year. In other words, the effect of these new traits might already have been counted. When people claim that, by developing drought- and stress-resistant varieties, crop yields can also be maintained in the face of the additional drought and heat caused by climate breakdown, they could be counting the same gains twice.

And then there is the soil, that fragile cushion between rock and air on which civilizations are built. One indication of how badly we have neglected the ecosystem that underpins our lives is that, while there are international treaties on telecommunication, civil aviation, investment guarantees, intellectual property, psychotropic substances, and doping in sports, there is no global treaty on soil. The implicit belief that this complex and scarcely understood system can withstand all we throw at it and continue to support us could be the most dangerous of our assumptions about the global food system.

The European Union did attempt a regional treaty—the Soil Framework Directive—which sought to reduce soil erosion and compaction, maintain organic material, prevent landslides, and stop soil from being contaminated with toxic substances.[178] But it was snuffed out, in 2014, by a fierce lobbying campaign, which had been waged for eight years by the continent's farming unions. It was the first legislative proposal in the EU's history to be withdrawn.

In the UK, the National Farmers' Union, which took partial credit for this demolition, celebrated the directive's demise.[179] It claimed that "soils in the UK, and across the EU, are already protected by a range of laws and other measures." But the only "laws and measures" it named—and indeed the only ones that existed—are the conditions attached to farm subsidies. Even on paper, these conditions are weak. Off paper, they scarcely exist at all.

Farmers who receive subsidies must fill out a form called the Soil Protection Review, on which they must say they are protecting their soil. The government is supposed to check that they do what they say, but only 1 percent of farms are inspected every year, which means that the average farm can expect a visit once a century.[180] Most inspectors have no expertise in soil erosion. Nor do they have powers to conduct an "invasive investigation," which is the government's term for using a spade to dig a hole. As a result, they can't tell whether or not the soil is being compacted by heavy machinery. But even if they somehow identify a problem, the most they tend to do is "offer guidance" on rectifying it. As long as the farmer has correctly filled out their Soil Protection Review, they cannot be deemed "non-compliant" and, crucially, cannot have their subsidies withheld.[181]

I came across this exchange between two farmers on the online Farming Forum.[182] One asked:

Is the Soil Protection Review the biggest load of red tape codswallop that Defra [the government's environment department] have ever written? Farmers do have common sense, so this should be scrapped.

The other responded:

[The Review] is an entirely paperwork based affair that Defra invented to satisfy the EU that they were "doing something" about soil management, without actually doing anything. . . . Defra only want to see that it's been filled in, that's it. They will fine you if you don't so they can say to their EU masters "Look we're enforcing the rules like you told us to." But beyond that they pretty much let the farmers get on with it. They know we fill the thing in at the end of the year with any old rubbish—they

don't care, as long as the farm doesn't look like a warzone. It's the ulti-
mate in "We pretend to abide by the rules, and you pretend to enforce
the rules, and everyone's happy" concepts. Take 10 mins to fill your form
in once a year and be very glad Defra have decided this is the way to go.

Since then, the UK has left the European Union and designed a new
subsidy system. On paper, it looks better than the EU's (it could
scarcely be worse). It creates stronger conditions for receiving public
money. But the government has also cut the regulators' budgets even
further. The few remaining staff can scarcely afford to leave their
desks. Without monitoring and enforcement, the new rules are a dead
letter. Sometimes it seems to me that we are almost willfully destroying the basis of our survival.

Instead of developing new policies to protect our soils, our governments are accelerating their destruction. In Europe, corn farming is
probably the greatest danger to soil health. The plants are slow to
develop in the spring, and are generally harvested too late to follow
with a winter crop. The stubble is widely spaced and sparse. As a
result, the soil in these fields tends to be exposed to the elements at the
times of year when rain and wind are most likely to strip it from the
land. While corn is grown in Europe to feed dairy cattle, the rapid
expansion of the crop in countries like mine is mostly being driven by
new, perverse incentives for making biogas,[183] a variety of that deca-
dent habit of turning food into fuel.

When incentives for biogas were first proposed, they were sold to
the public on the grounds that, instead of leaving discarded food, sewage, and animal manure to rot in the open air, releasing methane which
contributes to global heating, these wastes would be digested in tanks
and their effusions would be captured and used as a substitute for
fossil methane ("natural gas"). It was a sensible solution, applauded
by environmentalists. But we were conned. From the outset, several
EU governments encouraged farmers to boost their methane production by supplementing the waste with crops grown specially to feed
the digesters. Corn was the most profitable, and it soon became the
primary feedstock. It's a similar story to the European Union's biodiesel incentive, which was supposed to persuade manufacturers to
turn used cooking oil into car fuel, but, as some of us warned,[184] instead

accelerated the destruction of rainforests in Indonesia and Malaysia to plant oil palm.

According to the farming press, feeding a biogas digester with a capacity of 1 megawatt requires between 20,000 and 25,000 tons of corn a year.[185] This means that 450 or 500 hectares of land must be used to grow it. By comparison, wind turbines need one-third of a hectare for every megawatt of capacity,[186] or 1,500 times less land. Today you can witness annual disasters in several parts of Europe, caused by this "green" fuel. During winter storms, soil pours from harvested corn fields, slumping down hillsides, covering the roads, and washing into the rivers, where it destroys habitats and causes floods.

One scientific paper reports that the soil structure has been damaged in 75 percent of fields used for growing corn, sampled in southwest England.[187] Partly because of the loss of carbon as the soil is washed away, biogas made from corn is likely to cause more greenhouse gas emissions than burning fossil methane. An estimate in Germany suggests that in some cases its emissions are comparable to burning coal.[188]

Despite our best efforts to catch up, soil erosion tends to be greatest in poorer countries. This is partly because many of them are in the hotter regions of the world, where extreme rainfall, cyclones, and hurricanes can rip exposed earth from the land, and partly because hungry people are often driven to cultivate steep slopes and other fragile places. One paper finds that erosion rates in the world's poorest nations have risen by 12 percent in just eleven years.[189] In some countries, mostly in Central America, tropical Africa and South East Asia, over 70 percent of the arable land is now suffering severe erosion.[190] In Malawi, where half the population is undernourished,[191] erosion is so extreme that replacing the lost nutrients with artificial fertilizers would cost, every year, between 1 percent and 3 percent of the country's GDP.[192]

Already, as a result of drought, soil erosion, and the overuse of land, desertification affects one-third of the world's people.[193] By 2050, 4 billion people are expected to live in the world's drylands, which are especially prone to soil loss and degradation. Soil damage in dry places is one of the reasons why grain yields in sub-Saharan Africa have mostly failed to increase since 1960, even as they have boomed in the rest of the world.[194]

Climate breakdown will exacerbate this loss. More intense droughts and storms of wind and rain will rip into soils in North and Central Africa, the Arabian Peninsula, West Asia, Peru, and Bolivia, even at the lowest levels of expected heating.[195] Under the worst climate scenario, extreme weather would also help denude central and eastern parts of the U.S. and Canada, Mexico, southern Brazil, most of Africa, Europe, India, China, and Russia.

While the most obvious vulnerabilities are caused by plowing and by compaction with heavy machinery, there are subtler impacts that might, in the long term, be extremely damaging. For example, on the Global Standard Farm in many parts of the world, seeds are dressed with neonicotinoid pesticides, which harm a wide range of soil animals[196, 197] and microbes.[198, 199]

We are weakening the soil's capacity to renew itself, undermining its structure and making it more vulnerable, like other complex systems, to external shocks. The loss of a soil's resilience might happen incrementally and subtly. We might scarcely detect the flickering until a shock pushes it past its tipping point. When severe drought strikes, the erosion rate of fragile and degraded soil can rise 6,000-fold.[200] In other words, the soil collapses. Fertile lands turn to dustbowls.

We have become dependent for our survival on a food system whose emergent properties we barely begin to understand. This system depends for its own survival on other complex systems, such as financial markets and global governance, which interact in complex and often novel ways. Above all, it depends on the soil, the atmosphere, the cryosphere (the world's ice and snow). We are hammering these systems with environmental shocks. We know the global food network is losing redundancy, modularity, backups, and circuit breakers, but we have little idea of where its tipping points might lie, or which of many possible shocks—in isolation or together—might trigger its collapse.

Some of our knowledge gaps are so wide that humanity could fall through them. As systems become more connected, study paradoxically becomes more segregated: there is a trend in academic research of greater specialization over time. A tremendous body of knowledge accumulates, but the walls between disciplines are high. In science, modularity is dangerous.

Somehow we need not only to reduce the external pressures bearing upon the system—environmental breakdown and rising demand—but also to change the system itself. We need to increase the number of nodes, reduce their size, weaken their links, compartmentalize the network, increase its redundancy, increase its diversity. Unfortunately, the world's governments and enterprises seem to be attempting exactly the wrong solution to the problem of declining redundancy: expansion.

3

Agricultural Sprawl

When you first encounter a complex system, it can seem too big for the human brain to comprehend. In its entirety, it is.[1] To begin to understand it, at least in broad terms, we need to focus on certain critical elements and watch them as they flow through the system, like the barium meal a doctor uses to make a patient's digestive tract visible to X-rays. The elements I've found most useful in making the global agricultural system and its relationship to the living world visible are nutrients. By nutrients, I mean the minerals, such as nitrates and phosphates, some of which occur naturally in the soil, some of which are applied in manufactured formulas by people, that are essential to the survival of all life forms. As they pass through plants, animals, soil, and water, their flow illuminates the way the system works, and helps us to see how it is expanding across the living planet. For this reason, and with my apologies, I will begin this chapter by immersing you in the engrossing issue of excrement.

When I first saw it, several years ago, I couldn't work it out. I knew the river well, and what I spotted from the bridge at Glasbury made no sense. In dry weather, the River Wye ran clear. There had been no rain for a fortnight, yet it was brown and turbid. Had someone dumped a load of soil—or worse—into the water? But when I walked down to the water, I found no deposits on the stones. No sediments settled in the cup I dipped. I was mystified.

I saw it again four years later. But this time, I had expert tuition. I had come to survey the river for insects with the head of Afonydd

Cymru (the Welsh Rivers Trusts), Stephen Marsh-Smith.* When we were standing in the water, he showed me something I'd missed.

"You see those white stones? Look at how the color flickers as the water passes over them."

The brown stain shimmered like an underwater heat haze.

"If it was soil in the water, the color would be consistent. But when it flickers like that, you can tell it's algae."

Stephen explained that this was a "bloom." Encouraged by high temperatures, low flows, and a surfeit of phosphate, the population of diatoms—microscopic, single-celled plants—had exploded.

"There have always been blooms on the Wye," he told me, "but they used to be rare. When I first came here we never saw anything like this, except in the odd year like 1976. But now it happens every year."

As we started kick-sampling, I felt a gathering despair. Kick-sampling means pressing a net against the riverbed, then kicking the stones immediately upstream. The small animals sheltering beneath them are swept into the net by the current. The Wye, which runs through Wales and England, is famous for its river life, and, in theory, strictly protected. Conservationists had worked hard to rebuild its ecology, stopping acid flushes from the conifer plantations upstream, helping farmers to reduce soil erosion, removing weirs and dams. They had begun to restore the river's magnificent runs of migratory salmon, sea trout and shad.

But the net came up almost empty, again and again. I caught a couple of heptagenids—nymphs of the beautiful yellow dun and March brown flies, whose hydrofoil bodies pin them to the riverbed—a few larval baetid flies, some baby stoneflies, and a single mayfly grub, which flipped back and forth in the net. But now, in the early summer, when the river should have been swarming, they were strangely, frighteningly sparse.

The issue, Stephen explained, was chickens. In just five years, the county councils in the catchment had granted permission for massive barns—it would be more accurate to call them factories—containing 6 million birds.[2] There might have been many more installed in this

---

* Stephen, tragically, has since died.

period, as a farmer can build a steel barn big enough to hold 40,000 chickens without a pollution permit,[3] and neither the county councils in the catchment nor the English or Welsh governments have bothered to count.[4] Altogether, in the known factories, according to figures compiled by two citizen-scientists,* there are now at least 20 million chickens in the watershed.[5]

There are rules governing how the farmers who run these intensive chicken factories should dispose of the manure they produce. They should spread it on their fields only in certain conditions and at certain times of year. The rules are scarcely enforced.[6] But if they were followed to the letter, it would make little difference. So many nutrients are imported into the catchment in the chickens' feed, then excreted in their droppings, that the soil cannot absorb them,[7] even if they are spread exactly as the guidelines stipulate. Rain washes the surplus phosphates and nitrates into the river, fertilizing the microscopic algae. When the water is low and warm, the algal populations, fed by the nutrients, explode. Sometimes, during these blooms, the river runs brown, sometimes green, depending on the dominant species of diatom. These events are more frequent and persistent than they used to be. Since 2010, they have advanced 110 kilometers upstream.[8] The spread of the blooms charts the spread of the chicken farms.[9]

During the day, the diatoms increase the amount of oxygen in the water by photosynthesizing. But at night, as they respire, they suck it up again. By dawn, oxygen levels can fall so far that fish die of asphyxiation. The algae also reduce the amount of light reaching the riverbed, killing the river plants rooted there. Among them are water crowfoot,[†] the long fluttering weed whose white and yellow flowers once starred the surface, and which is to rivers like the Wye what mangroves are to tropical seas: it is the nursery in which young fish and invertebrates grow and adults hide and breed. In recent years the crowfoot has entirely disappeared from some stretches.[10] Mapping by the team that made an investigative documentary I presented, *Rivercide*, suggests a loss from the main stem of the Wye of between 90 and 97 percent.[11] Instead of crowfoot, the riverbed is now coated with

---

* Dr. Alison Caffyn and Dr. Christine High-Jones.
† *Ranunculus aquatilis.*

brown slime: sedimented manure and the filamentous algae that thrives on it.

We had been talking and sampling for half an hour when Stephen suddenly broke off. "You see those white stones?"

I looked down. While the water was murky before, now it was opaque. "No."

"That's because the sun's come out."

"Surely that would make them easier to see?"

"When the sun comes out, the algae reproduce faster. It's been shining for five minutes. That's why you can't see the riverbed."

I thought of the alien ocean in Stanisław Lem's novel *Solaris*. I felt we were standing in a living, responsive field, shifting faster than I thought was biologically possible. Like the *Solaris* ocean, it was messing with my head.

I asked Stephen why it didn't all wash away. It does, he told me, "but it's continually replaced by the algae coming downstream. Even as they're swept along, they're multiplying. It can take days for the bloom to pass."

The diatoms, which are suffocating the river, pour from a regulatory black hole. Permission to build the intensive chicken units in the Welsh part of the catchment was granted by Powys County Council behind closed doors, without full democratic scrutiny.[12] Almost without exception, the council decided they would have "no likely significant environmental impact," which meant that their effects on the river didn't need to be assessed. By the time I met Stephen, no application for a chicken farm had been refused. Worse still, the county councils in England and Wales and the government regulators treated every application as if it existed in isolation, making no attempt to determine what the extra increment of excrement would do to an overloaded river.

His organization lodged a complaint with the European Commission. A local petition gathered 75,000 signatures.[13] At the end of 2020, the Welsh government was forced to publish guidance on phosphate levels in the river.[14] It was weak and contradictory.[15] But in 2021, Powys County Council refused permission for a chicken unit for the first time.[16] The tide of manure seemed to be turning.

But it was too late. It is hard to see how this and many other beautiful rivers can be saved unless many, perhaps most, of the industrial livestock units in their catchments are closed, which would be both expensive and politically explosive. The Wye is being turned from a rich and complex ecosystem into a filthy gutter.

The summer after I met Stephen, I canoed down the river with my family during a spell of dry weather. We seldom saw the chicken units, which tend to be set back from the river. Though the algae were not in full bloom, the water was brown and slightly cloudy. The first time I canoed the river, twelve years before, I had seen shoals of bream and chub, trout holding their territory, packs of barbel pressed against the riverbed, almost as if they were swimming in air. Now, on the rare occasions when I spotted a fish, it was a nameless blur.

We pushed the canoe onto a stony beach a few miles upstream of Hay, and, through force of habit, got out to swim. The stones in the shallows were so slippery that I could scarcely stay on my feet. I slithered into a deep pool. But as soon as I began to swim—my nose just above the surface—I realized with a shock that the water stank of chicken shit. Almost gagging, I stumbled toward the bank, slipped on the stones and banged my knee. When I emerged from the river, I had another shock. My skin was slimy. It felt as if I had bathed in a solution of snot.

The transformation of our rivers by intensive farming has taken almost everyone by surprise. Like many people's, my mental picture of river pollution was, until a few years ago, stuck in the 1970s: industrial effluent, heavy metals, armadas of foam. My perceptions first began to shift when I traveled with an old friend to the River Culm in Devon. We had hoped to spend the day fishing and watching wildlife. But as soon as we stepped out of the car, I knew something was wrong. I could smell the river. When we peered into it, we saw that the bed was covered in feathery white growths. Filamentous sewage fungus is a sign of chronic and severe pollution. I followed it upstream.

It wasn't difficult to find the transition point. Above, the water was clear. I could see bullheads, minnows, and little trout darting between the stones. Below, it was cloudy and lifeless, but for the fluffy clumps of fungus.* It was harder to find the outfall. When rivers are polluted

---

* Sewage "fungus" is really a colonial bacterium.

with nutrients, the vegetation on their banks can become almost impenetrable. At length, behind a thick screen of plants, I discovered a concrete pipe from which brown filth was pouring.

I followed the pipe uphill. It took me to a dairy farm. Though the weather had been dry, shit sluiced from the lip of one of the slurry lagoons. It flowed down a trench then into the concrete pipe, which appeared to have been built for the purpose. Later, checking historic images on Google Earth, I found that the farm's dairy units (where the cows are housed and fed) had doubled in size the previous year, whereas the slurry lagoons were unaltered. Instead of enlarging them, the farm was tipping the extra waste into the river.

It looked like a clear case of illegal pollution. I reported it to the government regulator, the Environment Agency. It sent inspectors and claimed it was "taking this incident seriously."[17] But when I checked a few weeks later, it told me it would take no action against the farmer, on the grounds that "the long-term ecological impacts on the environment were fortunately low." How did it reach this conclusion? Because the inspectors had found "no evidence of a fish kill."

Of course there was no evidence of a "fish kill": there were no fish left to kill. A local man told me he had been trying to stop the pollution for six months: most life forms in this stretch of river had already died or been driven out. A "fish kill" (finding dead fish in the river) is caused by a sudden event, not chronic pollution. The agency also told me there were no problems with the farm, as the new dairy units met the legal requirements. I had explained to the inspectors that the problem was not the units but the deficient slurry lagoons. I found it hard to believe that this was ordinary incompetence. It seemed to me that the officials were looking the other way.

Though I told the story in *The Guardian*,[18] and it raised a stink commensurate with the smell of the river, still the agency took no action. I happened to meet the local man the following year. He told me the slurry was still pouring into the water. But at least I had an explanation. Two separate whistleblowers from the Environment Agency wrote to tell me they had been instructed not to enforce the law against dairy farms. The government had insisted that the agency take "a voluntary approach." This meant, in extreme and unavoidable cases, a quiet word, but most of the time no action at all.

In 2021 *The Guardian* used Freedom of Information laws to discover how many cases of farm pollution the Environment Agency had prosecuted.[19] Though it had "investigated" 243 incidents since new rules were introduced in 2018, it had sent just fourteen warning letters, but taken no further action. No fines had been issued, no farm subsidies withheld.

The regulators' budgets have been cut so hard that, on average, a farm can expect a pollution inspection once every 263 years.[20] Only once has the Environment Agency actively investigated the impact of farming across a river catchment. Its officials visited eighty-six dairy farms in the watershed of the River Axe, also in Devon, between 2016 and 2019.[21] They discovered that 95 percent of them were breaking the rules on storing farm waste, and 49 percent were polluting the river or its tributaries on the days when they visited. The farmers, the agency's report noted, "often admitted to taking a business risk by not investing in infrastructure," as they knew they were unlikely to be caught or fined. Even if they were, the fine was likely to be smaller than the cost of upgrading their plant.[22] As well as failing to contain their manure, many of the farms were spreading it when their fields were waterlogged or frozen, which meant that it washed straight into the nearest stream. But every farm the agency inspected had been awarded the Red Tractor logo, which assures us that our food is "farmed with care."[23] The assurance is as meaningless as the rules.

As I investigated further, I discovered something that shocked me. The major cause of river pollution in the UK is no longer industry, and no longer sewage outfalls, harmful and poorly regulated as these are. It's farming.[24, 25] Overwhelmingly, the worst farm pollution is caused by livestock, particularly dairy cows.[26] Where milk production is concentrated, our rivers are dying. In Wales, for example, the collapse of sea trout populations maps neatly onto the distribution of dairy farms.[27, 28]

There are similar situations throughout the rich world. Farm waste—especially from livestock farms—has in many wealthy countries become the biggest cause of water pollution. This holds true even in Ireland[29] and New Zealand,[30] which have invested heavily in marketing what they call their "pure" food. Dairy farming in New Zealand devastates river life,[31] threatens native fish with extinction, contaminates groundwater, and closes beaches by feeding blooms of toxic algae in the sea.[32] Even in China, blighted by industrial pollution, pig manure now seems to be the main cause of lake and river death.[33] The Global Standard Livestock Farm creates the Global Standard River: over-fertilized and hostile to all but a handful of species.

The River Culm exemplifies the problem. As prices of meat, milk, and eggs have fallen, farmers have increased the number of animals they keep. Regions have specialized in particular livestock species, often to meet the demands of factories and packing plants set up by the multinational companies that dominate the trade in commodities such as chicken meat, eggs, pork, and milk. The farmers are often little more than subcontractors to these firms. The Welsh borders are now the UK's chicken capital, for example, while south Devon concentrates on dairy. In continental Europe, dairy cows seldom graze: a study in Germany shows that pastures supply only 5 percent of the food they eat.[34] Instead, farmers apply nitrogen and phosphorus to their land, to grow corn or grass for silage, which is fed to cows indoors. In the UK, they tend to spend more time outdoors, though this is beginning to change.

Many dairy farms have replaced their arable fields, or plowed up their old meadows, and reseeded the land with a single grass species, which they now cut four[35] or five times a year.[36] People often assume that a green land is a pleasant one, but little survives in these fields. Birds have no time to nest, insects have nothing to eat, no flowers bloom. Pig and chicken farms tend to import their feed: soybeans and corn from the Americas, ground-up fish from the sea, wheat or barley from other parts of the country.

In all these cases, imported nutrients are funneled through the animals and concentrated in their manure. As the ratio of animals to acres increases, the problem I encountered on the River Wye is reproduced in many of the places where livestock are farmed: the land can no longer absorb the slurry they produce. Nor can it be trucked to arable regions. At least 95 percent of most farm slurries is water: they are heavy and their value is low. Shifting cow manure further than about 10 kilometers, for example, is uneconomic.[37] Nutrients that should be valuable to farmers have become a problem they must offload.

Muck-spreading by intensive livestock farms often looks more like waste dumping than fertilizing the fields. Even when the rules forbid it—when the ground is wet or frozen for example[38]—farmers must either spread the manure or allow their slurry stores to overflow. As an agricultural contractor explained to a Welsh pollution inquiry, some farmers deliberately spread slurry on their fields before high rainfall, to ensure it's washed away.[39] Whether the slurry is spread or kept in an overflowing lagoon, it ends up in the river.

It would be bad enough if the animals produced, so to speak, pure effluent. But in many parts of the world, manure is contaminated with metal salts, some of which are toxic to humans and other species. One paper suggests that the two worst regions for this metal contamination are the European Union and South East Asia. In these regions, the sheer volume of manure dumping ensures that mercury, copper, and zinc accumulate in the soil.[40] China is also badly afflicted: by spreading pig manure, it poisons its land with arsenic, mercury, chromium, copper, and manganese.[41]

Around 75 percent of the antibiotics sold in the European Union and the U.S. are used for treating not human beings but farm animals.[42] On industrial livestock farms, antibiotics are used prophylactically: in other words, before infection takes hold. In some countries, the drugs are used not only to treat or prevent disease, but also as growth promoters: animals fed with antibiotics gain weight more quickly. This practice was banned in Europe in the 1990s, and "discouraged" in the United States in 2017. What this means is that the U.S. Food and Drug Administration has asked drug companies kindly to refrain from selling antibiotics as "growth promoters," and to sell them instead, with a nod and a wink, for "new therapeutic indications."[43]

One estimate suggests that 58 percent of the antibiotics given to farm animals are excreted.[44] Some of the drugs break down quickly, others persist for hundreds or even thousands of days.[45] When manure is stored, the bacteria it contains face powerful selective pressures to resist the excreted antibiotics.[46] When it is spread, both the drugs and the microbes selected to survive them meet bacteria in the soil.

As I explained in chapter 1, soil bacteria engage in brutal warfare, using antibiotics to repel or destroy their competitors. They have developed what scientists call an "intrinsic resistome."[47] This means that they are pre-adapted to defend themselves against dangerous new chemicals.[48] When they come into contact with manure, one of two things tends to happen. The first is that bacteria with resistance genes proliferate at the expense of those without. The second is that the soil bacteria acquire resistance from manure bacteria. They do this through a remarkable process called horizontal gene transfer, which means that different species of microbe exchange genetic packages.[49] The result is that soil treated with manure quickly becomes a reservoir of antibiotic resistance. One study found 114 different resistance genes in soil on the North China Plain.[50]

These resistance genes can find various ways into other ecosystems and the human food chain. They wash off, with the slurry, into rivers.[51] They soak into groundwater,[52] they are blown into the air,[53] and they are absorbed by plants. Resistant bacteria or genetic packages, as well as antibiotics, are sucked up by the roots of crop plants and incorporated into the parts we eat.[54, 55]

Of the twenty-seven classes of antibiotics used in livestock farming, twenty are also used in human medicine.[56] Among them are some of the most valuable remaining drugs, described by doctors as "the last line of defense." The spread of antibiotic resistance through the food chain, according to one scientific paper, "may even be greater than hospital transmission."[57]

Already, 25,000 people a year are reckoned to die of antibiotic-resistant infections in Europe.[58] They kill hundreds of thousands worldwide.[59] Without antibiotics, modern medicine could scarcely function.[60] Scientists and doctors beg governments to address this crisis, but the use of these drugs on the world's livestock farms is expected to grow by two-thirds in fifteen years.[61] Intensive farming threatens human health, as well as the health of the living world.

Could this story get any worse? Perhaps. Farmland is also used to dispose of human sewage sludge. In principle, reusing the nutrients that pass through our bodies makes sense. Unfortunately, they are not the only compounds being recycled.

We have little idea of what is being spread on the land. For example, in the UK the testing rules haven't changed since 1989. Though 87 percent of our sewage sludge is sent to farms,[62] it's screened for just a handful of contaminants before delivery: heavy metals, fluoride, and dangerous bacteria. Since the rules were written, we've discovered that sludge commonly contains a much wider range of toxins, for which no tests are conducted.

Greenpeace used a Freedom of Information request to obtain a report that was commissioned by the government but then suppressed.[63] It warned that the sewage sludge being spread on farmland contains a remarkable cocktail of dangerous substances, including PFASs ("forever chemicals"),[64] benzo(a)pyrene (a Class 1 carcinogen), dioxins, furans, PCBs, and PAHs, all of which are persistent and potentially cumulative.*

Because sewage treatment works receive outflows from many sources, including rainwater washing off the roads, carrying oil and tire crumbs, effluents from building sites, workshops and beauty salons, offices and laundries, showers and washing machines, and the many synthetic chemicals tipped, legally or otherwise, down the drain, night soil† isn't what it used to be. In some cases, unscrupulous contractors paid to dispose safely of hazardous waste instead mix it with sewage sludge before spreading it on fields.[65] Fines for the few who are caught are pitiful: generally much less than the money to be made.[66] Without comprehensive testing, farmers have no idea what they are buying. The government keeps promising new rules, then postponing them.[67]

The situation in the United States is, if possible, even more dire.[68] In one case, in Georgia, sewage sludge spread on the land was so toxic that it killed hundreds of cows.[69] A dairy farm in Maine had to close after critical levels of PFASs were discovered in the soil. Two years later, the farmer's blood still contained twenty times the national average concentration of these cancer-causing chemicals.[70] Many other farms in the state, similarly poisoned,[71] remain in business. In other states, people report serious illnesses they believe were caused by

---

* These are all human-made organic compounds (organic in this case means containing carbon-hydrogen bonds). Some of them, like PFASs and PCBs, are or were deliberately manufactured. Others are by-products of industrial processes or combustion. These chemicals do not break down readily, or, in some cases, at all, but accumulate as they are passed up the food chain, becoming particularly concentrated in the tissues of long-lived animals at the top of the chain, such as orcas and human beings.

† A polite term for human sewage.

sludge-spreading.[72] Yet not only is sewage sludge still licensed as a farm manure without full testing, but it's also sold in the U.S. as "eco-friendly" and for use in gardens,[73] without a hint of what it is, where it comes from, or how it might spread insidiously into our lives.

Perhaps the strangest aspect of this story concerns the fate of microplastics. Many of them come from the synthetic clothes we wear, depilated in our washing machines. We know these fibers travel down rivers and accumulate in the sea, where they might threaten entire food chains. This is commonly described as a wicked problem, almost impossible to solve. In truth, almost all these fibers and other microplastics are removed from wastewater by modern sewage treatment works: one study reports a recovery rate of 99 percent.[74] So far, so good. But—and at this point you may decide whether to laugh or cry—having screened them out of the water supply, the treatment companies then release them back into the wild, in the sewage sludge they sell to farmers.[75]

As if to complete the scrambling of our minds, microplastics are sometimes spread deliberately on the soil, to make it more friable.[76] Across Europe, thousands of tons of plastic are added to fertilizers, to prevent them from caking,[77] or to delay the release of the nutrients they contain, ensuring that they seep into the soil slowly, matching the demands of the crop. In this case, fertilizer pellets are coated with plastic films—polyurethane, polystyrene, PVC, polyacrylamide, and other synthetic polymers[78]—some of which are known to be toxic,[79] all of which disintegrate into microplastics. It is almost unbelievable that, in the twenty-first century, we deliberately contaminate agricultural soils with persistent and cumulative pollutants. Sometimes it seems to me that in farming, which should be the most sensitive and careful of all industries, anything goes. I find myself asking the same question over and again: "Why is this legal?"

Some studies show microplastics steadily accumulating in soil as more sewage sludge is applied;[80] others show almost all of them washing off into rivers.[81] It's hard to decide which is worse. In either case, the contamination of food chains is effectively permanent. The plastics don't disappear, but disintegrate into ever smaller particles. They could be as damaging to the life of the soil as they are to the life of the sea.

Experiments show how microplastics cascade through soil food webs,[82] poisoning snails,[83] springtails,[84] mites, ants, and nematodes,[85] stunting earthworms,[86] halving the fertility of potworms.[87] When they decompose into nanoparticles, they can be absorbed by soil fungi[88] and accumulated by plants.[89] We currently have no idea what the consequences of eating contaminated crops might be. Nor do we know what the combined and cumulative effects of the cocktail of toxins spread in sewage sludge might be, either on soil ecology[90] or on our own health.* Governments must act quickly to prohibit the spreading of sewage sludge, before large tracts of farmland become unusable, and the damage to ecosystems, from soil to sea, irreversible.

A question I'm often asked is why so few rural people raise their voices against the poisoning of rivers and soils, the felling of trees and killing of wild animals, and other harms inflicted on their surroundings and quality of life. I think part of the answer is that in the countryside, where communities are small, people tend to refrain from antagonizing their powerful neighbors, and farmers are often the most powerful of all. As I've investigated these issues, I have been told repeatedly of intimidation and bullying: of dead animals left on people's doorsteps, of slurry sprayed over a garden wall, of a rampart of chicken excrement constructed by a farmer on three sides of the house of someone who had filed a planning objection to the new barns he wanted to build, of scratched cars and smashed windscreens, of the mafia-like power of dominant farming families.

I have met people who moved to the countryside in pursuit of the bucolic idyll, only to find themselves immersed in fear and loathing. In other nations, particularly Colombia, Mexico, and the Philippines, the consequences of objecting are not just stressful but extremely dangerous. Environmental activists, especially those who confront agricultural interests, are murdered in large numbers every year.[91]

In rich nations, intimidation, as I'm often reminded by those on the front line, is most keenly felt by government inspectors and staff working for conservation groups. As one official told me, "I have to

---

* It might be possible, one day, to decontaminate sewage sludge. But it hasn't happened yet. Some companies propose a technique called hydrothermal carbonization to preserve nutrients while destroying poisons. But it's not hot enough to decompose persistent organic pollutants, such as dioxins and furans.

live in this community. If I start enforcing the letter of the law, I'll be ostracized, my kids will be ostracized, life will get very unpleasant."

Local newspapers, whose editors and senior staff tend to socialize with other powerful members of the community, and whose revenues are dependent on local advertising, tend to take the farmers' side. Local councils are often dominated by landowners or those who defer to them. In some places, fealty to the lords of the manor seems to be almost as strong a force as it was before most people had the vote. There are parts of the countryside where, it sometimes seems, democracy scarcely functions.

Like many of aspects of farming, the impacts of the chicken production system resonate far beyond the valleys where the birds are kept. Though soy is just one ingredient of chicken feed, there can be more of it in chicken than in tofu: one report estimates that it takes 109 grams of soy to produce 100 grams of chicken breast.[92] Over three-quarters of the world's soy is fed to farm animals. Much of the rest is used by industry or to make cheap vegetable oil.[93]* Only 7 percent is turned into substitutes for meat and milk, like tofu, tempeh, soy mince, and soy milk. The amount grown for animal feed has risen, since 1990, three and half times faster than the amount grown for vegan alternatives.[94]

In South America, 200 times more land is used to grow soy today than in 1961.[95] The 57 million hectares the crop now covers there is bigger than Spain. Some indigenous peoples have been almost entirely dispossessed. Stunning ecosystems, especially the *cerrado* (savanna) of central Brazil, and the Gran Chaco forests in Paraguay and Argentina, the homes of maned wolves, giant anteaters, jaguars, tapirs, and armadillos, have been swept off the Earth on an unimaginable scale.[96] The farms look more like sea than land: gigantic fields, unmarked by any feature, stretch to the horizon. The remnant pockets of life resemble islands in the ocean of destruction. Still they are being cleared and burned.[97] One paper suggests that the *cerrado* is approaching a critical threshold.[98] When trees are felled, their long roots no longer draw

---

* Soybean meal for feeding livestock and soybean oil for human consumption can be co-products of the same crop.

water from the aquifers and release it into the air. As water vapor reduces while temperatures rise, dew, which is essential to the survival of many wild plants and animals, stops forming. This triggers a cycle of collapse that might lead to the tipping and hysteresis of the entire ecosystem by the middle of the century, to be replaced by desert.

To a lesser, but still great, extent,[99] rainforest in the southern Amazon is also razed to plant soy. But the biggest impact here is indirect: as cattle ranches in the *cerrado* give way to soy farms, ranching shifts northward, into the forest.[100] We remain, on the whole, blithely unconscious of the wreckage caused by our consumption of chicken, eggs, and pork. We might congratulate ourselves on buying local meat and eggs, or even on keeping our own birds, forgetting that the feed is likely to have been grown, at great ecological cost, thousands of miles away.

The pesticides used on these and other crops threaten much of life on Earth. One study suggests that farmland in the United States has become forty-eight times more toxic to bees across twenty-five years.[101] The main cause is the shift to neonicotinoids: insect poisons that are lethal at tiny concentrations. They are typically used to coat seeds before they are sown, so the toxins are absorbed by the crop as it grows, killing any insects that eat it. But only around 5 percent of the neonicotinoids in seed dressings are absorbed by the crop: the rest disperse.[102] They work through the soil, killing or stunting the animals that live there,[103, 104] or are washed into rivers, with similar effects on freshwater life.[105]

Pesticides, some of which spread far beyond the land on which they are sprayed,[106] are a major cause of Insectageddon: the collapse of insect life.[107] A famous study found that the weight of insects in German nature reserves fell by 76 percent in twenty-seven years.[108] A twenty-year record of insects hit by cars on two routes across Denmark recorded declines of 80 percent and 97 percent.[109] In the Netherlands, caddis flies, whose larvae live in freshwater, have dwindled by 9 percent a year.[110] While neonicotinoids are now banned in Europe, they are still used almost everywhere else. Pesticide companies have lobbied hard to ensure it stays that way.[111]

The Danish study discovered that the rate of insect decline could be used to predict the population of swallows. Similar effects are reported all over the world.[112] Birds that feed entirely on insects are hit hardest, but many of the species we call seed eaters rely on insects to feed their young. They are also declining fast.[113] A study in Lake Shinji in Japan linked the first use of neonicotinoids in the surrounding farmland to the crashing of the fish population.[114] In just one year the weight of animal plankton in the lake fell by 83 percent. The fishing community's catch then fell by more than 90 percent. The global use of pesticides is expected to triple during the first fifty years of this century.[115]

Just as the chickens along the river cause algal blooms in the Wye, growing their feed appears to trigger similar effects, but on a much greater scale. Until 2010, *Sargassum*, a floating seaweed, was mostly confined to the Sargasso Sea and a few small patches scattered around the tropical Atlantic and Caribbean. But from 2011, something astounding began to happen. In the first six months of every year, the *Sargassum* patches now expand to create a continuous belt. It extends, some years, all the way from the Gulf of Mexico, down the South American coast and across the Atlantic to the shores of West Africa: a blanket of floating weed nearly 9,000 kilometers long. The cause, scientists believe, is "increased deforestation and fertilizer use in Brazil."[116]

As Brazil has become a global supplier of animal feed, its use of fertilizer has risen by 3 or 4 percent a year.[117] When fertilizers wash from the fields and minerals are eroded from exposed soils, vast rivers like the Tapajós, Xingu, and Tocantins, which drain millions of hectares of agricultural land, pour these nutrients into the ocean, where they cause a bloom that girdles almost a quarter of the Earth. When it dies, in the second half of the year, the weed draws oxygen out of the water column. It might contribute to the spread of some of the world's many dead zones.

Dead zones are areas of sea, mostly close to estuaries and coasts, in which oxygen levels fall so far that scarcely any living creatures can survive. There are now several hundred of them around the world,[118] caused by the extravagant use of reactive minerals. Since 1960, the global production of phosphate fertilizers has risen almost fivefold,[119] and nitrate fertilizers almost tenfold.[120] As the barium meal of nutrients passes through the agricultural system, its dysfunctions are revealed in grisly detail.

One reason for the overuse of fertilizers is the payoff between the cost of materials and the cost of labor. Ideally, fertilizers would be applied to crops in tiny increments, ensuring that the plants receive only as much as they need at each stage of their growth, and absorb all the nutrients they're offered, so that none run off the land. But farmers would then have to bring out their tractors dozens of times while their crops are growing, using enormous amounts of fuel and ensuring their labor costs exceed their income from selling the crop. Most farmers are short of time, and many are tempted to do exactly the wrong thing, spreading artificial nitrogen* in one big slug, whereupon most of it runs off into the rivers or the groundwater, or escapes into the air as the potent greenhouse gas nitrous oxide.

Sometimes, disastrously, they spread it in the autumn or winter, partly because this is when they have the most time, and partly because fertilizer suppliers tend to offer discounts then.[121] In this case the great majority of it is likely to be lost, as there might be no living crops to mop up the minerals, so the rain washes them off the bare soil. Altogether, roughly two-thirds of the nitrogen that farmers apply to their fields,[122] and between 50 percent[123] and 80 percent[124] of the phosphorus, are wasted.

Instead of feeding crops, these lost minerals cause algal blooms in rivers, dead zones at sea, high costs for water users (as they must be extracted from drinking water), and global heating.[125] The more fertilizer farmers apply, the less effective every new increment is likely to be. In some cases, when artificial nitrogen feeds the weeds in a field, helping them to smother the crop, the greater the dose, the lower the yield.[126]

Herbicides and pesticides are also wasted on an astonishing scale. One paper reports that "there is no correlation between agrochemical use and productivity or profitability."[127] What this means is that the farmers in the study who used extra herbicides gained nothing as a result. So why do people keep slapping excessive fertilizers and toxins on their land? Partly because they are guided to a large extent by "consultants" working for the chemical companies,[128] some of whom are little more than pushers. Farmers who are advised by consultants

* By "artificial," I mean reactive nitrogen synthesized industrially, mostly through the Haber-Bosch process.

from private companies are more likely to use pesticides as their first line of defense than those who are advised by public bodies.[129] The farming press is dominated by the same interests. So are the farming unions, and government agricultural departments. While strong counter-movements are beginning to develop, a farmer must be curious, brave, and assertive to resist this industry.

The global damage caused by intensive livestock farming is exacerbated even by well-intentioned government policies. The destruction in South America, triggered in particular by chicken factories, is mirrored by deforestation in Estonia, Latvia, Poland, and Romania. In several European countries, chicken farmers receive a subsidy for "green heat."[130] In the UK, the payment is called the Renewable Heat Incentive. Farmers who install renewable heating systems in their chicken units receive guaranteed and untaxed payments for twenty years. The handouts are so generous that their investments tend to pay off within five years:[131] after that, it's free money. The most profitable system is burning wood pellets. A large chicken shed uses around 120 tons of dry pellets a year.[132] This means that each shed consumes a little over a hectare of forest every year.[133] Together, assuming they are all heated with pellets, the 590 known chicken units in the Wye Valley are likely to burn the equivalent of 600 hectares of forest a year.

These incentives have helped start a gold rush in Eastern Europe. In Estonia and Latvia, even nature reserves are being clear-cut for woodchip.[134] In Romania, lovely Carpathian woods are obliterated,[135] while in Poland logging companies keep trying to push their way into the ancient Białowieża Forest.[136] Since 2015, the area of European forest being felled has risen by a remarkable 49 percent.[137]

Burning wood to produce either heat or electricity releases more carbon dioxide than burning coal.[138, 139, 140] Eventually, new trees might reabsorb this carbon, but it takes decades. In preventing climate breakdown, a ton of carbon saved today is worth much more than a ton absorbed in thirty or forty years. These payments might help to explain the rush of applications for chicken units in the Wye Valley in 2019 and 2020. The incentive program closed in March 2021: if your barns were up and running by then, you qualified for twenty years of public money.

*

You might, by now, have decided that you want nothing more to do with intensive farming: from now on you will eat only meat, eggs, and milk from animals that can roam outdoors (free-range), or have been certified as organic (farmed according to an agreed set of environmental rules). If so, I can offer you little comfort.

There are three main benefits of organic production, and they are important. Organic farms tend to be more complex and diverse than conventional ones, allowing a slightly wider range of wildlife to persist.[141, 142] They use fewer antibiotics and fewer pesticides, and the pesticides they do use tend to be much less toxic and ecologically damaging than those spread by conventional farming.[143] And by using manure or plant material, rather than artificial fertilizer, they return carbon to the soil. This, according to the emerging Theory of Soil, helps to maintain its structure.[144]

But in other respects, organic farming can inflict more damage. This is mostly because, with lower average yields, it uses more land to produce the same amount of food. One calculation suggests that if England and Wales became entirely organic, our land footprint would grow by 40 percent.[145] The global average gap between organic and conventional yields is, according to different estimates, somewhere between 20 percent[146] and 36 percent.[147, 148] The greenhouse gas emissions from organic produce tend to be similar, or worse, per kilogram to those of conventional food.[149, 150, 151] Organic beef farms—as the animals take longer to raise and need more land—lose twice as much nitrogen per kilo of meat as conventional beef farms.[152, 153] This will come as a shock to many: there might be no more damaging farm product than organic, pasture-fed beef.

Disturbingly, organic farming creates at least as much nitrogen pollution as conventional farming:[154] one paper estimates it releases 37 percent more.[155] This is because the animal manure used to grow crops tends to be even leakier than artificial fertilizer, for reasons I will explore in chapter 4.

Similar problems afflict free-range farming, of all kinds. One of the upsetting, disorienting findings by the groups trying to protect the Wye is that free-range egg farms are just as damaging, and possibly more so, as the huge broiler units or laying factories in the catchment.[156] Why? For the same reason that people want to buy their eggs: the chickens are allowed outdoors. While this is better for their welfare, it means they lay a scorching carpet of reactive phosphate across the fields on which they roam, often at far greater concentrations than any farm would spread manure. When rain falls, their droppings are washed off the land and into the rivers even faster, and in a rawer state, than the excrement extracted from the factories and spread deliberately on the fields. Because they use more energy than confined birds, free-range chickens also need more feed, exerting even greater pressure on the ecosystems of South America.

But this is just the beginning of the problem. The chicken factories in the Wye Valley testify to the wider economic failure of free-range farming. Many of the farmers who have built these barns once made their living from pastured (free-range) sheep or cattle. Farms on the upper Wye might own or lease a few fields in the valleys, and larger tracts of land in the hills. Traditionally, sheep were kept in the low-lands (the *hendre*) during the winter, then herded into the mountains (the *hafod*) in the summer. This farming has not been economically viable for many years. Across Wales, where most of the land is used for sheep grazing, the average farm income from agriculture hovers around £0 a year.[157] Almost all the money farmers receive comes from public subsidies. This system has also generated a series of disasters.

Sheep farming is uneconomic partly because the land can sustain few animals. In some Welsh ranges, the maximum possible stocking density is less than one sheep per hectare, and sometimes as little as one sheep per ten hectares. For several decades in the second half of the twentieth century, farmers were paid a subsidy for every animal they kept. These perverse payments created an incentive to pack the hills with far more sheep than they could support. This could have helped to create a terrestrial dead zone in part of the Wye's watershed: the southern Cambrian Mountains.[158]

The dead zone, which covers around 300 square kilometers, is a remarkable, if dismal, phenomenon. It is dominated by a single spe-cies of coarse, inedible grass called *Molinia*, which remains brown for most of the year. Scarcely any other plants grow. *Molinia* forms tough and stable clumps that cover the ground, preventing other species from germinating. You can spend all day exploring it without encoun-tering a bird or even an insect.

For at least 2,000 years, the land here was covered by blanket mire (a kind of wetland), supporting a fairly wide range of species. But a switch in the twentieth century from cattle to sheep grazing, followed by a rise in sheep numbers, appears to have flipped the entire system into a new stable state.[159] Though sheep have not been kept in some parts of the zone for thirty or forty years, as there is nothing left for them to eat, the system has not flipped back. In other words, hysteresis has occurred.

Maps of temperature and rainfall suggest that the natural vegetation type in much of the Wye catchment, and across the western uplands of Britain, is likely to be temperate rainforest.[160] This is a fascinating ecosystem, whose natural architecture creates niches for a great diversity of life. But sheep and cattle have eradicated almost all of it. Tree seedlings are highly nutritious, so sheep, even when they are few in number, selectively browse them out.* Over the course of centuries, by preventing young trees from replacing the old ones as they die, sheep have turned forests into pastures or heather moorland, both of which sustain a smaller range of species.[161]

There are more trees per hectare in some parts of inner London than there are in the "wild" British hills where sheep graze.[162] Some of the barest regions are our national parks, most of which are glorified sheep ranches. Rainforest is confined to a few tiny patches, mainly in gorges too steep for livestock to enter.

Just as chickens in the valleys demand a profligate use of nutrients, grain, and wood, sheep in the hills demand a profligate use of land. My estimates suggest, conservatively, that some 4 million hectares of hill and mountain in the United Kingdom are used for sheep farming,[163] or 22 percent of the entire farmed area.[164] This is roughly equivalent to all the land used to grow arable crops in this country.[165] It is more than twice the size of the whole built environment (all the towns, cities, factories, warehouses, gardens, parks, roads, and airports),[166] and twenty-three times the area we use for growing fruit and vegetables.[167] But, in terms of calories, lamb and mutton, across both the uplands and the lowlands, supply just over 1 percent of our food.[168]

---

* In most parts of the British uplands, trees don't begin to return until sheep numbers fall below five per square kilometer, in other words, one per 20 hectares.

Sheep farming, in other words, is an example of agricultural sprawl. Environmental campaigners rail against urban sprawl: the profligate use of land for housing and infrastructure. But because most of us live in cities, we fail to see the bigger picture. Agricultural sprawl—using large amounts of land to produce small amounts of food—has transformed much greater areas. In the UK, a highly urbanized nation, humans occupy 7 percent of the land,[169] while livestock, their pastures, and rough grazing lands occupy 51 percent.[170] If aliens landed here, they might conclude that the dominant life form was the sheep.

The global figures are even starker. Just 1 percent of the world's land area is used for buildings and infrastructure.[171] Crops occupy 12 percent, while grazing, the most extensive kind of farming, uses 28 percent.[172] Only 15 percent of the world's land, by contrast, is protected for nature.[173] The rest of the Earth's land surface is either uninhabitable (glaciers, ice caps, deserts, rocks, mountaintops, salt flats) or mantled by unprotected forests. Yet the meat and milk from animals fed entirely by grazing provide just 1 percent of the world's protein.[174]

The more land that farming occupies, the less is available for forests and wetlands, savannas and wild grasslands, and the greater is the loss of wildlife and the rate of extinction. All farming, however kind and careful and complex, involves a radical simplification of natural ecosystems. This simplification is required to extract something that humans can eat. In other words, farming inflicts an ecological opportunity cost. Minimizing our impact means minimizing our use of land.

I have come to see land use as the most important of all environmental questions. I now believe it is the issue that makes the greatest difference to whether terrestrial ecosystems and Earth systems survive or perish. The more land we require, the less is available for other species and the habitats they need, and for sustaining the planetary equilibrium states on which our lives might depend. It is also among the most neglected of environmental issues. Like soil ecology, total land use is a subject that the great majority of us have unconsciously agreed to ignore; another fatal chasm in public understanding. We obsess about certain alarming topics, often with good reason. But I suspect the most dangerous issues of all are the ones we scarcely consider.

Thanks to the team at Oxford University who run the website Our

World in Data, we can make quick and easy comparisons between the foods we eat.[175] Its charts show that, to produce 100 grams of soy protein, eaten by humans in the form of tofu, requires just over two square meters of land. To raise 100 grams of egg protein requires just under six square meters. Chicken protein needs seven, and pork ten square meters. Chickens and pigs need more land than tofu does because they cannot turn everything they eat into meat, as they have to sustain themselves and build other body parts. Milk, the website reports, requires an average of 27 square meters, beef 163, and lamb 185. Lamb protein, in other words, requires 84 times as much land to grow as soy protein.

There are two reasons for the amazing profligacy of pasture-fed meat. Grazing animals with four-chambered stomachs, such as cattle, goats, and sheep, are less efficient converters of protein than chickens and pigs,[176] and grass is less digestible and contains less protein than grain or soy.[177] While cattle and sheep need a great deal of land, however they are raised, beef finished on grain in an intensive feedlot system needs (for both the grain growing and the cattle keeping) roughly one-twentieth of the area required by beef reared entirely on pasture.[178]

This is why the nation with the world's greatest hunger for land is New Zealand.[179] If everybody ate the average New Zealander's diet, which contains plenty of free-range lamb and beef, another planet almost the size of Earth would be needed to sustain us. If, on the other hand, we all stopped eating meat and dairy, and switched instead to entirely plant-based diets, we would reduce the amount of land used for farming by 76 percent.[180] We would need to eat more grains and beans to compensate for the loss of meat from our diets, but far less than the amount currently used for feeding livestock. So, as well as being able to return all the world's pastureland to nature, we could reduce the arable area by 19 percent.[181]

Even the pitiful levels of production on the world's pastures cause great damage. Around a quarter are grazed so heavily that their soil has been damaged.[182] This means they can now carry even fewer animals.[183] As an invasive species called cheatgrass sweeps across overgrazed ranches in North America, some regions appear to be crossing crucial thresholds.[184] They too are in danger of becoming terrestrial dead zones.

Livestock farmers often claim that grazing "mimics nature." If so, the mimicry is a crude caricature. Nature has no fences. Its food webs tend to be deeper and wider than those of any human-managed system. A review of evidence from over 100 studies found that when livestock are removed from the land, the abundance and diversity of almost all groups of wild animals increases.[185] There are more predators, more wild herbivores, more pollinators. The only broad group (or guild)* of species whose numbers fall when grazing by cattle or sheep ceases are those that eat dung and other kinds of detritus. Where there are cattle, there are fewer wild mammals, birds, reptiles, and insects on the land, and fewer fish in the rivers.[186]

Only when livestock numbers fall so far that their husbandry scarcely qualifies as food production is animal farming compatible with rich, functional ecosystems. For example, the Knepp Wildland project,[187] run by my friends Isabella Tree and Charlie Burrell, where small herds of cattle and pigs roam freely across a large estate, is often cited as an example of how meat and wildlife can be reconciled. But while it provides an excellent example of rewilding, it offers a terrible example of food production. If their system were to be rolled out across 10 percent of the UK's farmland, and if, as its champions propose, we obtained our meat this way, it would furnish each of the people of the United Kingdom with 420 grams per year, enough for around three meals.[188] This means a 99.5 percent cut in our consumption. I don't mean three steaks, by the way, but three meals containing meat of any kind, including the cheapest cuts. We could eat a prime steak roughly once every three years. If all the farmland in the UK were to be managed this way, it would provide us with 75 kcal a day (one-thirtieth of our requirement) in the form of meat, and nothing else.[189]

Needless to say, this is not how it would be distributed. If we ate meat of only this kind, its price would reflect its scarcity, ensuring that the very rich could consume it every week and other people not at all. Those who say we should buy our meat from farms like this, who often use the slogan "less and better," tend to present a rare and exclusive product as if it were available to everyone. It's a cruel elision, with

---

* A guild is a group of species that use resources in similar ways.

perverse consequences. I've heard several people use the Knepp Wild-land project as their justification for eating beef, though they might never have bought it from that source.

Ecologists talk about "wildlife-human conflicts." But when you explore them, you soon discover that many would be better labeled wildlife-livestock conflicts. Almost without exception, where live-stock are kept, large predators are killed. Animal farming is the major cause of their decline.[190] In some places, like the United King-dom, they have been wiped out altogether. In others, farmers either kill predators themselves or lobby the government to do so on their behalf.

In the United States, both federal and state governments wage war against wildlife, often with astonishing brutality. A federal agency called Wildlife Services uses poisoned baits, snares, and leghold traps, and shooting from planes and helicopters, to kill wolves, coyotes, bears, bobcats, and foxes. Its agents have incinerated pups in their dens, or dragged them out and clubbed them to death.[191] It operates even in public lands supposed to be set aside for wildlife, such as the Sawtooth National Recreation Area in Idaho.[192] It appears to kill with impunity. Its agents have allegedly tested their weapons on pet dogs, slaughtered endangered animals, and spread illegal poisons.[193]

But perhaps its most controversial killing tools are cyanide land-mines. These are spring-loaded canisters of sodium cyanide planted in the ground, designed to spray poison into the face of animals that trip them. They have killed a wide range of endangered animals, dozens of domestic dogs and at least one person (Dennis Slaugh, from Utah), while injuring or poisoning several others.[194] So powerful is Wildlife Services that it is seldom held to account even for the injuries inflicted on people.[195]

In a few places, livestock farmers seem able to live alongside large predators, mostly in East and Southern Africa. In the early 1990s, I spent several months working with a Maasai community, witnessing the end of its traditional, benign system of management, as the land was privatized and fenced at the behest of the Kenyan government and the World Bank.[196] Across most of the rangelands of East Africa, the old system has collapsed.

*

Here's a paradox that at first sight is hard to understand. As farming has intensified, the amount of land used for grazing has slightly but steadily shrunk.[197] But the expansion of grazing land remains the world's greatest cause of habitat loss. It's responsible for 40 percent of the deforestation caused by the food industry, making it almost three times more destructive than palm oil.[198] How could the land used for grazing simultaneously be shrinking and expanding?

The answer is that it's shrinking in some places and expanding in others. In continental Europe, for example, livestock farming has retreated from infertile hills, and forests have returned. But in Latin America, especially Brazil, it continues to expand. Unfortunately, some of the places into which grazing is expanding are among the richest ecosystems on Earth. Because 92 percent of the world's natural grasslands have already been occupied by livestock or crops,[199] most of this expansion destroys tropical forests, including some of the richest forests on Earth, in Madagascar, the Democratic Republic of Congo, Ecuador, Colombia, Brazil, Mexico, Australia, and Myanmar.[200] Meat production could swallow 3 million square kilometers of the world's most biodiverse places in just thirty-five years.[201] That's almost the size of India.

The total demand for new farmland—driven partly by human population growth, partly by biofuels, but mostly by the shift in diets toward meat and dairy—could amount to 10 million square kilometers by 2050.[202] This is the area of Canada. Most of the expansion, unless something changes, will happen in South America and sub-Saharan Africa, sweeping not only through tropical forests but also through wetlands, savannas, and seasonal woodlands. By eating farm animals, we arrogate to ourselves the right to a diet using more land than the planet can safely provide.

Centuries of experience show that in the face of this industry nothing is secure. Livestock farming has displaced millions of indigenous people and destroyed billions of hectares of wildlife habitat. Its political power is greater than its economic weight. As it rattles the gates of protected areas, governments in many nations have unlocked them.

In 2018, Brazil became the world's largest beef exporter.[203] Since 2020, when the United States lifted its ban on raw Brazilian beef (imposed not for ecological reasons, but for food safety),[204] it is highly likely to have imported large volumes from places where the rain-forest was illegally cleared by ranchers, including ecological and indigenous reserves they have felled and burned.[205] There is no require-ment to label the meat as produced in Brazil, let alone in the Amazon: customers are kept in a state of ignorance. Some of this beef is pro-moted as "pasture fed." People who believe this is a good thing either never discover, or tend to forget, that the pastures were carved out of forests or savannas.[206]

One paper looked at what would happen if everyone in the U.S. followed the advice of gastronomes and switched to entirely pasture-fed beef.[207] It found that, because they grow more slowly on grass, the number of cattle would rise by 30 percent, and the land area used to feed them would rise by 270 percent. Even if the U.S. felled all its forests, drained its wetlands, watered its deserts and annulled its national parks, it would still need to import most of its beef.

The climate costs of farming mirror its land costs. Raising a kilogram of beef protein releases 113 times more greenhouse gases than growing a kilogram of pea protein, and 190 times more than a kilogram of nut protein.[208] Again, pasture-fed beef and lamb have by far the worst impacts:[209, 210] three or four times worse, according to a scientific review, than beef raised intensively on grain,[211] harmful as this is. This is because of the lower efficiency of converting grass into protein, and the slower growth of pastured animals: the longer they live, the more methane is released from their stomachs and nitrous oxide from their dung. Both are powerful greenhouse gases.[212] Peas and nuts cause less global heating than chicken, which causes less than pork, which causes less than lamb and beef.[213] Switching from a diet that's high in meat to one entirely based on plants would cut the greenhouse gases from your food by 60 percent.[214]

Just over one-third of the world's greenhouse gas emissions are produced by the food system.[215] Of these, roughly 70 percent are released by farming, and the rest by processing, transport, selling and cooking. An analysis by Our World in Data shows that even if green-house gases from every other sector were eliminated today, by 2100, food production alone would bust the entire carbon budget two or three times over, if we want to avoid more than 1.5°C of global heating.[216] Even if we sought to hit the less ambitious, and far more dangerous, target of 2°C, the food sector would account for almost all of it, unless its impacts are drastically reduced.

These figures measure only the gases *released* by farming. But just as it imposes an ecological opportunity cost, farming also imposes a carbon opportunity cost. This means the carbon dioxide the land could absorb if it weren't used for production. While the gases that farming releases could be seen as its climate current account, the carbon the land would otherwise remove from the atmosphere could be seen as the capital account. This account is almost always in debt, because the ecosystems that would have occupied the land—such as forests, marshes, natural grasslands, and the intact soils that underlie them—tend to store more carbon than the fields and pastures that have taken their place. The debts in some cases are enormous.

A study of carbon opportunity costs published in *Nature* found that, while the global average cost of soybeans is 17 kilograms of carbon dioxide per kilo of protein, the average carbon opportunity cost of a kilogram of beef protein is an astounding 1,250 kilograms.[217] Beef is roughly one-quarter protein. So four kilograms of beef, if the current and capital accounts can be directly compared, has a carbon impact more or less equivalent to one passenger flying from London to New York and back.

Another paper calculates that if a magic switch were thrown, causing the entire world to shift to a plant-based diet, and the land now occupied by livestock were rewilded, the carbon drawn down from the atmosphere by recovering ecosystems would be equivalent to all the world's fossil fuel emissions from the previous sixteen years.[218] This drawdown could make the difference between our likely failure to prevent more than 1.5°C of global heating, and success.

Often, when I've raised this issue, I have been told that the global figures on carbon opportunity costs are inflated by the extreme impacts of cattle farming in Latin America. In other parts of the world, I'm assured, they are much less severe. When I've questioned this claim, I have been referred to a report by the UN's Food and Agriculture Organization.[219] It does indeed show much greater climate impacts from beef produced in Latin America than in most other regions. But this reflects a strange decision by the authors of the report. The climate impacts of land clearance, they state, "were quantified for Latin America only." This, they explain, is because in the period they studied, Latin America was the only region in which "significant" pasture expansions were taking place.

Their decision was wrong in two respects. The first is that the expansion of pasture for cattle farming, when their report was written in 2013 was—and remains—a powerful driver of destruction in other parts of the world.[220] The second is that they count only current land clearance: to provide a more balanced picture, the figures should seek to include the ongoing opportunity costs of land that has already been cleared. Overlooking those costs reveals, I feel, the kind of observation bias the scientific method is supposed to avoid. Just because, in regions like Europe and North America, land was converted from wild ecosystems to cattle pasture before scientists could witness the process does not mean it didn't happen. Nor does it mean that the carbon costs of continuing to keep cattle there can be ignored.

A more recent study, published in 2021, takes this Visibility Fallacy to a remarkable extreme, claiming that land usage by established cattle pasture (as long as it has not been "degraded") should be estimated as zero. It justifies this amazing claim on the grounds that the land occupied by extensive cattle pastures "could not be used for other human food production."[221] But the land could be used for restoring wild living systems, while drawing down carbon from the atmosphere.

Understandably, cattle and sheep farmers are not very happy about the idea that their animals are a major cause of climate breakdown. They have used another scientific reappraisal to push back. In 2018, scientists at Oxford University pointed out that our estimations of the effect of methane in the atmosphere were wrong. Methane is a powerful greenhouse gas. The biggest source for which humans are responsible is the digestive systems of the cattle, sheep, and goats we keep.[222] (Other major sources are oil and gas production, coal mines, waste dumps, paddy fields where rice is grown, and the thawing of permafrost.) While carbon dioxide accumulates in the atmosphere across many hundreds of years, methane quickly breaks down. Its impact on global temperatures is sharp and short.

The scientists explained that methane had mistakenly been treated in climate calculations as if it were an accumulating gas.[223] This means that its long-term impacts were overstated, and continued emissions at a constant level make little contribution to rising temperatures. Livestock farmers leaped on this finding, claiming it meant that the contribution of their animals to global heating had been over-estimated.[224] But the recalibration also means that the short-term impacts of reducing methane were *under-estimated*. Almost as soon as you stop releasing the gas, its contribution to global heating stops.[225] As effective action against climate breakdown has to be quick, to prevent temperatures and Earth systems from crossing crucial thresholds, this makes cutting methane not less important, but more important.[226, 227]

Another mistaken belief is that the best way to cut greenhouse gases is to eat food that is locally grown. There might be good social and cultural reasons to buy local food. Local markets might help to enhance the food system's modularity and resilience. But there are seldom good climate reasons. This is because the greenhouse gases emitted by moving food are tiny by comparison to those emitted by growing it.[228] For example, if you buy pasture-fed beef or lamb, the contribution of transport to its total climate costs, depending on where it comes from, is likely to be between 0.5 and 2 percent.[229, 230] Raising the meat accounts for roughly 95 percent of its emissions (the rest are caused by processing, packing, storing, and displaying it). You would have to ship a kilo of dried peas roughly one hundred times around the world before its greenhouse gases matched those of a kilo of local beef.

Air freight accounts for just 0.16 percent of our total food miles: planes tend to carry only the most perishable and expensive products, such as shellfish, French beans, sugar snaps, asparagus, and blueberries.[231] All other foods are trucked or shipped.* Generally, the carbon footprint of local, out-of-season fruit and vegetables is much bigger than that of fresh produce imported from other countries: keeping them in cold storage[232] or growing them in heated greenhouses[233] uses more fossil fuel than trucking them. If you make a dedicated round-trip

---

* It's worth noting, however, that shipping fuel is currently very dirty, an issue that could be addressed, but has so far received too little attention.

of over 6.7 kilometers to buy your vegetables directly from the farm that grows them,* you counteract, on average, all the emissions you save by avoiding the carbon costs of storage, packing, transport to a regional food hub, then delivery of the vegetables to your doorstep.[234] This is because much more fuel is used by individual car journeys than in mass transport. No change to the way we eat comes anywhere near the impact of reducing our consumption of animal products, especially beef and lamb.

So livestock, even when integrated into mixed farming systems (producing both animals and crops), are powerful drivers of ecological destruction. But in recent years a new front in the public relations war has opened, promoting a remarkable claim: if managed in a particular way, animal farming can restore the living world and reverse climate breakdown.

Such assertions had been made for three decades. But they remained obscure until, in 2013, TED published a 20-minute talk by the Zimbabwean rancher and ecologist Allan Savory.[235] He maintained that by raising the numbers of cattle, sheep, and goats kept on drylands—in one case by 400 percent—and using "planned grazing" in a "holistic management" system, he could reverse soil erosion and the spread of deserts, restore lush vegetation, bring back wildlife, and even undo climate change. He showed before-and-after photos that appeared to provide spectacular proof of his claims. His talk has now been watched 11 million times, between the TED site and YouTube.[236] His story was taken up by several documentaries, including the viral Netflix film *Kiss the Ground*, narrated by Woody Harrelson.[237]

I like Allan. When I was diagnosed with cancer, he sent me a kind and charming email. I know he's sincere and believes what he says. But, as soon as I watched his talk, I noticed that at least one of his claims could not possibly be correct. He stated that if we follow his prescription, "we can take enough carbon out of the atmosphere" to "take us back to pre-industrial levels."

Since 1750, roughly 490 billion tons of carbon have been released

---

* Visiting farm shops has become a major pastime in my country, where it is widely, and wrongly, believed to be a greener kind of shopping.

from fossil fuels, and around 190 billion tons by cutting forests, draining wetlands, plowing soils, and other kinds of land use.[238] So, to return atmospheric carbon to pre-industrial levels, grassland soils would need to absorb 680 billion tons.*

Since the dawn of agriculture, roughly 133 billion tons of carbon are reckoned to have been lost from the world's soils.[239] Of this, between 70 and 90 billion tons have been released from steppes, savannas, and grasslands, the ecosystems Allan is talking about.[240] As the great soil scientist Rattan Lal notes, the carbon lost from the world's living systems is roughly equivalent to the maximum amount they could, in a perfect world, absorb.[241] This means that grassland soils could draw down from the atmosphere a maximum of 13 percent of the carbon released in the industrial era.

This would still represent a massive contribution toward preventing climate breakdown. But unfortunately what could be done in theory is not the same as what can be done in practice. A study of the global potential for sucking up carbon by changing the way we farm suggests that, at most, 64 billion tons could be absorbed this century by agricultural soils.[242] If we assume, again being generous,† that two-thirds of this absorption happens on steppes, savannas, and grasslands, this brings the potential down to 43 billion tons, or about 6 percent of Allan's target of 680 billion.

Unfortunately, that's not the end of the matter. For even if Allan's system does cause carbon to be absorbed by the soil, that gain is counteracted by the greenhouse gases cattle, sheep, and goats and their manure release: methane and nitrous oxide. A global review drawing on 300 papers found that, in the very best cases, the carbon absorption on grazing lands amounted to 60 percent of the greenhouse gases the animals on the land release, through burping and defecating.[243] In other words, livestock grazing, even if we make the most generous possible assumptions, cannot wash its own face, let alone reverse historical emissions.

To make matters worse, more recent scientific findings challenge

---

* The paper I've cited is nine years old. So I've added to its total at the average emissions rate since then (roughly 9 billion tons of carbon a year).
† In reality, the study suggests that pastures could absorb just 40 percent of the total.

the very notion of storing carbon in soil.[244] The old belief that large, stable carbon molecules (collectively called humus) persist in the soil for long periods appears to have been debunked.[245] Most of these molecules can be broken down by soil bacteria. And, as temperatures rise, increasing the speed at which bacteria process it,[246] carbon is likely to be released from soil faster than scientists once reckoned.[247] It now seems wrong to treat any carbon as safely removed from the atmosphere, if it's lodged in soils in which air circulates. (The carbon in waterlogged soils, such as peat and the mud in marshes, is more stable.)

I phoned Allan to ask for evidence. I found his answers rambling and unconvincing.[248] He was unable to direct me to any scientific papers supporting the claims in his talk. But I wanted to be sure I wasn't missing anything. So, after *Kiss the Ground* was released, I set aside a month to read scientific papers about the "holistic" systems that he and other ranchers were promoting.

I discovered that there were similar problems with all his major claims. In a minority of cases, there were some improvements, by comparison to ordinary grazing, in soil quality and plant production, on ranches using his methods.[249, 250, 251, 252] But, as one paper notes, "the vast majority of experimental evidence does not support claims of enhanced ecological benefits," even by comparison to other kinds of grazing.[253] Even a review of the few scientific papers approved by Allan's organization, the Savory Institute, found that his system performed no better than conventional but well-managed grazing.[254] A global review of the scientific evidence for Allan's system of "holistic planned grazing" found that there is no difference, on average, in plant growth between ranches following his methods and ranches managed conventionally.[255] Instead, there seems to be plenty of evidence that his methods can inflict severe damage on ecosystems.

In his TED talk, Allan described the "crust of algae" that often grows on desert soils as "the cancer of desertification." Trampling by cattle destroys this "cancer," and allows a dense sward to grow in its place. In reality, the crust is a rich ecosystem of bacteria, fungi, algae, mosses, and lichens, which prevents erosion and absorbs carbon and moisture.[256, 257] These crusts are often extremely fragile, and are quickly destroyed by cattle, often with devastating and irreversible consequences for ecosystems, as invasive, exotic plants can then colonize the land, replacing native species.[258] Trampling by livestock, which Allan claims improves the soil and helps it to store carbon, in most cases has the opposite effect, compacting and eroding it.[259, 260, 261]

Intense grazing of the kind Allan promoted in his talk damages the vegetation on riverbanks, the crucial habitat for many species in drylands and deserts.[262] Drylands that livestock have never entered tend to have a greater range and abundance of native plants than those used for any kind of grazing.[263] As a general rule, the best way of ensuring that dryland ecologies recover is to remove the farm animals.[264, 265, 266]

So what do the famous photos in his talk show? They purport to show bare, eroded land miraculously springing back to life when his grazing regime begins: thick grass and shrubs surge from the naked ground and erosion gullies refill with soil. But is that really what they depict? Because they are either unlabeled or mislabeled,[267] it's hard to tell. But at least a couple of them appear to show the opposite of what he claims: the survival or recovery of the ecosystem was caused not by introducing livestock, but by taking them away.[268]

Sadly, scientific findings have not prevented some of the world's biggest meat companies from using false claims about the alleged benefits of pasture-fed beef in their advertising.[269, 270] Worse still, a new market has developed,[271] in which companies such as Microsoft buy carbon credits from ranches practicing holistic grazing,[272] on the mistaken grounds that this offsets their emissions.[273] By making ranching more economically viable, this money is likely to accelerate climate breakdown, as land that might otherwise be rewilded continues to be grazed. In other words, the companies investing in these programs ignore the opportunity costs of livestock farming. You might as well buy carbon credits from a coal mine.

While the biggest threat to habitats and wildlife is agricultural sprawl in the tropics, among the greatest threats to the climate is agricultural sprawl in the Far North. Peat soils in the boreal and Arctic regions hold more carbon than all the world's forests.[274] Much of it has, until now, been inaccessible to farming because the weather is too cold. But as the North warms, its governments have started to explore the potential for developing a new agricultural frontier. The temperature band suitable for farming could march 900 kilometers north across Alberta by 2100, and 1,200 kilometers through eastern Siberia.[275]

One paper mapped the land that global heating could make available for farming by the third quarter of this century. It discovered that roughly another 15 million square kilometers, the size of the U.S. and European Union combined, would become warm enough.[276] Most of it is in Canada, Alaska, Russia, Scandinavia, and northern China. Frighteningly, the agricultural possibilities cover land with some of the world's highest concentrations of soil carbon. In the unlikely event that all this land were used, 177 billion tons of carbon could be released: equivalent to 119 years of current CO_2 emissions from the United States.

Other scientists argue that while the climate on this frontier might become suitable for farming, much of the soil, being acid or rocky, will not be.[277] This may be true. But it's worth remembering that forty years ago, arable farming was considered impossible in most of the Amazon, and fifty years ago, the notion of growing oil palms in the deep peat of Indonesia would have been considered ridiculous. Researchers, unfortunately, discovered how to do it. In any case, the governments of Russia,[278] China,[279] the Northwest Territories,[280] and Newfoundland and Labrador[281] in Canada are cheerfully attempting to shift the frontier.

Campaigners, chefs, and food writers rail against "intensive farming," and the harm it does to us and our world. But the problem is not the adjective. It's the noun.

Farming is the world's greatest cause of habitat destruction,[282, 283] the greatest cause of the global loss of wildlife,[284, 285, 286] and the greatest cause of the global extinction crisis.[287, 288] It's responsible for around 80 percent of the deforestation that's happened this century.[289] Food production (including commercial fishing) is the main reason why the global population of wild vertebrate animals has fallen by 68 percent since 1970.[290] Of 28,000 species known to be at imminent risk of extinction, 24,000 are threatened by farming.[291] Only 29 percent of the weight of birds on Earth consists of wild species: all the rest are poultry.[292] Chickens alone weigh more than all other birds put together, including farmed ducks and turkeys. Just 4 percent of the world's mammals, by weight, are wild; humans account for 36 percent, and livestock for the remaining 60 percent.[293] This is caused not by intensive farming or extensive farming, but a disastrous combination of the two.

For years a debate has raged about whether it's better to farm more intensively, to reduce the amount of land required, or to farm less intensively, to allow wildlife to flourish within farmland. This dispute is known as sparing versus sharing.

The problem with sharing is well established. The great majority of the world's species cannot survive in farmed landscapes of any kind.[294, 295, 296] Many can persist only within very large areas of unexploited land.[297] The more land that can be spared for nature, without disturbance by farming or any other extractive industry, the fewer species are likely to become extinct. Sprawl threatens ten times more of the world's "biological hotspots"—the places with the greatest diversity of species—than intensification.[298] The continued expansion of farmland is forecast to endanger 30 percent of the world's species by 2050, while intensification endangers 7 percent.[299]

Case closed? No. One problem with the sparing solution is that even large protected areas might fail if they're completely isolated. If animals cannot travel between them, they are less likely to escape environmental change (such as global heating), and more likely to succumb to in-breeding and extinction. Populations that currently look healthy could be living on borrowed time.[300, 301] Their survival could depend on corridors and stepping stones: habitats on farms that let them travel from one nature reserve to another.[302, 303]

Sharing is necessary if we want to avoid the horrifying prospect of a tripling in the use of pesticides during the first half of this century. One alternative is to use biological control: encouraging predators to eat the pests attacking our crops. These predators cannot survive unless some farmland is shared with nature, leaving strips of grass and wild flowers, hedges, copses, and riverbanks in which they can safely shelter and breed.[304] The same applies to wild pollinators, which are often much more effective at fertilizing crops than honeybees.[305, 306, 307, 308] Honeybees are livestock: domestic animals reared by people. Like cows and chickens, they can inflict great damage to ecosystems, as they overwhelm wild species.[309, 310] A land of milk and honey is a land of ecological destruction. The great professor of entomology Pat Wilmer once told me that introducing beehives is like "moving a city into the countryside." While honeybees live in mobile homes, wild pollinators need wild places.[311, 312]

Another issue with pursuing only the most intensive forms of farming is the efficiency paradox. Just as improving the efficiency of irrigation leads to greater use of water, improving the efficiency of farming can cause a greater use of land.[313, 314] This is because efficient farming tends to be profitable, attracting capital. Capital wants to grow, so it seizes more land in which to reproduce itself.[315] The demand for animal feeds and biofuels could allow farming to keep expanding, long after everyone is adequately fed. This could lead to the worst of all worlds: intensive production across a vast area.[316] Intensification will spare wild places from farming only if it's coupled with a strong political commitment to protect or restore them.[317]

The fourth issue is the one I explained in chapter 2: intensive farming converges toward uniformity. It unwittingly begins to build a single, integrated system, whose nodes become bigger, whose links become stronger and whose behavior begins to synchronize. Efficiency threatens resilience.

In other words, we appear to be trapped between two dangerous forces: efficiency and sprawl. Farming is both too intensive and too extensive. It uses too many pesticides, too much fertilizer, too much water, and too much land.[318]

There is an obvious solution to this conundrum: for everyone to adopt a plant-based diet. If we stopped eating animal products, we could relax the intensity of production within the remaining arable area, reintroduce redundancy, modularity, backups, and circuit breakers into our farming systems, and create corridors and stepping stones for wildlife.

But while the number of vegans has been climbing, mainly in the rich world, this shift is much smaller than the global rise of meat-eating. Without better substitutes for animal products, which are cheaper than and scarcely distinguishable from the foods they seek to replace, and without deliberate policies to encourage it, the current drift away from meat-eating in a few small corners of the world offers no realistic prospect, in the first half of this century, of rolling back livestock farming. This transition could happen by design, as I'll suggest in later chapters. It will not happen by accident. Even if the shift can be catalyzed, we also need, for all the reasons explored in this and the preceding chapter, radically to change the way we grow plants.

So how do we resolve this dilemma? How can we ensure everyone is fed while farming becomes both less intensive and less extensive? How do we meet the demand for food as conditions in some parts of the world become hostile to agriculture? How do we do all this without causing systemic collapse?

In the rest of this book, I follow some of the remarkable people who are seeking to answer these questions. I meet the pioneer of a revolutionary system for growing fruit and vegetables, which manages to share land with wildlife and protect the soil without reducing yields: in other words, without causing agricultural sprawl. I follow farmers and researchers who are trying, with varying degrees of success, to grow grain and cereal crops that could feed everyone without destroying our life-support systems. And I meet scientists developing radical new ways of producing protein and fat with minimal environmental impacts, leading to new cuisines that might transform our relationship with the living world. Some aspects of their work raise as many issues as they resolve. But their experiments allow us to develop general principles that could help open the jaws of the trap. They also expose some massive gaps in research and practice that urgently need to be filled.

There are no perfect solutions in an imperfect world. As time runs down, our opportunities for action contract, and the possible answers become more conflicted. Even so, I think there are ways out. Improbable as it might sound, we can produce more food with less farming.

4

Fruitful

"You can make music with this soil," Tolly said.

As he tapped the flints with his trowel, the bright notes rang across the valley. "It sings when you work it. When you come down the field with the plow, it gets louder and louder."

I ran my hands over the eerie, zoomorphic flints, like bones and antlers, crooked fingers and bunched fists. Some had split, revealing their glassy interiors. They would cut me if I picked them up carelessly. On the ramparts of the ancient fort that crowned the hill above us, beside the weir on the river below, on the commanding ground where the great house stood, worked flints had been found, chipped into knives, arrowheads, scrapers, and spear points, dating from the Lower Paleolithic to the Upper Neolithic.[1] This land had sustained people for hundreds of thousands of years. Among the flints were umber lumps of quartzite that looked like old potatoes. They were dumped from the snouts of the glaciers that terminated north of here: the last cores, weathered and oxidized, of the hard white veins that once ran through the rocks of the north. They are used locally for cobblestones.

In the hedge above our heads, a blackcap warbler paid out its lovely song, then reeled it back on the in-breath.

"No conventional grower would even look at this ground. It's forty percent stone. They'd call it building rubble. During the war they plowed it and grew potatoes. By the time I got here, it was pretty much stripped. Almost everything had been taken and nothing put back. That's why I was able to get it. If it was better soil, it would have been snapped up."

A shadow fell across us. I looked up and saw a kite staring into my eyes, just a few meters above my head.

"Some days when I'm plowing, forty of them come to take the worms. They swoop down and pick them off the soil without landing. Once the kites get going, the goshawks hop out of the trees and join them. There's wagtails flitting around between them picking up wireworms, completely unfazed by all the raptors.

"This land is rough on machinery and rough on your hands. It isn't even classed as arable: an agronomist would say it's only good for grass or trees.* But over the past twelve months, we harvested 120 tons of vegetables and fruit."

For thirty-three years, Iain Tolhurst—Tolly—has been farming seven hectares of this rubble without pesticides, herbicides, mineral treatments, animal manure, or any other kind of fertilizer. He has pioneered a way of growing vegetables and fruit that he calls "stockfree organic." This means he uses no livestock or livestock products at any point in the farming cycle, yet he also uses no artificial inputs. Until he proved the model, this was believed to be a formula for sucking the fertility out of the land and destroying its productivity. Vegetables, in particular, are considered hungry crops, which require plenty of added nutrients to grow. Yet Tolly, while adding none, had raised his yields until they hit the lower bound of what intensive growers achieve with artificial fertilizers on good land: an achievement that was widely considered not merely remarkable, but impossible. As he did so, something even more astonishing happened: the fertility of his soil steadily climbed.

At the same time, a great diversity and abundance of wildlife returned to his farm. Almost single-handedly, across many years of trial and error, he developed a new and revolutionary model of horticulture.

I don't believe in magic. So I had come to learn the conjurer's tricks.

Tolly is a big, tough-looking man in his late sixties, with etched and weathered skin, a broad, heavy jaw, long blond hair, one gold earring, hands grained with earth and oil. He had started farming without training or instruction, without land or—as his origins were humble— any means to buy it. By trade, like his father, he was a woodworker.

* It's classed as Grade 3b, or Moderate Quality Agricultural Land, according to the Agricultural Land Classification System in England and Wales.

He was brought up in Bristol, and speaks with that city's long burred vowels and soft consonants.

"It was a tiny house, but it had a garden. My mum grew flowers. She thought only very poor people grew vegetables. She aspired to a better lifestyle than she got. A neighbor gave me a couple of strawberry plants, and I was fascinated by the way they reproduced themselves. It was my grandfather who encouraged my interest in gardening.

"I was bored to tears at school. I just wanted to be outside. The only thing I liked doing was woodwork. I was slightly disruptive, to be honest, and I don't think anyone was sorry when I left school at fifteen.

"I buggered off for a year, hitchhiked, went to Scotland, bummed around. Then I went to college in Falmouth, apprenticed in ships' joinery. But I always wanted an outdoor life. This was my problem with school."

After a year at college, when he was seventeen, Tolly met Lin, who, like him, was energetic and adventurous, ready to weather whatever life threw at her. She was a Scillonian, born on an open boat on the way to the island hospital.

"I jacked in the course and went back with her to Tresco. It's a small island—just 180 people. I learned to drive and worked on a farm there, daffodil picking and stuff. But I ended up back doing woodwork. That's what paid. I worked with Lin's father, who was a mason, renovating the old abbey on the island."

Tolly and Lin married when they were nineteen, and moved back to Bristol, where Tolly spent a year or so working with his father in construction. At twenty, Lin and Tolly had their first child, but shortly afterward Tolly lost his job when the company he worked for went bust. "We got into financial difficulty, and ended up squatting a farmhouse in Somerset. We got evicted, squatted another one, got evicted again."

Out of desperation, he took a job on a dairy farm near Milton Keynes.

"It wasn't long before I started to question the things I was doing. I wasn't happy with the health of the cows or the state of the soil. When I heard there was such a thing as organic farming, I realized it didn't have to be like that. We grew strawberries in the big garden we had, and sold them to a local health farm. We saved some money and bought some rubbish land in Cornwall to set up an organic vegetable farm: five acres for £6,500. It doesn't sound like much now, but in 1975 it was more than a year's wages. It was a pretty inhospitable part of Cornwall, up above St. Austell, in the china clay area.

"We were completely naive. We had no house, no infrastructure, no plans. Just a bit of land with a spring and a caravan. We lived in the caravan for seven years with two kids and no electric.

"We were on that land nearly ten years. We built a house out of granite and timber and set up a farm shop, mostly selling strawberries. The land was rubbish for vegetables. I supplemented my income with odd woodwork jobs. Eventually we sold the house and the farm as a going business, for good money.

"So in 1987 we bought some much better land and borrowed money to build a house there. We set up a good horticultural business, and we were selling strawberry plants all over Europe. But halfway through building the house, the economy crashed and we found ourselves in negative equity. We had to give the whole lot back to the bank.

"We came here with a couple of hundred quid and had to start all over again. But we were young and healthy. At the end of the day it's only money."

Tolly lives and works on the Hardwick Estate, overlooking the River Thames at the foot of the Chiltern escarpment. The estate is owned by Sir Julian Rose, an actor with a neat beard and an Errol Flynn dash. It had been bought by his great-grandfather, Sir Charles Rose, a horse breeder, yachtsman, motorcar enthusiast, adventurer, and philanderer, who is said to have been an inspiration for Toad in *The Wind in the Willows*. Its author, Kenneth Grahame, was an occasional guest at Sir Charles's house parties.[2] Sir Julian hadn't expected to inherit the estate, but his older brother died in a motor-racing accident. When he took on the property, in 1966, it was a mess. But he decided to use it to try to create a thriving rural community. He leased land only to people who were prepared to live there and work it themselves. He became a powerful advocate for organic farming, notorious for bringing a cow to London to protest against the government's plan to ban unpasteurized milk.[3]

Tolly and Lin rented two fields, the walled garden and one of the gatehouses. By then, Tolly told me, "Lin and I weren't getting on too well. I had a girlfriend. I was leading a double life. But even after we got divorced, we kept working together. She and the kids stayed in the gatehouse, I moved into the workshop behind it. We stayed very close, and worked together all the while. After I married Tamara, people would joke that I had two wives.

"I met Tamara at the gate to this field." He gestured toward the opening between the shaggy hedges. "She had come here from Moldova with fifteen farmers. I'd never heard of Moldova. She was a biology teacher, but when the Soviet Union collapsed, and Moldova became independent overnight, her job disappeared. She survived by smuggling vacuum cleaners into Romania. Then she got a job working for a British charity, bringing farmers together to find new ways of growing food.

"I gave them a farm walk. Tamara was heading the group, and she ended up translating, as the official translator didn't know the agricultural terms.

"Later I got a call: they liked what they'd seen here, and would pay me to give advice. Would I go to Moldova and tell them what I was doing?

"Tamara was amazing. She fixed for me to meet all the ministers and other senior people. We became good friends, but I didn't push my luck. Anyway, I got another invitation about three months later, to set up the national organic standards for Moldova. Toward the end of the week, she asked me whether I fancied extending my stay: we could go to Romania on the train. I knew what the invitation was. I changed my ticket. It was the most romantic thing I've ever done. When we got married, she moved in here.

"Lin welcomed her into our lives and made her feel at home. We made a great team. The three of us shared the running of the farm. Lin was a genius at propagation. Tamara, who was the first person in Moldova to win a scholarship to do an MBA in the U.S., ran the business side and makes preserves from the surplus veg. My daughter, Gena, runs the shop. I was unbelievably lucky to work with them all.

Eighteen months ago, Lin collapsed in the walled garden. I was with her when the light went out of her eyes. It was a devastating blow to all of us."

Tolly looked across the valley to where the river glittered in the distance.

"I've got a plumbing crisis to fix, so I'll leave you here for a bit. Come and meet me in the walled garden in a couple of hours."

Beside the plowed land where Tolly had left me was a strip of wild flowers, running the width of the field. Above it, hundreds of marbled white butterflies, with checkerboard wings, danced in the wind, trying to settle on the chicory flowers, the purple knapweed and yellow spires of agrimony. Grasshoppers sprang from the stems as I approached. To find out what was living on Tolly's land, I unbundled the folding sweep net in my bag and started working down the strip.

With the first swing of the net, I caught a shoal of little capsid bugs, whose backs were gorgeously grained in beech-brown and green.* There was a thistle gall fly with red eyes and zebra wings[†] and a wool carder bee as brightly marked as a wasp. In the next sweep, I caught a tiny sawfly,[‡] a red damselfly, and a tiny bronze furrow bee,[§] which excited me, as I'd never seen one before. (Yes, I know.) I found a wolf spider eating the beetle it had caught on an oxeye flower. I watched a pale, skeletal robber fly[¶] swing slowly through the plants like a spectral helicopter, picking off aphids and carrying them away. A massive ichneumon with yellow-striped legs and perpetually twitching antennae** stalked among the leaves, looking for caterpillars in which to lay her eggs. Red-tailed bumblebees blundered through the borage flowers.

A sudden flash of orange rose from the knapweed. I chased it down the row until I came close enough to identify it as a dark green fritillary, a butterfly I hadn't seen in these parts for a couple of years.

---

* *Stenotus binotatus.*
† *Urophora cardui.*
‡ *Arge ustulata.*
§ Either *Halictus tumulorum* or *Halictus confusus.*
¶ *Leptogaster cylindrica.*
** *Amblyteles amatorius.*

Everywhere, moths rose from the sward. I counted—I think—twelve species of bee. There were hoverflies and craneflies, wasps and soldier beetles, lacewings and shield bugs. I have rarely seen such a range or profusion of insects anywhere in Britain.

Flowerbanks like this ran across Tolly's two fields every 25 meters, frothing with ribwort plantain, birdsfoot trefoil, black medick, bloody cranesbill, bedstraw, corn marigold, self-heal, yarrow, cinquefoil, wild carrot, campion, and the strange cerise flowers of corncockle, whose soft petals unfold from a ring of daggers. The scientists studying Tolly's fields had recorded seventy-five species of wild flower, some of which he had sown, some of which had arrived by themselves.[4]

In the blocks of plowed land between the flower banks was an impressive range of crops. One plot was a blue haze of onion plants, another a patchwork of sea greens: young cauliflower plants, cabbages, and several kinds of kale. There were a few lines of rainbow chard, with gold, green, white, and crimson stems. Young carrots feathered up; old asparagus feathered down. Broadbean pods had begun to sprout from tight pillars of flower. Sweet corn grew in alternate rows with French beans. The potatoes were in full flower, filling the air with their honey scent, nightshade sinister, stamens like yellow stings. Lines of pumpkin plants were badged with the seed companies' corporate pastoral: Autumn Crown, Jack O'Lantern, Butternut Hunter, Blue Ballet.

The strawberry harvest was over: most of the remaining fruits were softening and bird-pecked. But I found a couple that were still sweet and sharp, as different from the strawberries in the supermarket as a peach is from a potato.

A group of kites tussled and screeched in the air above the copse at the top corner of the field, black against the sky, then ginger and gray against the trees when they swooped. As I watched them, a pigeon flew past. A sparrowhawk swung out, lunged at it, missed, then heeled back into the wood.

These fields seemed to me to exemplify the fruitful but hospitable landscape that resolves the sparing-sharing dilemma. Their high hedges and copses, flower banks and rows of trees connect the floodplains from which they rise to the beechwoods and chalk downlands that hang over them. The barn owls, goshawks, nightingales, kingfishers, polecats, water voles, orchids, and other unexpected wild flowers* that have returned to the land since Tolly started working here testify to the value of its corridors and stepping stones, which enable wild species to move over this ground and settle on it. Yet this abundance does not detract from the land's productivity. As Tolly would later explain to me, the wild animals and plants are essential to it.

I passed Sir Charles's stud farm, in which he bred a succession of famous racehorses in the late nineteenth and early twentieth centuries. Thoroughbreds in blankets still grazed the trim paddocks. When I rounded a bend in the long drive, the jumbled Tudor roofs and chimneys of Hardwick House came into view. As Kenneth Grahame wrote of Toad Hall, it's "a handsome, dignified old house of mellowed red brick, with well-kept lawns reaching down to the water's edge." In the first chapter of *The Portrait of a Lady*, Henry James, another visitor, describes its "long gabled front of red brick, with the complexion of which time and the weather had played all sorts of pictorial tricks, only, however, to improve and refine it."

The old kitchen garden was enclosed within a high wall of flint and biscuity brick, barely supported by vast buttresses, against which fig trees and apricots, a crusty old pear tree, grapevines, tayberries and loganberries, hollyhocks and rhubarb grew. In the garden, the crops were packed even more tightly than in the fields. Tolly later explained to me that the garden is worked by hand, and the fields by his ancient tractor. Machinery requires more space between the rows of plants, which is one of the reasons why manual labor is more productive. But manual labor is also more expensive. In any horticultural system, yield and profit are delicately balanced.

In the open beds, runner beans twisted up bamboo poles, courgettes extruded rudely behind their trumpet flowers, spring cabbages grew whorled and molluscan. There were bunching onions, kohlrabi and calabrese, carrots and kale. Lettuces had grown together until I couldn't see the ground between them. In the polytunnels, great trusses of tomatoes were ripening. It was hard to believe that they all belonged to the same species: there were flat-bottomed bruisers and tiny berries, peardrops and bombs, spheres and cylinders. Beside them

* Such as round-leaved fluellen and field madder.

grew parsley, green and purple basil, young celeriac and spring onions, several breeds of sweet pepper. The delicate tendrils of cucumbers, ogee and volute, wound up their strings. Like the crops in his fields, none of them showed signs of insect damage: the leaves were dark and wide, with scarcely a hole or a spot. Given that Tolly uses no pesticides and his farm supports such an abundance of insects, I found this astonishing.

In a propagation shed, the last of the modules were waiting to be planted out: blocks of broccoli, New Zealand spinach, Valmaine lettuce, and lollo rosso. Not an inch in the beds was wasted. The paths were edged with wild melilot, calendula, yarrow, poppies, and cornflowers, while nettles sprawled around the polytunnels. As Tolly would explain to me, the nettles were as valuable to him as the vegetables.

I had arrived in time for "crib." Crib is the mid-morning meeting, held under a canvas gazebo at the bottom of the garden. The name was brought by Lin from Cornwall, where it was used by tin miners to mean their morning break. It was during crib that she collapsed and died.

As Tolly spoke, I began to appreciate the complexity of his task. There were six other people round the table, but altogether, in these peak months, he employs twelve. On any summer's day, there might be twenty or thirty crops that need attention: propagating, planting out, watering, weeding, harvesting. Every year, Tolly's team must sow 350 times, as most vegetable crops can be picked for only three weeks. Multiple sowings, a few weeks apart, extend the harvest season. An arable farmer, by contrast, might sow just three times a year. He raises 100 varieties of vegetables, which he sells in his farm shop and to subscribers to his weekly veg box in nearby towns. "We grow green beans, yellow beans, purple beans, round beans, flat beans, so people seldom have the same box from one week to the next." But variety means work.

Tolly holds a map of the land in his head, which, as crops are planted or removed and the weather delivers surprises, is constantly adjusted. He raced through the day's business.

"We were going to plant out the broccoli, beans, and lettuce today, but the soil's too wet. We need to do something about the beans, otherwise they'll start growing round each other."

Tolly explained that the problem was making the hole. The tool they would normally use—a bulb planter—smears the soil when it's wet, creating a barrier that stops the roots of the beans from spreading.

The team discussed the problem for a moment, then decided to plant them with a trowel instead. "It takes more time, but it's only 150 plants. Not too bad. The lettuces can wait till Monday. They're big, but they'll hang on. There's a tray of celeriac down there, which nearly got planted as parsley."

"I'll put it out."

Tolly sketched a diagram on his clipboard, to show where it should go.

"Is there any kale to pick?"

"There's a bit."

"As much as you can get Chris, 'cos we've been really short in the shop this last couple of weeks. All the best stuff has gone in the boxes. What about spring onions?"

"I've done a few."

"The lettuce isn't selling because it's been raining. Cabbage and turnips would do better at the moment, if we had them. Are you still doing the rhubarb?"

"Nearly done. There are 300 to sow."

And so it went on, ranging across dozens of tasks. They talked about the plants as if they were customers, each of whom had to be served according to their whims.

I watched as the growers spread back into the garden. One woman harrowed a bed for planting, another ran compost through a rotary sieve for potting, one man pulled spring onions, releasing their sharp oily smell, which spread deliciously across the garden, another collected strawberry runners to propagate. Two workers set up a rack of poles for a new crop of beans to climb.

I had dozens of questions, but I could see that Tolly was preoccupied, so I suggested coming back at a less busy time.

"A less busy time doesn't exist. Hop in the van and I'll take you back to the fields."

When we were standing among his broad beans, I asked him why they weren't covered in blackfly (a species of aphid). When I've tried to grow them, they get slaughtered by these pests.

"Well, the important thing is to keep the pests alive in the winter."

"You mean the predators?"

"No, I mean the pests."

"Sorry. You're saying you want to keep the *pests* alive?"

Tolly smiled slowly.

"Well, if you don't keep the pests alive, how will the predators survive?"

The next thing he said surprised me even more.

"It's not just that I want to sustain the predators. We also need the pests. They're an essential part of the system. They show me that something has gone wrong. If your crop is being attacked, it's because there's an underlying problem. If I see a pest problem, I don't want to know what nature's done wrong to me. I want to know what I've done wrong. If you've got it right, your plants will defend themselves."

I thought of the way that plants use the bacteria in their external guts to prime their immune systems, and how they send signals to underground predators.

"In conventional horticulture," Tolly continued, "there are major problems with aphids, as the soil is often unhealthy. If you apply nitrogen—whether it's organic or chemical—you encourage them, because aphids love nitrogen. But if your soil's healthy, there should never be a problem, as so many species eat them. Hoverflies, lacewings, ladybirds, the aphid midge. If I was an aphid I would be paranoid, as everything is out to get you.

"Insect pests aren't an issue for us at all. Every so often a new species turns up and the industry panics. Leek moths arrived here eight or nine years ago, moving north with global warming. They did some damage to us in the first year, less in the second year, and by the fourth year they'd gone completely. One of our predators must have adapted to dealing with them. Perhaps a parasitic wasp. But for conventional growers, they never go away. Now we're told there's a leek miner moth on the way, and everyone's freaking out again. I just think, 'Oh yeah?' It never comes to anything. Not in a well-balanced system.

"Slugs aren't a problem, as we have loads of ground beetles. They eat the slug eggs and the baby slugs. You can never eliminate pests— it's impossible. So you work with them, not against them. The only issues we have is with pigeons and badgers. I've yet to find an insect that controls them."

The tremendous number of flower species in the beds between his crops might help to explain Tolly's success in controlling pests. Several studies show that the greater the diversity of plants in a wildflower bank, the greater the variety and abundance of beneficial insects they harbor.[5, 6, 7] The greater the diversity and abundance of predators, the wider the range of pests they attack.

When he tills the land, Tolly creates a desert. But because every plot he plows is bordered by flowerbeds, the insects have a refuge, and they can colonize the crops as soon as they begin to grow.

"Other farmers ask me 'How do you afford to give up so much land?' I reply 'How can I afford not to?' They see these banks as a production loss. But they're not. They're a gain. They raise our total yield.

"Your biodiversity has to be integral to the whole operation. It's not something you bolt on. Biodiversity is the driver of the farm. If I'm asked 'What do you grow?' I say 'Biodiversity.' The vegetables are a by-product.

"People look at the nettles round the polytunnels and think 'What a mess!' But those nettles keep the aphids alive, and the aphids keep the ladybirds alive. As soon as we need them, they move into our crops. If you look in the polytunnels now, you'll find a ladybird on every leaf."

I later learned that Tolly's decision to sow mostly the seeds of native wild plants in his flower strips made good horticultural sense. Native plants tend to support a wide range of predators and few crop pests, whereas non-native plants do the opposite.[8] For Tolly's system to work, it must be the opposite of the Global Standard Farm: it should be matched to the local ecology.

Now I understood how Tolly had managed to avoid pesticides. But I still didn't understand how he could survive without fertilizer. It seemed to defy everything I had learned about farming and growing. The need for fertilizer—whether in the form of manufactured products or animal manure—drives many of the disasters I discussed in chapter 3. But no one I had spoken to believed we could do without it. Like any farmer, Tolly was exporting nutrients from the land in the fruit and vegetables he was harvesting and selling. So surely, over time, the soil would get poorer?

"You'd think that, wouldn't you? I thought it myself for many years. But I came to see that the amount of minerals we export in food is actually quite low. It turns out you don't need a huge lot of nutrients to grow a crop—you just need them in the right place at the right time. And for that to happen, you need the right soil biology.

"The biggest losses in any farm system are caused by leakage from the soil. Not just soil erosion, but water moving sideways or downwards, taking the nutrients with it. This is a very leaky soil: water drains through it vertically. If you leave it bare, it won't produce anything the following spring. So first you have to make your system watertight—literally watertight. That means using plants to hold onto the nutrients over winter. Keeping the soil covered is the first principle of good management. So you always have something growing: either crops or green manure."

Green manure is a confusing term, as we think of manure as something applied to the land. But it means plants that are grown after a crop has been harvested, not to be eaten but to maintain or enhance the soil's fertility. Tolly's two fields are each divided into seven two-acre blocks, separated by his strips of flowers. Every two acres (a little less than a hectare) represents one-seventh of his rotation. He walked me around them.

The first block looked a bit like his flower banks. It was a thicket of color: blue chicory flowers, crimson clover, yellow melilot and trefoil, mauve *Phacelia* and lucerne, pink sainfoin. This was the green manure, the crop that covered the soil for the first two years of the rotation. Tolly explained that the long roots of these flowers, especially the chicory—which might reach one and a half meters into the ground—draw up nutrients from the subsoil.[9] The members of the pea family—clover, melilot, trefoil, lucerne, and sainfoin—harbor bacteria in nodules on their roots that convert nitrogen in the air into a reactive form the plants can use. Every so often, Tolly runs a mower over the flowers in this strip, chopping the plants into a coarse math that he leaves on the ground. The earthworms pull it down and incorporate it into the soil. "The idea is to let the plants put back at least as much carbon and minerals as we take out.

"Some green manures build fertility, some—the overwintering ones—are there to maintain it, to hold onto the nutrients. There's green manure under the green manure. You can't see it at the moment, but there's white clover under all this. As soon as we cut the bigger plants, it comes into flower, and the bees go crazy.

"Other growers look at what I'm doing and break into a cold sweat. That's because I don't mow these plants until the flowers are over and their seeds have set. The birds eat some of them. The rest fall into the soil. People have nightmares about the seedbanks in their soil. They think weeds are bad and want to eliminate them. But at many points in a crop's life, if the soil is fertile and healthy, weed competition is not a problem at all. They are the crop's understory. They keep the soil covered. Eliminating weeds isn't possible and isn't desirable.

"The green manure is key to the whole system. It ties up nutrients, fixes nitrogen, adds carbon, and enhances the diversity of the soil. The more plant species you sow, the more bacteria and fungi you encourage. Every plant has its own associations. So it's not just the biodiversity above ground that you want to cultivate. It's also the biodiversity below ground. Roots are the glue that holds and builds the soil biology.

"After the green manure here, we'll plant potatoes. As soon as they're finished, we sow the ground with green manure again, to keep the soil covered through the winter. Then cabbages, broccoli, kale, cauliflower, sprouts. This is the point of the rotation when the soil's in really good condition, and the cabbage family likes a lot of fertility. We grow green manures under them too: cereal rye or white clover and trefoil.

"Year five is onions and leeks, with cereal rye, oats, and vetch growing through them, to keep the soil covered and build fertility at the same time as growing the crop. In year six, as the soil fertility begins to drop again, we put in carrots, beetroot, parsnips, and celeriac. Root crops don't need a lot of nutrients. We let some weeds grow through them.

"And this is year seven."

We had arrived at the pumpkin patch.

"You can see how the green-manure seedlings are coming up underneath the squash plants. We let the squashes establish for a couple of

weeks and then sow the green-manure seeds. I pioneered this technique, and again people thought I was crazy, introducing weeds into
my crop. But if you get the timing right, the leaves of the squash shade
out the weeds and keep them small. By late August, as the squashes
start to die off, the green manure is well established, and spreads out
to cover the soil before winter comes. Then you're back to year one."

Tolly, as always, had done the sums. Out of 364 weeks (seven
years), green manure, often growing through or under crops, covers
the soil for 234. Across the whole rotation, there are only a few days
in which his soil is bare.

The wide variety of Tolly's crops is likely to be a further defense
against pests, as no insect that specializes in attacking a particular
vegetable can proliferate for long or spread very far. His method also
hampers invasive weeds: one scientific review suggests that weeds are
reduced by 49 percent in complex rotations like his.[10] Because different crops are sown and harvested at different times, and compete with
wild plants in different ways, weed species cannot easily hop from one
stage of the rotation to the next.[11]

"From the soil's point of view, our growing is not ideal. When we
till it, we damage the soil structure and harm its diversity. Horticulture is much more demanding of the soil than arable farming. We
might take the tractor over it ten or fifteen times a year, each time
doing some damage. So it's a cycle of exploitation and regeneration:
break it, mend it, break it, mend it. We don't plow deep, and we don't
turn over the soil, and we do it only three years out of seven. If I could
do it all without plowing, I would."

Tolly took me down to the potato patch and put a fork under one
of the plants. As he lifted, he exposed its clutch of yellow eggs.

"Look at that. Twelve good ones from one plant. We've got an
incredible crop this year. High temperatures, lots of sunshine. I don't
think we've ever had this much sunshine.

"It's completely the wrong soil for potatoes: stony, alkaline, dry.
But we can get it to work. I mean look at this."

He held out a handful of brown earth, which looked as rich as
chocolate cake. The soil in his palm was so ingrained that it was hard
to see where his hand ended and the earth began.

"Just twelve weeks ago, we turned in the green manure. You can't see it at all now. It's completely broken down to humus. That means we have a lot of biological activity."

Tolly started farming like this before he realized what he was doing. He told me that he didn't do it to change the agricultural system: it was a practical consideration.

"When I came here, I looked round for sources of manure. But there was nothing nearby that was acceptable. Just factory farms. The only possibility was the stable manure from the stud. But I wasn't happy with the medication going into the horses.

"I heard that the Chinese had sustained their land for centuries with green-manure crops. So I began doing something similar, through lack of choice.

"It seemed to work. We weren't losing yields or fertility. In fact our fertility was rising a little, though very slowly. Then twelve years ago I started doing something else, and that made all the difference."

Tolly led me past the open-fronted wooden shop he had built from his own milled wood, where people paid for his produce with an honesty box, to two stacked rows of something brown and fragrant and granular, about 30 meters long and roughly my height.

"Compost?" I asked.

"No. Woodchip."

At the front of one row, closest to the little road that cut between his fields, the chip was fresh and pale. It filled the air with its spicy scent. I worked my fingers into it to feel its heat.

"I wouldn't stick your hand in too far. We've had these heaps up to eighty-three degrees. Sometimes I push potatoes in, and cook them over twenty-four hours."

At the other end of the row, the chip was darker and cooler: Tolly told me the temperature here, caused by the bacteria breaking down the wood, was down to 55°C. I shoved my hand in. It reminded me of the warmth you feel when you take eggs from under a sitting hen.

"This is about ready now. We don't want it completely broken down."

He explained that the chip is left here by the local tree surgeon, who lives on a boat on the river beyond his fields. It all comes from people's gardens, within the five-mile radius in which the surgeon works. If he didn't bring it to Tolly, he might have to pay to dispose of it, so the arrangement suits them both. Tolly turns it with a digger every so often, to ensure it keeps cooking. Then, when it reaches the dark, slightly cooler state of the chip at the end of the row, he spreads it over his green-manure crops, twice every seven years, and leaves the worms to pull it into the earth. This, apart from seed, is the only substance applied to the fields that comes from outside the farm. Altogether, he lays down a depth of just seven millimeters of chip during the rotation: an average of one millimeter per year.

In other words, it's a very light dressing of a material that contains scarcely any nutrients. It's hard to believe that this thin gruel could make much difference. But it's the revolutionary shift that has transformed his business. Tolly has kept meticulous records throughout his thirty-three years on the farm. They show that soon after he started adding the chip, the fertility of his soil soared. In the past five years, his yields had roughly doubled. How?

No one really knows. A team of scientists is studying his system, but the results so far are inconclusive.[12] What they do show, however, is an extraordinary abundance of earthworms. Their initial report notes that cropped soils usually contain between 150 and 350 earthworms per square meter. If there are 400 or more, that's a good sign, as these numbers are associated with healthy soil and high production. Tolly's soil, their sampling discovered, contains "just under 800 earthworms per square meter."

While it's hard to measure the diversity and abundance of microbes and small soil animals,[13] and the extent to which, at any point in the farming year, they lock up or release the minerals in the soil, earthworms can be used as a rough proxy: an indicator of the health of the soil system.

Because of the phenomenally complex nature of soil biology, it could be several years before we grasp exactly how Tolly's method works. But he believes he's adding enough carbon to stimulate the activity of fungi and bacteria, but not so much that it causes them to lock up the nitrogen in the soil, which is what happens if you give microbes more carbon than they need.[14]

"The woodchip isn't fertilizer. It's an inoculant, that stimulates microbes. The carbon in the wood encourages the bacteria and fungi, that bring the soil back to life.

"We aren't feeding the crop, we're feeding the soil. Most farmers have a big hang-up about nutrients, especially nitrogen. They douse their land with nitrogen, and half of it washes off and gets in the rivers. If you get the biology right, the nutrient demand is much less than people imagine.

"I'm maintaining fertility by adjusting the carbon levels. Once you find the right carbon balance, the bacteria and fungi will make nutrients accessible at the right time of year, and make sure the plants get what they need. We should let the plant choose what it needs to meet its own demands, instead of loading it with the minerals we think it wants.

"You smelled that soil when I dug the potatoes? What did you make of it?"

"Well, it was pretty rich."

"It's a particular smell, isn't it? It's the smell of the soil you get in the woods. What I'm trying to do here is to copy the forest. The woody litter from trees rains down slowly, not in one great dump. That's what we're doing, applying the chip very lightly, when the plants are best able to use it."

I realized as he spoke that because this land was forested until it was cleared for farming in the Neolithic, these "farm soils" are actually forest soils, which we happen to be using for agriculture. Perhaps, as Tolly suggests, to make them work we need to restore a forest soil ecology, aligning farming with the original ecosystem.

"The way farming has gone, removing old trees from the landscape, is seriously damaging. Trees are really good for the soil. They bring back what the soil needs. I've made this connection between my two biggest interests: trees and soil. I've always loved wood; now I can combine it with the land. If I didn't have to farm veg to make a living, I'd just grow trees."

Tolly had started planting trees in his wildflower banks—some fruit trees, some forest trees—to help push his soil even closer to that of the surrounding woods. He had also planted a boggy patch at the bottom of his lower field with willow trees, which he would use to produce his own wood. Eventually, from the trees in this copse, in the hedges and between his crops, he hoped to produce much of his own woodchip.

Researchers in Canada have discovered that chip made from the thinnest branches* doesn't have to be composted before it's spread on the soil, as it has a higher ratio of nitrogen to carbon than the chip made from large branches and trunks. So it doesn't prompt bacteria to lock up the nitrogen.[15, 16] If you use chip made from twigs, you need only half as much material, as wood shrinks and loses nutrients when it's composted. Tolly has started using this research to change his system yet again: his experiments never end.

Listening to him explain how he thinks his method works, it seemed to me that, through patient observation, intuition, and experience, he had almost anticipated the emerging Theory of Soil: the way in which bacteria turn carbon into the structures that stabilize it and make air and water and nutrients accessible to the roots of plants.

Tolly's farm was, I realized, a genuinely regenerative, organic system. I'm sorry to say that it stands in contrast to certain other kinds of organic farming, where something seems to have gone wrong.

In the last chapter, I mentioned the leakage of nitrogen from organic farms using animal manure, which, according to one paper, is 37 percent worse than the leakage from conventional farming.[17] After stumbling across this issue, I became slightly obsessed. Though big and troubling, it is seldom discussed. But it should alarm all those who grow and eat organic food. The fundamental principle of organic farming is that it closes the nutrient loop. In other words, instead of leaking minerals, it recycles them, generally between animals and plants. If, in reality, it leaks even more nitrogen than conventional farming does, how does it sustain itself?

Some nitrogen is replaced by planting clover, beans, and other members of the legume family, which have formed a remarkable relationship with bacteria species that turn atmospheric nitrogen into the nitrate minerals that plants and animals need. These plants form nodules on their roots in which the bacteria live and grow, exchanging the nitrates they produce for sugars the plants create. But legumes like clover and beans cannot compensate for losses on this scale. Surely organic soils, in a system that claims to rely on recycling but loses so much, must

---

* They call it ramial woodchip. It comes from branches of less than 7 centimeters in diameter.

gradually become starved of nutrients, and deliver ever-lower yields? I was mystified, until I downloaded the standards published by the bodies that approve organic farms. I read them with growing incredulity.

It turns out that manure used on organic farms doesn't have to be organic. Take the UK's organic standards, for example, which are published by the Soil Association.[18] These allow organic farms to apply animal feces at the rate of 170 kilograms of nitrogen per hectare every year. While the standards urge farmers to "minimize the need for brought-in nutrients," and state that the nutrients they buy should "preferably" come from organic sources, there is no rule enforcing these "preferences": in principle, farmers could buy all their manure from non-organic farms and still qualify as organic. Cow manure contains roughly 6 kilograms of nitrogen per ton, which means that organic farmers could apply up to 28,000 kilograms of manure from conventional farms to every hectare every year. That's quite a lot. In other words, organic farmers can grow their crops with the aid of artificial nitrogen, as long as it has first passed through someone else's animals.

The only restrictions on this supply set by the standards are that the manure must have been fermented or diluted, and must not contain genetically modified ingredients or arise from "factory farming." The Soil Association's definition of factory farming seems quite loose. The manure can come from chickens kept indoors, as long as their weight does not exceed 30 kilograms per square meter. This sounds quite close to factory farming to me, given that the maximum legal stocking rate, unless the farmer can meet extra government requirements, is 33 kilograms.[19] Manure can be bought from indoor pig units, as long as they have straw on the floor. Or from cattle that have "access to pasture for at least part of the year."

Astoundingly, while the standards list maximum concentrations of heavy metals in the household waste that organic farmers might use, they set no limits on the amount of heavy metals in the animal manure they can buy. Nor do they mention persistent organic pollutants in manure or—even more remarkably—antibiotic residues and antibiotic-resistant bacteria. This is particularly surprising, coming from an organization which insists, when treating their own livestock, that organic farmers should use "homeopathic products" in preference to antibiotics.[20] In fact, while there is a general requirement in the standards to identify "unauthorized or prohibited substances" and "reduce the risk of contamination," they set no limits for residues of any kind in the manure that organic farmers can buy from other businesses.

When I asked the Soil Association about this, it pointed out that non-organic manure should be used only "when a clear need can be demonstrated" and that organic farmers "are obliged to maximize the retention and recycling of nutrients within their farming system." Even so, the rules allow them to use conventional farming to plug their nutrient gaps, which to my mind makes a mockery of what organic farming claims to be. Because there are no limits on residues, they could accidentally expose their customers to the very hazards people seek to avoid when they pay extra for organic food. Mainstream organic farming is sometimes lampooned as "all muck and magic." It begins to look like all muck and no magic.

What this shows is not that organic farming is evil, or that organic farmers are trying to rip us off. It shows that, while animal manure might return carbon to the soil, in other respects, contrary to the claims of many practitioners, it is a problematic soil additive. The problem is timing.[21] Crops absorb most of the nitrogen and other minerals they need during a short growth spurt. Before they are sown, they obviously don't need any minerals. After the growth spurt they need far fewer, and after they've been harvested, none at all.

But, while artificial fertilizer often releases its nutrients too quickly, manure releases its nutrients too slowly. If the crop is not to starve, the dung needs to be spread long before the maximum growth spurt. Even then, the plants are unlikely to receive all the nutrients they need to reach their full potential. This is especially true of modern, high-yielding crops, whose growth phase is particularly fast. Long after the plant has matured and been harvested, manure continues to release its minerals. If minerals are delivered when plants aren't growing, they tend to leach away, either into the groundwater or into the rivers.

Some of the nitrogen and other nutrients can be mopped up by sowing temporary "catch crops," which are later plowed back into the soil, but there will always be gaps and disparities, when animal dung is applied, between the demand for minerals and their supply. While older crop varieties, whose growth is slower, might be slightly better matched to the way in which manure releases nutrients, there will still be significant leakage at either end of the growth phase. And, of course, their yields are smaller.

For the first few years after a conventional farm has converted to organic growing using animal manure, crop production will be low, as little free nitrogen is available. As manure is added in later years, yields will rise, but so will the amount of nitrogen that leaches away. One study that modeled this relationship showed that even with massive applications of manure from beef cattle (600 kilograms of nitrogen per hectare per year—more than three times the amount permitted by the Soil Association), it would take twenty-five years before corn grown on that land reached its maximum yields.[22] If any less manure were applied, the crops would never achieve full production, as there would not be a sufficient accumulation of nitrogen in the soil. But because much of this nitrogen continues to be released when plants can't absorb it, the losses rise accordingly. There's a fundamental principle that should apply to all farming: release nutrients when they are needed. Lock them up when they are not.

Those who use animal manure argue that the way they farm is how nature works: animals excrete on the ground, plants suck it up, and the cycle sustains itself indefinitely. But there are few natural systems that look anything like agricultural ones. The vast herds of wild herbivores that Europeans encountered when they first arrived in Africa and the Americas are likely to have been an artifact of the suppression of predators by the people who already lived there. Paleontological evidence suggests that, before humans began competing with them and killing them, large carnivores existed in far greater concentrations than they do in any ecosystem today.[23, 24] Rarely, if ever, would dung have been deposited at agricultural rates.

In countries like mine, most of the places we now farm would have been either forests or wood pastures (a mixture of trees, shrubs, grass, and flowers), whose minerals tend to cycle more slowly and to be better conserved than those used by crops. Natural systems lose some nutrients, but far fewer than farmland does. Their nitrogen is replenished by bacteria and lightning strikes; other minerals by the weathering of rocks.

*

Unlike either conventional or mainstream organic growers, Tolly neither applies nutrients, nor, he believes, loses them. He doesn't need to close the nutrient loop, because he's not opening it in the first place. What he seems to have done, in line with the fundamental principle, is to induce the microbes in the soil to deliver minerals when they are required, and to hold on to them when they are not.* By striking the right balance between carbon and nitrogen in the soil, which plays a major role in the behavior of bacteria and fungi, he appears to have created a self-regulating system.

As he explained to me, his system lets "the plant choose what it needs to meet its own demands." What this might mean at the microscopic scale is that, in a healthy soil, crops can regulate their relationship with the bacteria in the rhizosphere (the plant's external gut), using their chemical signaling to switch microbes on and off as they desire, unlocking exactly the amount of nutrients they need to support their different growth phases.

Tolly's success forces us to consider what fertility means. It is not just about the quantity of nutrients the soil contains. It's also a function of whether or not they're available to plants at the right moments, and safely immobilized when plants don't need them. In other words, fertility is a property of a functioning ecosystem.

Tolly, who, I've discovered, is greatly admired by other organic growers, is widely employed as a consultant. When he looks at someone else's farming system, one of the first questions he asks is how many "ghost acres" it uses. Through his influence, this concept† has come to haunt my thinking. A ghost acre is land in another place, on which a farm depends. The chicken farmers in the Wye Valley harvest ghost acres in Brazil and Argentina: the land from which their feed comes. Farmers who buy their manure are importing fertility from someone else's land. Tolly calculates that if he were a mainstream organic grower, using animal dung, his land would be ghosted by an area between two and three times as big as the one he farms.

He also questions the amount of green material that people might

* The scientific terms for releasing and holding on to nutrients are mineralization and immobilization.
† It was originally formulated by Georg Borgstrom.

import onto their land. "No-dig" systems have become popular among some small growers, who bring in very large amounts of manure or composted green waste or woodchip and pile it onto their soil without tilling. It suppresses weeds and, for a few years, produces massive yields. But Tolly has been called in by several of these growers to explain a sudden collapse in their production.

"You get this build-up of phosphate and potassium to ridiculous levels, and that causes big problems. I call it soil obesity. Overfeeding the soil reduces the activity of fungi and bacteria, at least on some soil types. And you can't keep building up carbon for ever. You have to use it. Otherwise you end up with a peat bog, which doesn't grow food. So these 'no-dig' growers eventually have to do some digging.

"The other issue is where do all these nutrients come from? It's an ethical issue: how big a share should you take? You might call this plant material 'waste,' but it's still fertility that you're taking from somewhere else. When you add fertilizer to the soil, you should ask yourself: Is it needed? Is it affordable? Is it ethical? I've opened a can of worms by drawing attention to this problem. I've pissed people off. But nothing changes if you don't piss people off. I expect you know all about that."

He has not managed to exorcize all his ghost acres. Though he is gradually replacing it, some of Tolly's woodchip still comes from people's gardens. He buys most of the seed he uses, and calculates that growing it requires a little over half a hectare. But in other respects he is self-sufficient, unspooked by other people's land.

I returned several times over the following year, following Tolly around the fields and the garden, or sitting in his tiny house, poring over spreadsheets. He had built it in the shell of his old workshop, cunningly designed to fool the planning authority. "We got away with it for fourteen years, then the fucking council caught up with us, didn't they?"

Ironically, as a farmer, he could have constructed a massive steel barn without an environmental permit, as long as it was designed to house livestock, not people, and was 400 meters from human habitation. He had used the small space beautifully, tucking the tiny bathroom under the stairs, and the kitchen into a nook beside them. All the heating was provided by the wood he grew on the farm, which he sawed himself. Somehow it didn't feel cramped. It felt like an old ship's cabin, and it was no surprise to discover that he had also built a stunning wooden sailing boat in his backyard, *The Naida*, starting with the oak trees thrown down in a great gale: a feat commemorated in a celebrated photo-essay.[25] He didn't get much time to sail it, however: over the year in which I followed him, he had two days off.

As we talked, he would tap away on his calculator. "At school, I was really bad at maths. I got dumped in the dumbos' class. We had to make cardboard models to do trigonometry. But now I love it. There's a lot of mathematics in running a farm." One of Tolly's achievements is to make everything visible. As well as helping to bring the issue of ghost acres to light, he was one of the first farmers in the UK to calculate his carbon footprint. In contrast to the wild assertions made by the cattle ranchers who claim to be saving the planet, his sums are meticulous and unspectacular.

Tolly's rigorous accounts show how difficult it is to balance the carbon budget on a working farm.* While the carbon in waterlogged soils (in salt marshes, peat bogs, under mangroves and seagrass beds, for example) is stable, as long as the habitat is not destroyed or drained, and while trees store it quite reliably, the carbon in farmed soils tends to be less secure, and harder to measure.[26] An apparent gain one year can be lost in another part of the rotation. Small increments or reductions are difficult to assess:[27] the smallest detectable change is about 5 percent.[28] As a result, in most cases at least ten years must elapse before you can measure any gain or loss.[29, 30] Techniques for holding carbon differ from soil to soil.[31] And a working farm, almost by definition, is one that produces greenhouse gases.

"We try to keep the carbon footprint as low as possible. That's what my passion is. All the hard standing round the shop is recycled: it comes from demolished buildings and planings from the road. The shop is mostly reclaimed timber. The joinery and crates come from two old trees on the farm that had to be cut down. We have a fair bit of sequestration, through planting hedgerows and building the organic matter in the soil.

"The biggest effect is how much money we spend. If we buy a new piece of kit, it hits our carbon budget for a long time. Some years, we

---

* He uses an accounting tool called the Farm Carbon Toolkit.

save more carbon than we use. But as soon as you start laying concrete or buying machinery, your impact goes through the roof. Last year I bought a power harrow. It'll take years to clear the carbon debt."

Many of the apparent carbon savings on farms are an artifact of false accounting. When organic material is incorporated into the soil, this raises its carbon content. But if the material has been imported from ghost acres elsewhere—in the form of manure or animal feed or green waste—the farm is robbing Peter to pay Paul: moving carbon from one place to another.[32, 33]

Like all complex systems, soil seeks equilibrium. It tends to stabilize around a carbon-to-nitrogen ratio of 12:1.[34] If you add too much carbon, nitrogen, because of the complicated ways in which microbes interact with minerals,[35] can become less available to plants. If you try to balance the extra carbon by adding more nitrogen, you could release nitrous oxide, undermining any improvements in the farm's climate impact.[36]

Some people claim that we can escape the constraints of this equilibrium by adding biochar to soils. Biochar is fine-grained charcoal, which can be made from forestry trimmings, green waste, sewage, and manure. It appears to remain stable once it has been buried, and it can improve the texture and fertility of some soils.[37]

Given the high excitement surrounding this magic powder a few years ago, you might wonder why biochar hasn't saved the planet yet. But it's not hard to see the problem. At the time of writing, the cheapest source I can find in my country costs £1,300 a ton.[38] Compare this to agricultural lime, which is sometimes described by farmers as prohibitively expensive, and is generally applied more sparsely than the recommendations for biochar. My enquiries suggest that the average cost for a full load of lime delivered across 80 kilometers is roughly £50 a ton.

Not only is biochar comically expensive as a soil amendment, but it's also an extremely dear means—by comparison to protecting forests or peat bogs for example—of saving carbon. There might be scope for reducing the price a little, but the technologies involved and the quantities of raw materials required don't lend themselves to the cost curves enjoyed by digital industries. The only cheap way of obtaining biochar is to make it yourself in a glorified dustbin.[39] But unless you get the burn exactly right, you're likely to cancel any possible savings by releasing methane, nitrous oxide, and black carbon, while the toxic smoke shaves years off your life.[40]

The most secure and effective means of removing carbon from the atmosphere is to reduce the amount of land we need for farming, and rewild the land we spare, restoring wetlands[41] and forests.[42] This is one of the reasons why it's important for farms to be like Tolly's: highly productive, even as they provide passages and habitats for wildlife.

"I've never bet on anything in my life," Tolly told me. "Yet every year I wager hundreds of thousands of pounds on this business. I'm always gambling with the weather, and I have to accept the cards it deals me. A lot can go wrong. But by doing loads of small plantings, I reduce the risks.

"I get up at five to do the maths. I have to make constant decisions about things I don't want to think about. We have to keep reinvesting in the business just to stay in the same place. I'm not in this for the money. But we do have to pay our way."

Tolly told me that he qualifies for half the state pension, as he hadn't made his full contributions, "and that effectively doubles my income." This suggests his income from the farm is about £70 a week: a pittance, though he also earns money from consulting. "I have almost no costs. I live out of the business. I get paid much less than my staff, but I get more benefits. We pay soft rents on the houses here as Julian [Sir Julian Rose] believes there should be people working on the land. Commercial rents would be more than I earned every month. You can't pay commercial rents from agricultural incomes."

Because farm subsidies are paid by the hectare, and Tolly works only seven, they account for just 0.3 percent of his farm income: a stark contrast to the big arable and livestock farms, many of which are kept afloat by state support. He sells 60 percent of his produce through his subscription program, which delivers boxes of vegetables and fruit to Pangbourne (on the other side of the river), Reading, and Oxford. The rest is sold through his farm shop. "My ideal would be just to feed people in a two-mile radius. There are one thousand homes in Pangbourne and Whitchurch. We could feed half of them on basic veg."

He won't sell to restaurants. "You have to be a masochist to deal direct with them. Temperamental chefs, high staff turnover. They want everything yesterday, want it all washed, want it all tiny. I can't be arsed to pamper to their whims. And they throw a lot of their food out: it's far more wasteful than home cooking. I don't grow veg to be dumped in the bin."

In most years, demand and supply are tragically mismatched. Just as production on the farm starts to boom, in mid-July, half his customers go on holiday. The best thing that ever happened to his business was COVID-19: when the shelves emptied at the beginning of the pandemic, demand for his produce rocketed. During the UK's multiple lockdowns, he had a captive market. Subscribers to his delivery program doubled, and he had to ration the amount that anyone could buy from the shop, to ensure everyone got something. "It put us in the tax bracket, which has never happened before."

Agricultural science has devoted a great deal of attention to soil chemistry. But the more we understand, the more important the biology appears to be. Developing an advanced science of the soil is, I think, part of the solution to the dilemma we currently face: the need to produce more food with less farming. This new agronomy would draw on emerging knowledge about the rhizosphere and the developing Theory of Soil, to devise specific organic treatments, precisely tuned to the world's many soil ecologies.

If we can discover how to mediate and enhance the relationship between crop plants and bacteria and fungi, as Tolly has done, in a wide range of soils and climates, it might be possible greatly to reduce the need for both artificial fertilizer and manure, while raising yields. The new science would deliver, in other words, a Greener Revolution. Its precision could help reverse the growth of the Global Standard Farm, as techniques and materials would everywhere need to be adapted to local ecologies.

Small farmers around the world are seeking such solutions, and have come together to build a global agroecology movement.[43] But they lack the government support and funding that Big Farmer has enjoyed.[44] One paper found that the UK, while spending £6 billion of foreign aid across seven years on conventional farming projects, provided no funds at all for projects whose main focus was the development or promotion of agroecology.[45] In fact, not a penny was spent on organic farming of any kind: it was all poured into the sort of agriculture the private sector already promotes. This, the researchers found, was typical of the aid disbursements by rich nations.

The full development of a new agronomy would require many billions of dollars. It would doubtless cost less, however, than exploring the surface of Mars, a project inspired, in part, by the preposterous ambition of terraforming that planet to make it habitable to humans. Ensuring that Earth—whose happy denizens enjoy the luxuries of oxygen, radiation-screening, atmospheric pressure, and one g of gravity—remains habitable seems to me a more urgent and plausible ambition.

In other words, we need an Earth Rover Program,* to explore thoroughly the surface of our own planet, whose intimacies are as little known to us as the surface of Mars. Scientists working under this program would seek to map the world's agricultural soils at much finer resolution than has yet been achieved,[46] understand their varied ecologies,[47, 48] and work out the means by which large amounts of food can be grown with the lowest possible supplements and the smallest possible impacts. This science is likely to be of little interest to agrochemical companies, which could explain why it has received so little money and attention.[49, 50] Government funding for research often follows commercial agendas. I feel it should do the opposite, exploring techniques that cannot be monopolized by corporations and used by them to dominate the global food system.

No technology is entirely safe from capital's tendency to concentrate and centralize. Even a method like Tolly's could go bad. For example, farmers might find it cheaper or more convenient to import woodchip than to grow their own or to use local waste. Tolly uses less woodchip in seven years than a single chicken shed does—for heating—in one. But if it suddenly became a popular inoculant, the demand for mashed-up tree could contribute to the unbearable pressure on the world's forests. It seems to me that the standards for "stockfree organic farming" that Tolly has helped to develop should

---

* This should not be confused with a brand of agricultural robot, sold by a company called Earth Rover.

include strict limits on the amount of plant materials a farm can import, which currently don't exist.[51]

There's a tendency among well-meaning companies and their customers to treat green materials as infinite. Bioethanol and biodiesel can replace the transport fuels we use. Biokerosene can take the guilt out of flying. Heating oil and coal can be replaced with wood. Disposable cups can be made from corn starch, and plastic bags out of potatoes. But everything is finite. Everything we use, we take from someone or something else.

All these substitutions have proved to be disastrous. Bioethanol and biodiesel have launched a deadly competition between cars and people, raising food prices and spreading hunger,[52] while helping to catalyze the destruction of tropical forests, as they're converted to oil palm and other industrial crops.[53, 54] Biokerosene, if it's widely used, will compound these catastrophes. Corn and potato starch are likely—when the use of pesticides, fertilizers, irrigation water, and diesel has been fully accounted for, as well as the soil erosion for which these two crops are notorious—to prove more damaging than fossil plastic.[55] Unless we are to complete the destruction of the world's forests and other wild places, we need to minimize our use of plant materials. In other words, we should stop treating them as a guilt-free alternative to our excessive consumption of hydrocarbons.

Even in a system like Tolly's, we should use plant materials as sparingly as possible. Tolly estimates that if 20 percent of the land now used for farming were used to grow trees, that could provide enough material to sustain the farm carbon cycle and raise soil fertility. This sounds like a lot. But if the 20 percent helps to achieve a transition out of livestock farming, it will save far more land than it uses. And growing these trees, especially if a wide variety of native species were used, could create important corridors and habitats for wildlife.

On every visit, I bought some vegetables from Tolly's farm shop. They were always fresh and firm. Even in winter, the variety was wide enough to create hundreds of possible meals. In early summer, I bought broad beans, new potatoes, and spring onions. The bean pods were still lined with deep white down. Evicting the small seeds that nestled within felt almost cruel. I boiled them for less than a minute, until the skins puffed out, then sprinkled them with a little lemon juice. They were sweet and fleshy. I cooked the potatoes until they were just soft enough to fall off a knife, and tossed them in vegan butter.

As I ate Tolly's produce throughout the year, I marveled at its depth of flavor. He was producing quality as well as quantity. In late summer, I bought three varieties of tomatoes, which I ate roughly chopped with a little salt, olive oil, and basil that I grow on the windowsill. Sharp and sweet, I could have told the tomatoes apart blindfold, as their flavors were so distinctive.

In the autumn, I bought an uchiki kuri squash: the bright orange Japanese variety that looks, with its delicate neck and perfect curves, as if it has been thrown on a potter's wheel. I roasted it, skin on, until it blistered and browned, adding a head of garlic cloves halfway through. When it had cooled a little, I blended it and the shucked cloves with tahini, lemon juice, and pepper. The result was a smooth, sweet paste, speckled with toffee-squash skin and caramelized garlic.

In the winter, I bought carrots, kale, leeks, and onions and cooked a ribollita, based on Hugh Fearnley-Whittingstall's recipe. A ribollita is an Italian bean and vegetable stew. This makes it sound simple and crude, but it's a delicate meal that's easy to spoil. As I cooked it, the kitchen filled with the sharp aroma of herb oils. I scrubbed the big knobbly carrots, and tore the leathery kale from its stems. I served the ribollita in bowls, sprinkled with pieces of toast and olive oil. It was thick and glistening, but tasted light, fragrant, and buttery. Even in this dense stew, the flavor of every vegetable stood out.

I last visited Tolly on a bright day in late January. Low sunlight illuminated the columns of figures on the paper strewn across his kitchen table. On the flooded horse pastures below his house, there was a thin skein of ice, around which black-headed gulls whirled and yowled.

We were talking about the prospects of his system being widely adopted.

"I've had three ministers of agriculture from different countries visit this farm."

"Including this country?"

"No, of course not. Why would they?"

Tolly's phone rang. He spoke for a moment, then stood up.

"The trailer's got a flat tire, so we can't move the veg.

"This is the reality. You go to a conference and come back with loads of great ideas. Then you spend half the day mending a puncture."

I followed him to the lower field, where the stricken trailer was parked. Fooled by the early sunshine, a robin sang in an old ash tree.

"It's a very nice day. I'm just wondering what we'll have to do to pay for it."

He deftly braced the jack under the trailer, fitted the wrench to the nuts, lengthened the handle with a piece of scaffolding pipe, and loosened each one with a clang that echoed across the fields and startled the dunnocks in the hedge.

"I've had this trailer for thirty-five years. I've built new sides and floor, fitted a new tipping ram, put new wheels on. There's not much left of the original. I get the maximum life out of everything. I've had my tractor for thirty-three years, and it was seventeen when I bought it. A whole generation is growing up that knows nothing about machines. That is a worry."

He spun out the nuts and jacked up the trailer.

"Something like this can completely fuck your day up. There's a whole trailer full of veg that has to be moved to the shop."

He flipped the wheel into the back of his van, and we drove to a little industrial estate that seemed to have been left behind when everything else changed. Outside his workshop, a big man with a mild pink face sat on a chair in the sunshine, wearing a boiler suit and an old suede cap like a shark's nose, its brim shiny with handling. Radio 2 was playing in the shop.

"Not got much on then?" Tolly asked as we got out of the van.

"Well, it's better than sitting in there."

They talked for a while about the gaffer's horse and carriage, which he hires out for weddings.

"You ought to do funerals. You'd get a lot of trade that way."

"It's a dying trade, that's what they say. I'd have to do weekdays, because there's no telling when people will die. Weddings is all at weekends."

The gaffer pushed his cap up, so the sun fell full on his face.

"I'm glad you came. I've been wanting to ask you about those giant black slugs. You don't normally get them in the winter, do you?"

"No," said Tolly, "it's the unseasonal weather."

"Some of them are this long. What I want to know is, what are they for?"

"What are they for? Oh they're very useful, they clean everything up." Even as Tolly answered, the gaffer seemed to have moved on.

"What's been puzzling me is all those orange and black caterpillars you see on the ragwort. How do they get there?"

Tolly nods at me. "He'll tell you."

"They're cinnabar moth larvae." I felt, beside these practical men, like the earnest, annoying boy I was, raising my hand while my classmates rolled their eyes.

"Moths, is it? You're saying that's where they come from?"

"Yes."

"So these moths, are they stripey too?"

"Sort of."

"Orange and black are they?"

"No. Pink and gray."

"Ah. So it's the moths that bring the caterpillars?"

"Yes."

"So what about wasps? What are they for?"

On the way back, Tolly said, "I once had a fantasy about running the farm on horses. I looked into it, did a study. I worked out that we currently use 4.5 liters of diesel a year for each family we sell veg to. That covers the tractor movements, the deliveries, and everything else. Farming does use a lot of energy. What if we used horses instead?

"We would need two of them, which means using 4 acres of our 17, to feed them with oats, grass, and hay. To plow an acre would take one day. You would walk about 16 miles. Ten acres would take ten days. It takes me one in the tractor. If it rained, you'd have to stop. Delivering the veg would take two days, instead of four hours. On the other hand, horses would be much better for the land. They don't compact it like tractors do.

"Two horses means more people, to look after them and the grazing. We would have to charge three and a half times more for the vegetables.

"There'd be a similar problem if we used livestock to manure the ground. I could keep either one and a half cows here, or three sheep. That's the issue with livestock. They use a hell of a lot of land."

This was one of the many things I had come to appreciate about Tolly: his constant investigations and conjectures. He is a natural scientist, endlessly curious, questioning everything, challenging his own ideas, recording all his data. "What we're doing looks like it might be sustainable," he once told me. "But we won't know until we've done it for a hundred years."

I doubt I'm the only one in this field exasperated by people who claim that their variety of farming is the best of all possible systems in the best of all possible worlds. I've seen farmers who set out with high ideals gradually becoming hucksters, overlooking the drawbacks of their practice, exaggerating the advantages, subordinating their intellects to their interests. The people I'm drawn to are those, like Tolly, with a capacity for self-correction, who recognize the flaws in what they do and seek to address them. One of the most difficult life balances is summoning enough self-confidence to carry on, however many times you're knocked back, but sustaining enough self-doubt to listen to criticism and change course when necessary.

Tolly's method, as he readily acknowledges, is not perfect. Like most vegetable growers, he uses plenty of irrigation water. He needs plastic covers for his polytunnels, and bags and punnets to deliver his vegetables and fruit.[56] He would love to generate his own electricity, and replace his diesel-powered machines with electric versions, but he can't afford it.

He has worked hard to make his system as easy as possible to copy, using simple measures like earthworm counts to provide rough metrics for success. Other growers are beginning to adopt his techniques. But while his designs might be replicable, he isn't. As he told me, "It's not just a full-time job. It's a full-time life." How many people are prepared to live it?

While he jacked the trailer higher and fitted the wheel, I asked him whether, as we cannot survive only on vegetables and fruit, he thought his methods could be transferred to arable farming.

"Technically, yes, you could do it. You would have to design a rotation of three or four crops, using beans, hemp as a filler, maybe incorporating vegetables grown at scale. You could integrate arable crops with trees, using alley cropping.* The yield would probably be lower than from conventional cropping, but as livestock wouldn't be needed, there would be much less waste of land. A few people are experimenting with it already. It would completely change how agriculture works."

He tightened the last nut.

"The issue is finance. Can you make money from it? Arable margins are extremely tight, which is why farmers crop their land 100 percent, all the time. Adopting this system means one third of their farm would not be growing food. If an arable farmer took one third of their land out of production, they'd go bankrupt. That's the problem. The price of cereal is so low.

"OK, that's job done. We'll go and find a bit of grub."

* Planting rows of trees and growing the annual crops between them.

5

The Number of the Feast

The day after our *Rivercide* documentary was broadcast live from the banks of the Wye,[1] I was walking through Hay, one of the small market towns in the valley. I was accosted by a smartly dressed woman of about seventy.

"I saw your film last night," she announced in a voice that could have carried across the border. "I agreed with most of it. But *why* didn't you mention the real solution?"

"The real solution?"

"Yes. That everyone should spend 30 percent of their income on food."

"But—"

"That's the issue, isn't it? Food is too cheap. That's why it's all gone wrong. But you didn't say it."

"Well—"

"So it was a waste of time. Well-meaning, but a waste of time."

She marched away before I had a chance to answer.

This is a claim I've often heard: food is too cheap. In some ways it's true. Food is too cheap to provide sufficient income for small farmers, among whom number many of the world's hungriest people. It is too cheap, as Tolly pointed out, to enable large farmers to farm well. It is too cheap to reflect the damage it inflicts on the living world, the unpaid costs that economists call "externalities." But for the poor, the problem is not that food is too cheap, but that good food is too expensive. Never mind 30 percent; 3 billion of the world's people couldn't buy a healthy diet even if they spent 63 percent of their income on food.[2]

Before moving on to the critical question of how our staple

crops—mostly cereals, oilseeds, and other grains—might be grown, I think it's essential to explore the needs they have to meet. Sometimes it seems to me that the economics of food production and consumption are as poorly understood as the biology. Because the agricultural crisis is a health and social crisis as well as an environmental one, we should be constantly aware of the danger of exacerbating one problem while solving another.

Sometimes environmental protection and food justice align; sometimes they clash. To address the crucial dilemma this book explores—how to feed everyone without destroying our life-support systems—we need, as well as we are able, to resolve these issues. Food, while produced within environmental limits, must be healthy and affordable. This is a massive challenge. While it may be true that rich people should spend more of their income on food, the claim that everyone should do so seems ignorant and callous.

I live on a street of the middling sort, on the border of a large housing estate. It was built to accommodate workers in a car plant that's now mostly automated. On the outskirts of one of the UK's richest cities, this project has one of the country's highest rates of child poverty. Half a mile from our home, the estate wraps around an ancient village, incorporating it into the city. Once inhabited by professors, its lovely limestone and red-brick mansions are now populated by bankers, tech entrepreneurs, and other mysterious millionaires.

If you stand on the edge of the village on a Friday afternoon and gaze across the recreation ground that separates it from the housing project, you might see a queue of people arriving at the glass door of the community center. Each of them spends a couple of minutes there, fills their bags, then walks away. It's the food bank. Crossing the recreation ground is like passing through a portal between two worlds.

During one of my visits, I asked the customers what they thought of the claim that food is too cheap. Their responses ranged from mild puzzlement—as if they couldn't have heard me right—to outright incredulity. Some responded pithily.

"If food was too cheap, I wouldn't be here."

"The place I'm at right now, if it's not free, I can't afford it."

"There's no such thing as cheap these days."

"They're taking the mick."

Others sighed or shook their heads. As I spoke to them from within the community center, I kept glimpsing over their shoulders, on the other side of the recreation ground, a row of £3 million homes.

My journey to the food bank began on Tolly's farm. He told me that when he has vegetables he can't sell—generally potatoes and onions at the end of the season—he gives them to a charity that collects and distributes surplus food. I decided to follow them, to discover why, in nations as rich as this, so many go hungry, and what needs to change. I tracked Tolly's vegetables first to a warehouse on the outskirts of Didcot, a bland commuter town in Oxfordshire. It's run by school leavers employed by a local charity (Sofea),[3] and handles the goods gathered by Britain's biggest redistributor of food, an organization called FareShare.[4]

It's a small warehouse but a big logistical challenge. Every day food that would otherwise have gone to waste arrives from dozens of businesses. Sometimes it comes by the pallet load, sometimes in a carrier bag. As it enters, it's listed in the inventory, then stacked and mapped so that it can be quickly found. The updated stocklist is then sent to more than 100 charities, and they submit their orders. As some of this surplus food is approaching its sell-by date, it must be ordered, docketed, picked, and dispatched almost immediately.

On steel shelves rising to the ceiling sits the "ambient food": products that last a long time and don't need to be chilled. Surplus cans and packages are often hard to obtain, as shops and manufacturers needn't rush to clear them. When I visited, the shelves were full, albeit with an idiosyncratic range: canned mushrooms, stock cubes, noodles, tuna, chili sauce, cans of readymade latte, doughnut mix, breakfast cereal. At the beginning of the pandemic, when the shops emptied, spare ambient food was almost impossible to find, so FareShare had to press the government for a grant to buy it. The last of that stock—canned tomatoes, baked beans, chickpeas, and sweet corn—was now running down, and there was a slight air of panic about how it might be replaced.

The most frantic activity surrounded the fresh food. This tends to arrive in smaller and more varied batches than the ambient stock, and it has to be shifted faster. There were sacks of gnarly parsnips bigger than the supermarkets had specified, carrots that weren't straight enough, mauve potatoes whose baroque shapes had offended quality control. Onions had come from the packing plant in huge net sacks: "These are the ones that dropped off the production line," Adèle from FareShare told me. "They do an extra sweep for us and bag them up."

There was a builders' bag full of pears, boxes of cucumbers, apples, peppers, watermelons, lemons, sweet potatoes, kiwifruit, kohlrabi, a crate of Polish bread, a few hundred eggs. Half a pallet of flowers had arrived in cellophane bouquets. "They'll send them out with the food. Why not?"

A young man was unpacking an entire trolley-load of powdered coconut milk.

"I don't know how this got here," Adèle said. "There's only so much powdered coconut milk you can shift."

In the cavernous industrial chillers were crates of radishes, cabbage, broccoli, mushrooms, cauliflower, French beans.

"These beans were probably too long for the packing plant. Packing machines are very particular about what they can use. If the veg is too big it splits the bags. The broccoli's a bit overgrown. Good to eat, but doesn't fit the specs."

A walk-in freezer was filled with sausages and cuts of meat, fish whose eyes were glazed with icy cataracts, veggie burgers, frozen pasta, cakes, and pies.

"We never have enough fruit or enough meat."

This is a big and common issue for food banks. It's the fresh produce that people on low incomes struggle most to afford. But it is also the fresh produce that food banks find hardest to supply. This is partly because of its short shelf life, and partly, as I would soon discover, because of difficulties in extracting it from the food chain.

The supervisor in charge of collecting and dispatching the orders was a tiny, slight woman of twenty-one, her face almost entirely covered by her mask. She had left school without passing any exams. Now Soph runs a team of six.

"Before I started here, I wasn't actually doing anything, because I couldn't get an apprenticeship or any sort of education because of my grades from school. I was a really socially awkward person, like if I was sat in a group, I'd be the one that didn't say anything, and everyone else would just be talking.

"It all changed for me when I got the opportunity to actually do some work. Then I started to see you can be accepted by people that you wouldn't think would accept you, and given a chance."

In the packing area, Soph had met the young man who became her partner, and they had recently put down a deposit on a house.

FareShare's figures show that a ton of food, averaged across all lines, would cost around £1,500 to buy. It can find and distribute the equivalent in surplus food for £210.[5] It performs crucial work. But it knows that reducing food waste, while essential, is one of the many solutions to our agricultural crisis whose potential has been greatly exaggerated.

The charity told me that while the food industry in the UK discards around 2 million tons a year, the standard estimate is that only 250,000 tons can be rescued.[6] FareShare challenges this figure—it believes that, with a little creativity, hitherto inaccessible parts of the food chain could be induced to surrender their surpluses. But some businesses are more persuadable than others.

Twisting the arms of supermarkets is relatively easy, as they worry about their public image. The food they handle can be shifted straight to warehouses like Sofea's because it has already been packaged for final use. Much of the time, it seems, the supermarkets aren't giving their own produce away, but someone else's. They enjoy such exploitative relationships with their suppliers that if they don't sell the produce they've commissioned, the suppliers don't get paid. So they tend to over-order.[7, 8] Then they pose as heroes when they give other people's goods to charity.

But the processors and packhouses further up the food chain have no direct relationship with shoppers, and little to lose in terms of corporate reputation if they dump much of the food they handle. Sometimes there is a payoff between labor efficiency and material efficiency: it can cost a company more to prevent food from going to waste than it would save by rescuing it. A lot of food is lost long before it takes a form that can be easily packaged or consumed.

FareShare has made some clever interventions. For example, it discovered that when sausage factories switch from making one kind of sausage to another (Lincolnshire to Cumberland, for example), the sausage meat stranded in the factory's pipes at the time of the switchover is pumped out and goes to waste. FareShare persuaded one factory to turn this meat into generic sausages and give them to the charity. It found that by paying the extra costs incurred by growers of fruit and veg, it could prompt them to harvest, pack, and deliver food that would otherwise have been plowed into the soil.

But much of the time, there's no way in. Fruit and vegetables are unusual in that they are more or less final products by the time they leave the farm. The same does not apply to grain or livestock: the charity is not equipped to accept a herd of surplus pigs. In the rich world, much of our food waste occurs at the other end of the chain:[9] dumped half-eaten or uneaten by people whose eyes are bigger than their stomachs.[10, 11] It cannot be redistributed, and this loss can be reduced only through moral suasion, which is notoriously difficult.

So when people claim that, because roughly one-third of the world's food is wasted,[12] salvaging it could feed all those who go hungry today, while saving vast tracts of farmland and much of the fertilizer, pesticide, and water that farmers use, don't believe them. The majority of that food is not recoverable.[13, 14]

There's a related mistake that even scientists sometimes make. Some calculations assume that any reduction in environmental impacts caused by reducing food waste is a net saving.[15] But in poorer nations, where a great deal of food is lost as a result of slow and unreliable transport and spoiling by high temperatures, rot, pests, or bruising, the solutions, such as more and better roads, more refrigeration,[16] and more packaging can accidentally reverse some or all of the environmental savings.

For example, while improving roads in existing agricultural areas might concentrate production in those places, expanding road networks is often the major driver of habitat destruction. New or better roads can trigger a goldrush of people seeking farmland in areas that were previously hard to exploit. In the Amazon and Congo Basins, the process blandly described as "infrastructure improvement" is ripping the rainforests apart.[17] Scientists describe deforestation as highly "spatially contagious": in other words, any expansion of the agricultural frontier triggers further expansion, largely through road-building.[18] There's a tight relationship between proximity to a road and the number of forest fires.[19] There's a similar relationship between roads and the hunting of wild animals.[20] The assumption that better infrastructure translates into reduced environmental impacts is unsafe.

Any savings we might achieve by reducing food loss are small by comparison to the savings we could make by changing our diets. One paper compared the reduction of greenhouse gases through halving food waste to the reduction achieved by switching to a plant-based diet.[21] The difference was enormous: 5 percent versus 80 percent. Yet it seems to me that more effort has been spent on trying to persuade people to eat what they buy than to change what they eat. Both forms of persuasion are weak instruments. But, given that people's readiness to listen to exhortation is limited, perhaps we should focus on the kind that could make the biggest difference.

None of this is to disparage the service that FareShare and its partners provide. The charity sees food redistribution not only as essential in its own right, but also as a means to other ends: bringing people together, building community, helping to meet a range of unmet needs. When I asked for good examples of the projects it works with, Fare-Share named the food bank and youth club half a mile from my home. They turned out to be run by someone I knew, who had worked with my partner when she was in charge of community development on the estate. A 60-mile journey, through Tolly's farm and a warehouse in Didcot, took me almost to my doorstep.

Fran Gardner is a very young- and fit-looking sixty-nine. She has brown skin and short gray hair, big sad eyes, a fine aquiline profile, and strong knuckly hands. She too had come here the long way round, after an extraordinary life elsewhere, which, among many other lessons, had taught her the unparalleled social power of food.

"I married young and divorced a few years later. Then I just wanted to get out. I flew to New York and joined a minibus tour around the U.S. I was the only one who could cook, so I fed them for six weeks on four gas rings. When we got to Los Angeles, the driver offered me a job. I worked on a Trailways bus, cooking for thirty-nine people.

When I returned to the UK, I was hired as cook on a ninety-five-foot yacht, which was about to sail to Miami."

She had never sailed before, and began throwing up as soon as they hit the English Channel. Thus began a series of remarkable adventures. Thanks to an appalling blunder by the captain, the boat ran out of fresh water a day out of the Cape Verde islands. Stowaways emerged midway across the Atlantic and triggered a massive row when half the crew wanted to throw them overboard. A horrifying gale smashed the boat's fittings and forced them to land in Cuba. There they were boarded by soldiers and held for ten days while the government demanded a ransom.

But Fran is a tough and determined woman. She stayed with the boat, working the Miami to Bahamas run for most of the year. When she returned to Britain, she rented a cottage on the Sussex Downs. She found she wanted only to work outdoors so, through sheer persistence and hard labor, she managed to become first an apprentice keeper on the local pheasant shoot, then full gamekeeper. At the time, she was the only female keeper in Britain. Her description of what the job involved testifies to one of the many aspects of rural life we prefer to overlook, in constructing our bucolic reveries.

"The killing was relentless. Every day I had to kill: foxes, stoats, weasels, rabbits, pigeons, everything, all the time. I had to sit and wait to see where fox cubs had been born and then lay down cyanide. I saw things that horrify me to this day. The keeper I worked with was old-school. He set pole traps on the posts round the pheasant enclosure, to catch the owls that came down to eat the young birds. They'd be caught by the legs and would be left hanging and fluttering there all night until he came to kill them in the morning. When I challenged him, he said 'They're cruel to my birds, so I can be cruel to them.'"

The killing sickened her, and she was never accepted by the men, however hard she worked. So when the shoot closed, she decided she'd had enough. A friend introduced her to computers, and she became fascinated by IT. "That's the way it's always been for me. I get hooked on something and give it everything." She taught IT at a college, then was taken on by a small charity. The chief executive taught her to raise funds. "So that became my next obsession: finding money. I became pretty good at it."

After a few more career shifts, she was offered a job on the Rose Hill estate in Oxford, working for a housing association, mostly as a fundraiser. One of the city councilors pointed out that there was nothing for primary-school kids on the estate after school hours. "So I raised the money and set up a junior youth club in 2010.

"It was very popular: after five years we had a hundred and fifty kids in the club. The level of need was unbelievable. Many of them were hungry, and you could see the way it affected their behavior. If they haven't eaten, they're going to feel rubbish, and there will be issues. So as well as doing sports and games and music and crafts, we gave them a hot meal, perhaps pasta with homemade sauce, or pizza, as well as fruit and veg. It was the only dinner some of them would have.

"Now it's the first thing we do with any group of children—offer them food. Food is the way in. Some of the older boys only come here for the meal, eat it, then go. But even in that short time you can talk to them and find out what's going on. That's the power of food. It creates links that wouldn't otherwise exist."

I remembered the etymology of the word "companion." It derives from the Latin *com panis*: with bread.

"Once we've won people's trust, we start to learn about their lives. We notice kids who're obviously hungry, and stuff their pockets with food to take home. We look out for bruising, for signs of neglect, for children whose clothes are always dirty and never changed, whose hair hasn't been washed or brushed for weeks. We might notice that a child is being picked up by an 'uncle' who they're terrified of. One girl told us a friend of hers was about to run away to London with a middle-aged man. If it hadn't been for the meals we provide, no one would have known. Young people in difficulty are more likely to talk to us than go to the police. Because there's trust here."

The youth club teaches children to cook and introduces them to new kinds of food. It has taken on a plot at the local allotments, where the children grow their own vegetables. "The kids love it. Growing their own food, even just a little, gives them a massive sense of achievement.

"One day, when I was unloading the food from Didcot, I got chatting to two women who worked in the kitchen of the primary school.

They asked me what all the food was. When I told them, one of them said 'I wish we could afford food like that.' It turned out that both of them were in full-time work, and so were their partners, but they were really struggling. They were on the minimum wage, and couldn't afford to feed their kids well. They told me how many other people were in the same boat. The rents round here are astronomical. Often 60 percent of people's income.

"So those two women set up the food bank and ran it for us for the first four years. It's got bigger and bigger.

"We now have seventy people a week using it. We can't solve the fundamental problems here. We're just doing what we can, on the budget we have. We're sticking a thumb in the dyke, but we can see the wall is crumbling."

She told me that one of the main reasons why people have no choice but to use the food bank is the cruel and unnecessary rule that, if you fall on hard times, you won't receive financial help from the government until five weeks have passed.[22]

"People have literally been facing starvation. One woman has a six-week-old baby and no food in the cupboards. But she has to wait five weeks before she gets any help. Another woman went for three days without gas or electricity. There are people with severe learning difficulties or mental-health conditions who're just left to fend for themselves, without a penny.

"If I thought about it too hard, I reckon I would fall apart. But the one thing I can do is write those funding applications with all the skill and passion I've got. You need to put the passion in to get the money."

As she spoke, the first customers started arriving. Most of them were prepared to talk to me, though only a few were happy for me to use their names, and no one wanted to hang around for long at the food-bank door. But what I heard then and on my later visits seemed to confirm a familiar story. It doesn't much matter where you land on the wheel of misfortune: before long you will complete the cycle. You might fall into destitution because you lose your job, or because your relationship collapses, or because you're evicted from your home, or you have a mental or physical health crisis, or your debts become unpayable.[23] But eventually, as you hit the skids, you might experience all these disasters. And hunger, which many people in prosperous countries find almost unimaginable, starts to walk in your tracks.

One woman told me how her life had fallen apart.

"I was the manager of a fashion department. I worked hard all my life, then I lost my job and everything went to shit. Now I've got anxiety and depression and I've got a, a drug problem. I don't have any family or friends anymore. So to reach this stage in my life and feel I'm in the gutter, that's really hard.

"Why do I come here? Because everything else gets priority. You have to pay the bills, don't you, or they come after you. You've got to keep a roof over your head. So if you can't afford everything, it's the food you stop buying. Coming here, that's one less problem to deal with. Then you can pay the bills.

"It's a lifeline. This is one place where no one's looking down on me. I don't have to feel embarrassed coming here. Everywhere else, I'm made to feel shit."

A man began to cry as he told me how his crisis began.

"I'm a full-time carer for my disabled daughter. I broke up with my partner, so I had to leave our home. I ended up in temporary housing. Since then, I've been a hell of a lot worse off. I had a really tough time last month, paying off arrears. I literally survived off the food they gave me here. Everything has changed for me. It's really difficult."

He told me he has uncontrolled type 1 diabetes. He is supposed to eat every two hours, otherwise he suffers from severe hypoglycemic attacks. "I've had to go to hospital three times in the past eighteen months with hypos. The food here helps me fill in those gaps when I've got nothing. I don't know what I'd do without it, to be honest.

"Things would be a lot, lot worse, that's all I can say. I'm trying to get back on my feet and this is really helping. I can't praise these people enough."

A woman in her forties told me she had "always coped, however hard it gets. But I couldn't cope without this now.

"I got attacked, and they damaged this eye. I had a few operations to try to save it, but they didn't work. So now they've given me a glass one. I've been in quite a bad way. You know, struggling to hold it all together. I'm starting from scratch again. Picking myself up."

A man told me that he takes all the work he can get through agencies, but he's a single parent, and doesn't earn enough to support him and his daughter.

"When I first started using the food bank, it was very humbling. There was a bit of a stigma attached to it. People I know can be quite judgmental, they treated it like it was tantamount to begging. At first, it hurt me, but not so much now: I've got more used to it. It's needs must, isn't it?

"Would we go hungry if it wasn't for the food bank—is that what you're asking? My daughter wouldn't, but I would. Yeah, it's been a lifesaver for us."

Everyone had a different story, but every story had a similar trajectory. One woman began to starve during the panic-buying at the beginning of the COVID-19 pandemic when people pushed past her mobility scooter and cleared the shelves before she could reach them. Then everything else fell to pieces. Another had come on behalf of her mum, who is registered blind and struggles to survive on pitiful benefits. Another supports her family and her housebound neighbor, who would otherwise likely starve. She couldn't do it without the food bank.

Two of the people I met were very overweight. This is an issue some commentators find confusing. If someone is obese, they must be eating too much. So how come they need free food?

This reflects a misunderstanding of the causes of obesity. Obesity is highest among the very poor,[24, 25] because the cheapest food is obesogenic.[26] As Fran says, "When people live in poverty, they buy the things with the highest calorific value, which is normally really bad food—processed carbs, high fat. It's not that people don't want good food: it's that they can't afford it."

According to the UN Food and Agriculture Organization, a good diet costs five times as much as one that's merely adequate in terms of calories.[27] Obesity, paradoxically, is often associated with malnutrition.[28] Some people who consume too many calories are deficient in vitamins, minerals, and fiber.

A combination of poverty and the stress, anxiety, and depression associated with low social status appears to make people especially vulnerable to bad diets.[29, 30] But a survey published in *The Lancet* discovered, astonishingly, that over 90 percent of policy makers believe "personal motivation" is "a strong or very strong influence on the rise of obesity."[31] I've heard plenty of pundits insist, with an apparent thrill of disapproval, that the problem is "willpower": people are "failing to take responsibility" for their diets. I have yet to see anyone who makes this claim provide a convincing explanation of why, in countries like England, almost two-thirds of the population has rapidly and simultaneously lost its willpower.[32]

Their explanation overlooks rising poverty and precarity. It ignores time poverty: Fran tells me there are people on the housing estate who work, across two or three jobs, twelve hours a day. They spend so long putting food on the table that they don't have time to cook it. So they rely on the takeout restaurants that dominate the estate's shopping parade. In many poor neighborhoods, there's little else on offer.

It ignores the way junk food has been progressively tailored by scientists and technicians, through precise combinations of sugar, salt, fat, and flavor enhancers, to bypass our natural mechanisms of appetite control.[33, 34] It ignores the way the advertisers they engage use psychologists and neuroscientists to unlock our weaknesses and discover ingenious ways of persuading us to buy unhealthy food.[35] The same food companies then employ biddable scientists[36] and think tanks[37] to argue that weight is a question of "personal responsibility." In other words, after spending billions overriding our self-control, they blame us for failing to exercise it. Obesity is a communicable disease. Its vectors are corporations.

In truth, there *is* a lack of willpower at work: willpower on the part of politicians to improve the distribution of wealth, to ensure no one falls through the cracks, and to restrain the companies that prey on our stomachs and our minds. And, perhaps, to make good food more affordable.

Fran told me, "I've never really understood why the government doesn't subsidize fruit and vegetables. It could massively reduce obesity. You go to the supermarket, and it's £2 for four apples. If you're on the minimum wage and you've got two or three kids, how could you afford that? We'd love to give every child a bag of fruit to take home every week. But it would cost £3 a bag—£450 a week—and we don't have that money."

Almost everyone I spoke to told me how much they liked cooking. There were some ready or half-ready meals in the boxes lined up in the community center—sausage casserole in a sachet, hoisin duck noodles, cans of beef Bolognese and vegetable soup, pasta sauce, baked beans—but also a lot of raw ingredients. There were packages of rice and pasta and boxes of eggs (but no powdered coconut milk, I noticed). There were big bags of vegetables: onions and potatoes—which might or might not have been Tolly's—corn on the cob, courgettes, eggplant, Chinese cabbage. There were trays of purple sprouting broccoli, rutabagas, and leeks for people who wanted extra.

When I wondered what the clients would do with them, most explained they would either look up recipes online or invent something. As they examined the food they were putting in their bags, they told me they might cook a curry or a stew with it, or a stir fry or a soup. If they had an oven, they might try a vegetable bake. No one seemed daunted by cooking from fresh.

"I'm not a good cook," one woman told me, "but getting this box is like *Ready Steady Cook*. You've got these ingredients and you have to think of what to make from them. Sometimes it works. I made a chickpea and potato curry with last week's food, and that was good."

"You can always create something out of what you get here," another customer said. "I don't let anything go to waste. One way or another, I'll find a way of using it."

"Any stuff the kids don't like," one woman told me, "I chop up very small and put it in a sauce or a stew, so they can't tell. 'Has it got mushrooms in, Mum?' 'Of course not, love.'"

Almost everyone agrees that, in the words of one campaigning group, "We can't foodbank our way out of hunger."[38] Food banks in a wealthy country are a last, desperate resort, a symptom of deep political and economic dysfunction,[39] of an obscene maldistribution of wealth. Charity is what happens when government fails. Yet, across much of the world, including some of its richest nations, they have become essential.[40] In the UK, government figures suggest that just under 2 percent of adults now rely on them.[41] Many would otherwise face starvation.

Redistributing surplus food is controversial. The celebrated cook and community organizer Dee Woods notes that "leftover people get leftover food."[42] She makes an important point: it would be abominable if hungry people were treated as a kind of waste-disposal system, assuaging social guilt about the food we squander. But because charities' budgets are so strained and surplus food costs much less than purchased food, there are strong pragmatic reasons to use it. No one who cares about poverty and hunger wants things to be as they are. But these are structural problems, which require structural solutions.

Just as we cannot foodbank our way out of hunger, we cannot solve the dilemmas faced by farmers simply by raising the price of food, unless we are prepared to see many more people go hungry. According to one analysis,[43] global food prices in 2021 were already higher in real terms than at any point in the preceding sixty years, except for 1974 and 1975. Environmentalists often call for "the internalization of the externalities" (one of our many thrilling and uplifting slogans). What this means is that the price of goods and services should incorporate the cost of the harm done to people, places, and living systems. The famous *EAT Lancet* study states, "We believe that food prices should fully reflect the true costs of food."[44]

But even if it were possible to produce figures for the damage to wildlife and soil and water and atmosphere that could be meaningfully expressed in currency and incorporated into the price of food (a proposition I find entirely implausible),[45] would we want to do this? If, through the magical transubstantiation of our life-support systems into dollars and pounds, we discovered that wheat should cost not $200 per ton but $500, is that really what we would want people to pay? If so, vast numbers would starve.

A large and growing movement insists that the answer to all these conundrums is "food sovereignty." This is defined by its leading thinkers as the right "to healthy and culturally appropriate food," grown in ways that are "ecologically sound," and the ability of local people to control "their own food and agriculture systems."[46] It seeks to break the corporate grip on the food chain, uphold the rights of women in food production, improve the distribution of land and the condition of rural workers, guarantee fair incomes, and stop coercive trade agreements and the privatization of nature. All these, I believe, are essential steps toward a fairer world. But there are, I think, two problems.

The first is that the movement's aims say nothing about *not* farming. There is no mention in its founding declaration[47] of limiting the land area occupied by farming, in other words of controlling agricultural sprawl. When I've raised this issue with some food-sovereignty campaigners, they have fiercely rejected the idea that even land used in the most profligate ways—vast areas producing tiny amounts of meat—should be withdrawn from production and returned to nature.[48] Yet without such restraints, it's hard to see how we can stop the accelerating collapse of the diversity and abundance of wildlife, most of which depends for its survival on undisturbed natural systems. There's a fundamental clash of visions here: the declaration states that "all . . . biodiversity" should be conserved through "ecologically sustainable management." But the extractive management involved in any kind of farming appears to present an existential threat to most species.[49, 50]

The right to farm is sometimes incompatible with the right to a thriving planet. To pretend this conflict does not exist is to ensure that it cannot be resolved.

The second problem is that the vision tends to emphasize supplanting long-distance trade with local food production.[51] Let's explore this for a moment.

It is an important and understandable aim. Local networks create an opportunity for small producers to connect directly with the people who eat their food, cutting out the intermediaries who otherwise capture most of the money. They build the kind of trust and accountability that's often missing from international food chains. A local food economy is more likely to keep people on the land, favor small-scale and diverse production, and generate employment than a transnational one.

We should resist the urge to romanticize all forms of local production. The lives of small farmers range from the rich autonomy portrayed in Chinua Achebe's *Things Fall Apart* and John Berger's *Pig Earth* to the squalid hell in Anton Chekhov's *Peasants* and Jung Chang's *Wild Swans*. Not every aspect of traditional rural culture deserves to be celebrated. Consider the treatment of women and the power of landlords in parts of rural Pakistan and northern India, or the *caciquismo* (tyrannical control by local leaders) in some regions of Mexico, or the racism that disfigures some rural communities in the southern states of the U.S.

The food-sovereignty movement performs crucial work in fighting traditional injustices, as well as new forms of oppression. It contests authoritarian control, racism, and cultural and religious oppression.[52] It promotes feminism[53] and the redistribution of land, wealth, and power.[54] Its emphasis on local markets is, in principle, consistent with these aims. There should be greater scope for justice and the distribution of wealth in a local food chain than in a global one.

So what prevents a general transition to local production from happening? There are plenty of explanations: the concentration of land ownership, the structure of markets, the political power of big corporations, unfair subsidies, the dumping of food at less than the cost of production by rich nations, which undercuts farmers in poorer ones. All these hold true, and great injustices are done. But there is another, deeper problem, less often recognized and scarcely discussed: math. In most cases, sufficient growing areas close to our centers of population simply do not exist.

A paper in the journal *Nature Food* sought to discover how many of the world's people could be fed with staple crops grown within 100 kilometers of where they live.[55] It discovered that wheat, rice, barley, rye, beans, millet, and sorghum grown within this radius could feed only a quarter of the world's people. Corn and cassava grown within 100 kilometers could supply a maximum of 16 percent of those who need them. The average minimum distance at which the world's people can be fed is 2,200 kilometers. For those who depend on wheat and similar cereals, it's 3,800 kilometers. A quarter of the global population that consumes these crops requires food grown at least 5,200 kilometers away.

Why? Because most of the world's people live in big cities or populous valleys, whose hinterland is too small (and often too dry, too hot, or too cold) to feed them. Much of the world's food is grown in vast, lightly habited lands—the Canadian prairies, the U.S. plains, the Russian and Ukrainian steppes, the Brazilian interior—and shipped to tight, densely populated places. There is some scope for reducing these distances by increasing yields, but this is often limited, and runs into the constraints I discussed in chapter 2. In fact, as climate breakdown and other disasters are likely to render more places unsuitable for farming, trade distances might need to increase.

You can negotiate with politics and economics, market structure, and corporate power. But you can't negotiate with arithmetic. Given the distribution of the world's population and of the regions suitable for farming, the abandonment of long-distance trade would be a recipe for mass starvation.

But, as so often in this field, passionate debates about how we should grow our food take place in a numerical vacuum. Nowhere is this innumeracy more obvious than in our fantasies about the most extreme form of localism: urban farming.

Growing food in cities is, I believe, of great benefit to mental health. Without our orchard, especially during the UK's repeated lockdowns, I'm not sure how I would cope. City farms, allotments, and guerrilla gardens[56] help us to feel a sense of connection to the land and engage our minds and hands in satisfying work. But, with one or two exceptions (principally Cuba during the blockade), it's unlikely to satisfy more than a tiny fraction of demand. The reason should be obvious: land in cities is scarce and expensive.

This doesn't appear to stop some people from making bold claims about its potential. For example, one report asserts that "some 25% of the world's small livestock and fruit and vegetable consumption" could be supplied by urban farms and households.[57] Even if we were to ignore the fact that livestock kept in cities have to be fed on grain grown elsewhere, and present a high risk of transmitting zoonotic diseases, it's impossible to see, in the great majority of cities, where the land could be found. Cities occupy 1 percent of the planet's surface, and most of that land has other uses. When I challenged one of the authors of the report, an old friend, he responded amicably, but he couldn't provide the evidence required to justify the claim.

Some people seek to overcome this constraint by promoting indoor and vertical farming: growing food in buildings with multiple stories or tiers. The only financial advice I would offer anyone is "don't ask me for financial advice." But when this idea first became popular, in 2010, even I could see that the sums didn't add up. I wrote an article warning that a combination of vast embedded costs (in land and buildings) and the need to grow crops in artificial light made this model financially unviable.[58] Vertical farms in cities would have to compete with horizontal farms in the countryside, whose land and infrastructure costs are lower, and which receive their sunlight free of charge. Even when indoor farmers grow the kind of crops that do better under glass, their investments in load-bearing structures would go head-to-head with greenhouses, which cost much less.

I'm not saying it's impossible. There are some thriving farms in my city, neatly integrated into people's homes, expensively equipped with lights, pumps, and temperature controls, growing crops to precise specifications. Every so often they're busted, and the farmers are led away in handcuffs.

Weed can be grown this way because it's among the few farm products—all of which are illegal in most jurisdictions—whose price justifies the outlay. But thirty-story hydroponic weed towers are difficult to conceal.

Needless to say, my article did nothing to dampen the enthusiasm. I watched with incredulity as one start-up after another published their plans for impossible programs, raised millions from venture capitalists, then collapsed.[59, 60] Whenever it happened, the chastened entrepreneurs explained that their project was "ahead of its time."[61] But they weren't ahead of time. They were ahead of physics. These failures did nothing to discourage the next wave of investors, by comparison to whom I seem to be some kind of financial genius.

In most sectors, when a commercial model fails, that's the end of it: capital seeks to avoid making the same mistake twice. But, almost uniquely, in this field repeated failure appears to be no deterrent. At first sight, it's one of life's great mysteries, comparable to the Tardislike properties of Tupperware or the current whereabouts of my phone. But I think it reflects two things: how little most tech entrepreneurs understand farming, and the determined belief in magic that food production inspires.

Perhaps it's unsurprising that so many biblical miracles involve the supernatural delivery of food and drink: manna from Heaven, Elijah fed by ravens, the widow's self-replenishing flour and oil,* water into wine, feeding the five thousand. Secular folk tales, like the magic porridge pot and the magic pudding, have similar themes. There appears to be a deep and ancient appetite for improbable solutions to the question of how to feed ourselves. But there are no miracles: just good or bad ideas, and the compromises they must make when they collide with material reality.

So, within the grand dilemma that frames this book, we encounter some subsidiary but equally difficult dilemmas. Most of our food has to be grown, for simple mathematical reasons, far from where we live, and shifted in bulk. Long-distance trade and mass production favor transnational corporations and accelerate the homogenization of the Global Standard Farm. But this consolidation makes the food system less resilient, destroys the livelihoods of small farmers, and undermines food sovereignty. Somehow, we need to arrest and reverse it, but without causing mass starvation. Food has to be cheap enough to feed people in poverty, yet expensive enough to support those who grow it. It needs to be grown at low cost, but without the corner-cutting that destroys the living world.

If you are looking for easy answers, you've come to the wrong place. If you want miracles, other books are available. But, though the problems are wicked[62] and paradoxical, I believe there are some counterintuitive solutions.

---

* 1 Kings 17:12–17.

6

Putting Down Roots

It looked, in this quiet corner of Shropshire, as if a tank battle had been fought. As we traveled from the station to his house, Tim Ashton explained what had happened. Three terrible storms in February and early March, and the resultant floods that broke all records, had destroyed the crops sown the previous autumn. Now, in March, the fields we passed were being frantically plowed and sown again, in the hope of setting a spring crop in time for the growing season. But the soil was so wet that the heavy machinery had ripped it apart.

"That guy," Tim pointed to a field scarred by great ruts, "was so embarrassed about plowing in these conditions that he did it at night."

A few minutes later, he parked in the lane that led to his house, and we stepped into a field belonging to his neighbor.

"He lets contractors run his farm, and it's a disaster. They've been growing corn here all my adult life, without a break. I can't remember ever seeing a different crop in this field."

It was a remarkable sight. Though the field hadn't yet been plowed (corn is planted late in the UK), the storms had stripped the bare soil from the land, then sifted and winnowed it into its mineral components. Drifts of red sand filled the tractor ruts. Black silt had collected at the bottom of the field. In some places—especially where the machinery had turned—the deep soil had disappeared entirely, leaving bare glacial gravel. In compacted depressions, stinking ponds had formed, dyed cobalt green by cyanobacteria.

Soils in the UK, by comparison to those in many other parts of the world, tend to be quite robust. But the contractors, with the help of the extreme weather, had managed to turn a productive field into a war zone.

In trying to discover how we might produce more food with less farming, I've divided the foods we eat into three categories: vegetables and fruit; cereals; and protein and fats.* Through Tolly, I found some of the answers to the question of how we might reduce the environmental impact of growing vegetables and fruits while maintaining high yields. I was visiting Tim in the hope of finding answers to an even bigger challenge: how to secure and improve the arable farming that produces cereals and other grains. Over 50 percent of the world's current diet is supplied directly by cereal crops, such as wheat, rice, and corn (by "directly," I mean it doesn't first go through an animal).[1] Not only does arable farming—through plowing, pesticides, weedkillers, and fertilizers—impose massive impacts on the living planet, but it might also be highly vulnerable to environmental shifts and systemic failure. Faced with both climate chaos and a gathering loss of resilience, changing the way our staple crops are grown should number among humanity's most urgent missions.

We could argue about the merits of obtaining so many of our calories from cereals. There are good nutritional reasons for changing the balance of our diet. But whatever farming system we adopt, it needs to provide food that people are prepared to eat and can afford. If minimizing the ecological cost of our diets means not only changing the way they're produced, but also changing their fundamental nature, the task becomes harder. I could produce an excellent argument, based on both human and planetary health, for switching to a diet dominated by beans, lentils, and nuts. But beyond a particular social circle, it's unlikely, at the moment, to gain much traction. This could change, as more people become aware of the health and environmental impacts of the Global Standard Diet, and if governments alter subsidies, taxes, and the rules that determine how food is marketed. But the less we need to rely on moral suasion, the more successful a shift is likely to be.

---

* In reality, the categories don't divide quite so neatly. We obtain some of our protein from cereals, for example, and no one seems able to decide whether potatoes are an arable crop or a vegetable.

There's a useful comparison to be made with government efforts to reduce greenhouse gas emissions. The most successful reductions have so far been those that require the least consumer effort. The lights still come on when we flick the switch, though the power might now be supplied by wind turbines and solar panels, rather than coal and gas. The much harder transformations, at which many governments balk,[2] involve persuading people to change their habits: driving less and walking, cycling or using public transport more, flying less, changing our heating systems, renovating our homes to make them more energy-efficient. All these things also need to happen to prevent climate breakdown, but they are harder to effect than the remote and scarcely perceived replacement of our power sources.

So a large part of the shift needs to take place in one of the most difficult of all arenas: the broad fields in which our staple crops are grown, where prices are low, and—because cereals are easily stored and transported—competition is global. To make matters harder still, this shift needs to happen while the environmental conditions in which these crops are raised become more hostile.

Tim has a permanent half smile and a glint in his eye, as if there's a joke he can't wait to share. He's blond with hazel eyes, thirty-three years old, not very tall. He walks with quick, jerky movements and talks in the same style. He's ebullient, smart, impossible not to like.

His house, Soulton Hall, is one of the strangest buildings I've seen in this country. It's a high, square, flat-roofed seventeenth-century block of red brick, with limestone quoins and mullioned windows, a front door framed by pillars and overhung by a gigantic pediment bearing the ancestral arms. It has an enormous square chimney, architraved and corniced, on each of its corners.

"My friends are cruel enough to compare it to Battersea Power Station," Tim told me.

It's built around a medieval core, which was itself constructed on the site of a Saxon manor, whose owner was slain by King Cnut in 1017. This, in turn, might have been built on top of a Bronze Age village. In the grounds are the remains of an illegal twelfth-century castle, erected without the king's permission.

Tim introduced me to his father: a vigorous, slightly stooped old man with massive forearms. Remarkably, for the lord of a Shropshire manor, he has a broad Lancashire accent. Tim told me that his great-great-grandfather was "one of those legendary Victorians who managed radically to transform their circumstances." He began his working life as a messenger boy, and ended it as the owner of several cotton mills. In mid-life, his frenetic labor almost killed him. He came from Lancashire to recuperate at Soulton Hall, where he fell for the daughter of the house. Until the late 1940s, when the cotton mills, whose survival had depended on the Empire's coercive trading relations, closed, the family lived between Lancashire and the Hall.

The land is almost as strange as the house. They own 500 acres (200 hectares)—Tim calls it "a big small farm"—covered by a wider range of soils than you'll find in many parts of the world across a thousand square kilometers. Between the ridges of sandstone laid down in the Triassic are, in some fields, strips of clay deposited by glaciers; in others, sandy, silty, and peaty soils, some of which must have accumulated in post-glacial mires. A crop that thrives in one field might curl up and die in the next. On the escarpment overhanging the farm, the young Charles Darwin found his first fossils.

Tim is the twenty-first-century equivalent of the Victorian farmer-scientists, drawing on the latest findings and conducting his own experiments to improve his practice. Waiting for us at the Hall were his collaborators: the soil ecologist Simon Jeffery, who supervised Tim's Master's thesis, and the agronomist Paul Cawood. In character and appearance, they are polar opposites. Paul is confident, boisterous, deep-chested, with spiky dark hair and a wide-boy swagger. Simon is self-effacing, slender, with blond dreadlocks and an apologetic manner. But what they have in common with each other and with Tim is an understanding of the land and its uses that, in my view, approaches genius. I have come to see them as the Three Wise Men.

Tim picked up a spade and we followed him to the first of his fields. If I had witnessed nothing before I arrived, it would have looked like any other arable field. But today, it seemed as if I'd traveled to a different country. The disaster that had ravaged the neighboring farms had somehow passed his by. There was no visible damage at all. Tim hadn't yet sown his spring crops. The surface was covered with the straw from last year's harvest, moss, and worm casts, through which a few rapeseed plants* that had seeded themselves now grew. But the delicate sensory organs best attuned to the difference were my size 11 feet. The soil was as firm and springy as a rubber mat. The contrast with his neighbor's torn cornfield and the disasters we had seen on the way from the station couldn't have been greater.

Why the difference? Because Tim does not plow his land. He is among the fairly small number of cultivators in Europe (there are far more in the Americas and Asia) using "no-till" farming to grow grain.† He is trying to overcome the problem that Tolly identified in his own system: the need to keep breaking the soil by plowing.

Tim had plowed one strip across this field, to provide a point of comparison. I could easily have found it with my feet: the soil here was as soft as whipped cream. When I walked across it, I sank to my ankles. Looking at it closely, I could see that the surface was streaked with pale lines of sand, separated from the silt by the rain. On a much smaller scale, it had suffered the winnowing effect I had seen in the cornfield.

Tim struck his spade into the ground and extracted a spit of earth from the plowed strip, then one from the unplowed land. He laid them side by side. The plowed soil immediately fell apart, while the unplowed spit held together. I could see another difference: the unbroken soil was crawling with worms. Except when digging in the compost heap, I don't believe I've ever seen so many in a spadeful. They ranged from tiny purple and yellow specimens to giant pink lobworms. In the plowed soil, by contrast, I could see none.

Simon explained that it wasn't so much the roots of the rapeseed seedlings that were holding it together as the polymers secreted by bacteria, worms, and tiny arthropods, and the glomalin molecules extruded by fungi. These build the aggregates that harbor life.

"Plowing rips up the fungal hyphae and breaks open the aggregates. The bacteria then mineralize the carbon, turning it into CO_2, and it's released into the air. Then the cellular structure of the soil

---

* Known in other parts of the world as canola.
† No-till is not the same as "no-dig." The no-dig technique is largely used for vegetable growing. It involves piling organic material onto the surface of the soil. I briefly discussed its drawbacks in chapter 4.

collapses, the pores fill up with particles, and less air and water can pass through it. It becomes quite hostile to life."

Simon told me that since Tim had stopped plowing, the number of worms had increased sixfold: their populations here were even higher than Tolly's.

"That means better infiltration of water, better aeration, better root growth. A big earthworm can burrow down to two meters, and roots can follow the burrow. Then the virtuous circles start turning. The worms create habitat for other organisms, that improve the soil structure, which is good for earthworms, and so on."

Without worms and the many other creatures keeping the soil porous, water can scarcely penetrate. In some fields, the problem becomes so extreme that even steady rain flashes off the surface: while the top five centimeters of soil become saturated, underneath, it remains bone-dry. As the agricultural researcher and educator Niels Corfield points out, the aggregates—or crumb structure—created by living creatures greatly increase the internal surface area of the soil.[3] They're highly hygroscopic, which means they're structured in such a way that water clings to them.

Remarkably, even when healthy soil is air-dried, the relative humidity within the aggregates stays at 98 percent.[4] In other words, they are more or less impervious to desiccation: a property that at first sight looks like magic. It's not magic, but neither is it accidental. The vast internal surface area that makes this possible is a feature of biological construction, as the life of the soil creates the environment that suits it: in this case a saturated atmosphere. Just as carbon in the soil is held in the form of biological cements, water is held in the form of biological films and trapped vapor. Water is life and life is water.

"On the day after we plowed this strip," Tim said, "the diversity of springtail species fell by thirty percent, and their abundance by seventy percent. If I did this amount of harm to a rainforest, people would be up in arms. But every autumn we sing songs about it."

Paul showed me something I wouldn't have noticed: a horizontal fracture splitting the soil about 12 centimeters from the top surface of the untilled spit. A root had bent at right angles and was running along it.

"Until five years ago, this field was plowed. This line represents the deepest point the plow reached. Plowing creates a horizon here, for two reasons. One of them is that the finest particles, which are shaken through the soil by plowing, end up at the lowest level of cultivation. It's like shaking a jar of beans and rice: the smaller grains move to the bottom. When you do this repeatedly, the fine particles form a hard layer. It costs the crop a lot of energy to punch through it. As the roots try to get through, they become abraded, which exposes them to infection. So they often find it easier not to bother, and divert sideways instead. But when they're stuck in the top few inches of soil, they find less nutrients, and they're more susceptible to drought.

"The other issue is that plowing turns the soil through ninety degrees, leaving the straw that was on the surface under the ground, along the fracture line, making an anoxic layer.* Then the straw rots,† creating acidity. Plant roots hate it. It's like licking a battery."

Eventually, after several more years without plowing, the worms and roots in Tim's fields will split up the hard, acidic pan, and the fracture zone will disappear. But it takes a long time. It's easy to break the soil, harder to mend it.

"It's not just the carbon cycle you mess up when you plow," Paul said. "You also liberate nitrogen. You release a biologically weird portion of it at the wrong stage of the plant's life. The crop grows fast just after germinating in the autumn, and it looks vigorous and healthy. But in the spring, when it should be hitting maximum growth, it all goes wrong. The plants have taken on too much too soon, and they get stressed. That's when the aphids attack."

Paul reinforced something that Tolly had told me: aphids love stressed plants. "If you've got aphids, it means your crop is malnourished and struggling. What they want is protein, not carbohydrate, which is why they excrete honeydew.‡ Stressed plants have higher

* Anoxic means without oxygen.
† Rot is sometimes used to mean breaking down anaerobically (without oxygen), while decomposition, by contrast, sometimes means breaking down in the presence of oxygen.
‡ The sticky, sugary secretion that ants harvest.

levels of amino acids:* they're aphid heaven. You can spray them all you want, but if your plants are unhealthy, it won't do any good.†

"Insecticides have short half-lives. But pests come in surges. So what are you going to do? Keep spraying every day as new ones turn up? Once you start, you can't stop, as you kill all the predators too. And the more you spray, the more you select for resistant pests. Don't pick a fight with aphids. You will lose.

"But a well-nourished, healthy plant doesn't need to be sprayed. It deposits silicates on its leaves that protect it. The aphids won't touch it.

"If you don't plow, in the autumn your crops don't look like they do in the textbooks," he continued. "They don't stand upright, but sprawl on the ground. They look pretty sad, to be honest. Other farmers come round and suck their teeth, and you feel like you've failed. But come the spring, they bounce up. Before long the boot is on the other foot."

Giving up plowing, Tim told me, has not only allowed him almost to eliminate pesticides (he hasn't used any on his crops for four years), but also to reduce his use of fertilizer by 15 percent, as the soil is healthier and fewer nutrients are lost through erosion. As more organic matter accumulates, he hopes he will need ever less.

"And my fuel use has fallen off a cliff. After I switched to no-till, the man who sells me diesel phoned to ask if we'd changed suppliers."

Paul had the figures in his head—he always does. The usual arable package, he told me—using a plow, power harrow, combination drill, and roller—burns between 29 and 31 liters of diesel per hectare per year. But you need to perform only two operations on the soil in a no-till system. The first is drilling the seed into the ground, with an ingenious device pulled behind a tractor, a bit like a giant sewing machine with dozens of needles. The second is rolling the soil to force the weeds to germinate (I'd learn more about this later). Together, they need just 4.5 liters per hectare.

There's another saving, which might be greater than any of these. As Tim explained, "When farmers use loads of machinery to prep their fields, about forty or fifty percent of the cost of growing a crop is spent

---

* The building blocks of proteins.
† This explanation is contested by some crop scientists.

upfront. So they take a big hit when a harvest fails. But if a crop of mine gets ruined, it costs me very little. All I lose is the seed we grow ourselves and a small amount of time and diesel for the drilling.

"There's more time to get things right, and a lot less strain. If conditions are wrong in the autumn, you can plant in the spring. You exhaust the land less and you exhaust yourself less."

"A farmer who's always in their tractor is under massive stress," Paul added. "If they need to do lots of operations, they have to exploit every window. So often farmers start cultivating too early, damaging the soil. Or they have to leave it too late. The standard system is all about maximizing work, rather than optimizing conditions for crops. It drives people to farm badly."

"I could drill my fields this week without harming them," Tim said. "A plowed field would have to wait another fortnight. Otherwise, when the soil surface is this wet, you'd make a total mess of it."

So this makes Tim's system doubly resistant to climate breakdown. The soil, being firm and well-structured, is less susceptible to storm damage and more resilient to drought. And if extreme weather trashes his crop, his financial losses are smaller, so he is more likely to survive.

As they talked, the Three Wise Men kept using a phrase I hadn't heard before: soil armor. I asked them what it meant.

"Oh, sorry," said Tim. "It's a slightly romantic term for the crop residues we leave on the surface. When you don't plow, all the cut straw and other plant material stays on top. If you plow it in, it's like you're force-feeding the life of the soil. But if you leave it where it is, the microbes will digest it and the worms can pull it into their burrows in their own time. In the meantime, it protects the soil against the rain and sun."

"You should never let the sun see the soil," said Simon.

Useful as these changes are, the transition wasn't easy. Tim had to buy new equipment, and learn to use it. "There was some painful experience, and we had to sell some assets. Luckily, the bank manager was understanding. It took four years to get the gross margin back to where we wanted. We weren't helped by the crazy weather we had all through that time."

Now, thanks to the improvement in the health of his soil, Tim gets a slightly better yield than he did when he plowed: an extra 0.2

tons of wheat, which means he harvests roughly 7 tons per hectare: a decent weight for Shropshire's climate.* Because his costs are lower, Tim told me, he could produce less than he did before and still make more money. So, after a bumpy start, the shift is beginning to pay off.

Tim is likely to owe much of his success to his earthworms. In systems like his, earthworms seem to do the plow's work, without the destruction it involves. They aerate the soil by burrowing through it and, by digesting plant material, make nutrients available to crops. On average, crop yields are 25 percent higher in fields with earthworms than in fields without.[5] In other words, if all the crops that farmers grew were used to feed human beings, roughly a quarter of us would owe our survival to worms.

We walked through a few fields that in normal times would have been unremarkable, but in these abnormal times were remarkably intact. Tim showed me the habitat he had allowed to develop along the little stream that ran through his land: sedges, willows, and rough grass. Two American wood duck burst up from it and sped away downstream. A couple of blue tits moved through the little trees. It makes sense not to farm the land closest to rivers. Strips of thick vegetation catch fertilizers and farm chemicals washing off the fields before they reach the water.[6, 7] Wildlife corridors work best around rivers, as the interface between land and water tends to be the richest part of an ecosystem.[8]

We found ourselves back at the little road on which we had traveled to the Hall. We crossed into the neighbor's cornfield Tim had already shown me. Now the contrast with his land seemed even starker.

"In my professional opinion," said Paul, "that field is shagged."

So is no-till cultivation a complete remedy to the problems caused by arable farming? No. Because Tim does not plow, he relies to an even greater extent than most non-organic farmers on herbicide (weedkiller).

---

* This is unusual, however. In most cases, in temperate regions, the yields from no-till farming tend to be slightly lower than the yields from tilled systems (https://www.sciencedirect.com/science/article/pii/S0378429015300228).

This doesn't mean he uses more of it than other farmers do: he generally uses less. It means that he currently has no choice but to deploy it. He can't destroy his weeds and cover crops by plowing, as Tolly does, so he must kill them chemically.

The herbicide he uses is almost universal in large-scale farming: glyphosate, most often sold as the formulation Roundup. For many years, this chemical was marketed as harmless to everything except the plants on which it's sprayed. It is generally less harmful to other life forms than the herbicides with which it competes. But after it became the world's most popular weedkiller, scientists started reporting damage to a wide range of living creatures. For example, one experiment suggests that glyphosate affects honeybees' ability to navigate: bees fed on sugar containing the herbicide took longer to get home.[9] Another experiment found that it damages the bacteria in honeybee guts, reducing the number of larvae that survive in the hive.[10]

Glyphosate has been reported as toxic to many freshwater creatures—including frogs,[11, 12, 13] fish,[14, 15] and crayfish[16]—and to marine species such as mussels[17] and plant plankton.[18] But some of these studies used concentrations much greater than farmers apply, so the results are unlikely to reflect real conditions. The effect of the herbicide in freshwater might be exacerbated by phosphate pollution, which seems to prevent it from being rapidly broken down.[19] Glyphosate is extremely persistent in seawater,[20] as salt also appears to stop it from degrading. We don't yet know what its cumulative effect might be when it runs off the land and into the oceans.[21]

The herbicide works by inhibiting the action of an enzyme that plants produce, called EPSPS,* without which they cannot make a number of essential chemicals.[22] Alarmingly, many bacteria and fungi produce the same variety of EPSPS, and glyphosate appears to affect them in the same way.[23]

Among them are many of the crucial bacteria that inhabit the rhizosphere (the plants' external gut) and the guts of animals. Some studies suggest that it damages microbes that promote plant growth,[24, 25] though others indicate that the microbe community as a whole seems to recover quite quickly after spraying.[26] There's evidence that the

---

* 5-enolpyruvylshikimate-3-phosphate synthase to its friends.

chemical tips the balance between beneficial microbes and those that cause disease in plants, which could explain why at least one deadly crop disease seems to be more prevalent in soil where the herbicide has been used.[27] In most circumstances, glyphosate is quickly broken down in the soil by fungi, but in some places, depending on the soil type, it can persist for up to a year, and accumulate.[28]

You may remember that in chapter 3 I mentioned the "intrinsic resistome" of soil bacteria:[29] their readiness to defend themselves against hostile chemicals. One of the weird effects of this resistome is that exposure to one toxin can make bacteria less susceptible to others.[30] Some research suggests that, in trying to defend themselves against glyphosate, bacteria become better equipped to defend themselves against antibiotics.[31] This could, in principle, accelerate the crisis of antibiotic resistance. We might expect to see similar effects in our gut bacteria, which develop the same kind of generalized defense.

A recent study shows some minor effects of glyphosate on microbes in the human gut, potentially affecting our immune system.[32] One survey found that a third of people tested in Germany had traces of glyphosate or one of the chemicals into which it decomposes in their urine, though at levels that are unlikely to present a risk to human health.[33] The researchers were able to link these residues to particular foods the people had eaten.

Glyphosate might soon become useless in some places anyway, as so many weeds are now resistant to it. Already, soybean farmers in parts of the United States are beginning to give up no-till, as superweeds, which can be controlled only through plowing, make it impossible.[34]

There seems to be a standard trajectory in the development and sale of farm chemicals. They are marketed as safe to everything except the species they're intended to kill. Only after they are widely used do we discover that the manufacturers had failed to conduct sufficient tests.

While a standard arable farm might use glyphosate two or three times in the crop cycle, Tim uses it only once. Because he doesn't plow, the seeds shed by weeds remain on the surface of the field. Many of them are eaten by birds. He deals with the rest by running a heavy roller over the ground just after he drills his crop. This presses the weed seeds onto the soil, which wakes them up (soil contact is a trigger for germination). They sprout two or three days before the crop raises its head, which gives Tim a moment in which to spray them without harming the plants he has sown.

This technique ensures there's little chance of the chemical getting into the human food chain. His process, he says, makes his land less susceptible to a weed now causing massive problems for farmers in Europe: blackgrass.* It likes disturbed soil, so it's favored by plowing.

Simon argues that the damage inflicted on soil life by glyphosate is much slighter than the damage caused by plowing, as we could see from the remarkable difference in earthworm numbers between the plowed strip and the rest of Tim's land. The amazing rise in earthworm numbers on Tim's farm is confirmed by studies from around the world: though one study suggests that glyphosate harms earthworms,[35] the effects of plowing appear to be worse.[36] It slices them up, exposes them to predators and destroys the burrows which, in stable soils, are passed from one generation of worms to the next.[37] The cut worm does not forgive the plow.

Animal life of many kinds is massively reduced by plowing: there are fewer strands in the food web, fewer species of earthworms, springtails, and mites, and the surviving animals are generally the smaller and faster-breeding species.[38] Some groups vanish altogether. Plowing affects bacteria in the same way: there are more of them in no-till fields than in plowed lands.[39]

But, as always, it's hard to compare different kinds of environmental damage. How do we weigh the protection of the soil on farms like Tim's against the damage glyphosate might inflict on marine ecosystems? How do we compare the value of earthworms to the value of antibiotics? While some people have tried to create a single metric, often expressed in pounds or dollars,[40] I see these impacts as incommensurable: you cannot meaningfully trade one against the other.

Can these issues be resolved? In some parts of the world, you can use a machine called a crimper roller to kill weeds or cover crops without either plowing or herbicides.[41] It's a crude but effective technique: a giant mangling machine that smashes and frays the stems

---

* *Alopecurus myosuroides.*

of the plants, exposing them to frost and infection. It works in Canada and other continental farming zones. But the UK is no longer cold enough. You can't pull such heavy machinery over your fields in the winter unless the ground is frozen solid. Otherwise you mash and smear the soil, as some of Tim's neighbors had done while using other kinds of machinery. It can take many years to recover. And without deep frosts, the mangled plants might survive.

But a contraption that could be deployed in any climate is quickly gaining a foothold: the robot weeder. This is a fully automated machine that can roam up and down the fields by day and night, identifying selected weeds and destroying them individually. The most promising models, I think, are those that use bolts of electricity,[42] rather than flames or tiny jets of herbicide, as this appears to have the lowest environmental impact. As usual, there's a downside: the technology is expensive. Farmers might not need to buy their own robots: some companies sell the service, rather than the machine, charging for weed removal by the hectare.[43] Even so, it provides a further advantage to money over labor, giving large or well-capitalized farms yet another edge over small, cash-poor farmers. Every answer has its counterpoint.

But if Tim can afford such a thing, he might, he says, one day "find a sweet spot, bringing together no-till, better engineering, and new techniques which could let me make a transition to organic farming. That's where the future belongs."

Tim's results appear to be better than most farmers'. On average, no-till slightly reduces individual yields.[44] But in some places, it can nevertheless greatly increase the overall harvest.

How can this be true? The reason is that it makes switching from one crop to the next much quicker, which means that, in central Brazil and some other hot parts of the world, two crops can be grown every year, where only one could be raised before.[45] In India, Pakistan, and Bangladesh, no-till allows farmers to flip between rice (in the monsoon months) and wheat (in the drier season) without the devastations the traditional system requires.[46] Partly to suppress the weeds that would otherwise overwhelm it, rice is grown in waterlogged fields. Turning these fields back into the light tilth required for wheat requires weeks of repeated plowing,[47] which often causes massive soil degradation and erosion. To quickly dispose of straw and weeds between the two harvests, farmers traditionally burn them. The smoke, pouring into nearby cities, shortens the lives of millions.[48] No-till allows farmers to move from one crop to another without flooding, churning, or burning.

No-till farming seems to be better for keeping the soil damp[49] and reducing erosion[50] and compaction.[51] But, for complicated reasons, it's no better at storing carbon in the soil,[52] or preventing fertilizers[53, 54] and pesticides[55] escaping from the fields. Broadly speaking, it works well when people farm as Tim does: rotating their crops, keeping the soil covered with catch crops in the winter, and leaving straw and dead weeds (the soil armor) on the surface.[56, 57] But it works badly when farmers, as Paul puts it, "spray and pray": dose the land with weedkiller, drill the seed, but otherwise carry on as usual.[58] No technology is idiot-proof.

Does no-till further tilt the balance against small farmers? Seed drills of the kind Tim uses are fantastically expensive. Any system that favors capital over labor tends toward concentration. But in Bolivia, a seed drill that can be pulled by animals costs less than $400.[59] This is still quite expensive in a nation where the average smallholder earns $4.30 per day,[60] but it's many times cheaper than the standard machinery, and the cost could be shared between farms. It reduces the time taken to plant a hectare from twelve days to ten hours.

I returned to Tim's farm on a strangely wintry day in mid-July, a couple of weeks before harvest. As we crossed the land, lapwings dived and twisted above the fields, with cries that sounded like the rubber dog toys that squeak when they're bitten. They must have had chicks in the crops.

The spring storms and floods had been followed, soon after my last visit, by a scorching drought, which killed germinating crops across the country. If I wanted to assess the system's resilience to climate chaos, I couldn't have chosen a better year.

Thanks to the violent weather, Tim's rotation had been knocked out of shape. In the past, he had grown wheat, followed by a cover crop, then broad beans, which he sells to North Africa, then wheat, then rapeseed. But, in the early years of this century, a commodity supercycle—which means a sustained rise in the price of many basic products—tempted many farmers into growing nothing but wheat and rapeseed, to feed the buying frenzy. Pests love the continued cultivation of just one or two crops, as this allows them to proliferate without interruption. By repeatedly planting rapeseed, these farmers inadvertently triggered a plague of flea beetles, whose larvae drill into the rapeseed stems and kill the plants. At roughly the same time, the European Union banned the use of neonicotinoid pesticides, rightly I think, because of their terrible ecological impacts. As the excessive planting of rapeseed depended on these chemicals to kill the beetles, the system collapsed almost overnight. Now, in many parts of the country, the crop is almost impossible to raise.[61] Bad farmers hurt good ones.

So Tim replaced the rapeseed in his rotation with linseed. Some of his fields shimmered with the weird silvery globes of the seedheads, which seemed to float in mid-air. He is among the first farmers in the country to grow quinoa. Tim told me that as soon as new varieties of chickpeas, lentils, soybeans, and other pulses are bred to thrive in our cold climate, he'll sow them. Like many good farmers, he's desperate for new crops to patch into his rotation: the greater the variety you plant, the better the health of the soil and the less of a problem pests become.[62, 63]

But this year, there were so many disasters that he had to drill one field with seed five times before a crop took root. In some places he'd had to follow wheat with wheat and beans with beans. You could see the difference: where broad beans had been planted two years in a row, their stems and leaves were covered with flecks of brown rust (a kind of fungus). This disease scarcely affected the beans that had followed a cover crop. But the biggest difference in his bean fields was between the plants growing in the experimental strip he had plowed, and those on the rest of his land. The beans on the plowed strip were overrun with weeds: redleg, groundsel, mayweed, and, towering above the crop, fat hen. Perhaps as a result, as dense weeds tend to suppress growth, the beans on the tilled land carried fewer flowers. The plowed soil had hardened in the drought, while the unbroken ground was just as it had been before: sprung like a dancefloor.

Tim's wheat was looking pretty good. It was a fiercely armored variety called Skyfall, with a heavy, angular head and a sharp spine growing from every grain. It was at full height (a little above my knee) but still green: the soft grains burst easily between my finger and thumb. They needed only to swell, harden, and dry before he reaped them. A crop, at last, was on the brink of success.

Tim explained the constraints he faces. He sells to a flour mill, which seeks "total consistency. Every loaf has to be the same, so every load of grain must fit the specs. They want a protein content of over twelve percent, a Hagberg Falling Number* of over three hundred, and a water content of fifteen percent. If we miss any of that, it's disqualified as milling wheat and goes for animal feed instead, and we get a lower price. They don't make concessions. They don't accept suggestions.

"We want diversity. They want conformity. We get no points for wildlife, no points for soil quality, no points for a complex rotation. It just has to be cheap. This is the world we work in."

Tim's wheat is used to make bread through the Chorleywood Baking Process, a weird British invention that has spread around the world, whose story is both a classic case of unintended consequences and a reminder that local doesn't always equal good.

Until the middle of the twentieth century, baking was a local business. In some cities, there was a bakery on almost every corner. Here, people not only bought their bread, but also—as the baker whose family had worked at the end of my street for 170 years before he had to close the business explained to me—used the ovens to cook their Sunday roasts.

In 1932, a Canadian entrepreneur called Garfield Weston started importing wheat into the UK from his home country. As Canadian wheat, thanks to a more favorable climate, has a higher protein content than British wheat, he was able to undercut his local competitors. More protein means the bread rises better, which means more loaf from the same amount of flour. The company he founded, Allied Bakeries, threatened to put small bakers out of business.

In the 1950s, a group of small British bakers appealed to the British Baking Industries Research Association, based in the town of Chorleywood. It was a classic institution of the era: a state body that

---

* This measures an enzyme called alpha-amylase.

sought to make national industries more competitive. Could it devise a baking process that would use British wheat cheaply enough to topple Mr. Weston?[64] By 1961, the Association had cracked it: a quick and economical technique that could work with low-protein flour. Soon, we were eating British bread made from British wheat. But pride was all we got.

As Felicity Lawrence explains in her excellent book, *Not on the Label*, the Chorleywood Baking Process is a brutal business.[65] Instead of gently kneading the dough then letting it rise for three hours, as traditional bakers do, the process uses high-speed mechanical mixers to stretch the dough and lengthen its protein chains in just three minutes. After that it proves in precisely fifty-four minutes, and cooks in twenty-one. To make this work, it needs twice the quantity of yeast, as well as chemical oxidants, hardened fat, extra salt, enzymes, and emulsifiers that tend to be made from petrochemicals. This concoction produces the great majority of the bread we now eat. The loaves it makes appear to be indestructible. Molds and rots scarcely afflict them, they take ages to stale, they are the undead of bread. Long after civilization has collapsed, and wild dogs forage among the ruins, they will sniff out bags of Chorleywood bread with years left on the sell-by date.

It was inevitable that a highly automated, capital-intensive process like this would favor large corporations over small local enterprises. Within a couple of decades, most of the remaining small bakers had disappeared, and bread belonged to big business.[66]

The fact that bread seems to have become an unhealthy product cannot be blamed entirely on this innovation. Similar processes, using higher protein grains, have ruined this staple just as effectively in North America. Much of the damage is done before the brutal mixing, stretching, and adulteration begins, as flour today tends to be milled and refined in ways that remove most of those indigestible elements— the fiber—that turn out to be essential for health, and many of the minerals, vitamins, natural oils, and other components.[67, 68] But, in principle at least, it would be easier to make a cheap but healthy mass-produced loaf from Canadian wheat than from British wheat, which, because of our climate, is poorly suited to most kinds of breadmaking. Self-reliance doesn't always align with food security.

*

Even a commercial farm like Tim's struggles to pay the bills. Like many farmers, he and his dad have had to diversify to survive, turning their grand house and outbuildings into a hotel, conference center, and wedding venue. So far, so conventional. But after we had inspected his crops, Tim took me to see the craziest and cleverest farm diversification I've encountered. It's what every modern farmer needs: a long barrow.

He had built a full-scale Neolithic monument, brand-new but designed 5,500 years ago. The Neolithic, or new stone age, is the period in which farming began, and Tim's reconstruction closed the circle in a fascinating way. It's not a folly, but an essential part of his farm's economy. From the outside, it looks like a conical heap of earth with grass growing on top. But as you duck through the doorway, you see that it has been built around three massive yellow stones, two of which form the entrance, and the third the lintel and part of the roof. You step into a high, domed chamber built from small stone blocks. It feels cool and hushed and hallowed. Beyond the chamber is a long gallery, built on the same principles. The front door aligns with the rising sun of the summer solstice, the back door with the rising sun of the winter solstice.

"I was slightly nervous about getting it right. The first solstice was an anxious moment. But the sun rose where it was meant to, and shone straight through the door."

Built into the walls of the chambers are hundreds of loculi: alcoves where funeral urns are placed, in which Tim's customers pay, in some cases, several thousand pounds to be left in peace. Even before it was finished, the spaces began to fill. "We have pagans, Roman Catholics, evangelical atheists. It seems to be a place where people put aside their differences."

Tim calls his barrow "neo-Neolithic." He describes his farming the same way.

"In the Neolithic, they couldn't plow, as they didn't have draft animals. So they must have drilled their seed directly into the ground, perhaps with a tool made from an antler. It's taken five and a half thousand years to realize we made a mistake, and to relearn what they knew."

Soon after my visit, massive thunderstorms broke across the country, beating down the ripening crops and completing the worst farming year in living memory. Like everyone else, Tim suffered great losses. There are ways of enhancing farming's resilience to climate chaos. There are no means of making it impervious.

Tim is supplying the mainstream cereals market, and trying, within the extreme constraints it sets, to minimize his impacts. But the next farmer whose work I've followed, Ian Wilkinson, is approaching the problem from the other end: he is seeking to change the market. He practices a kind of agroecology. Agroecology means not only farming more sensitively, with fewer chemicals, less use of machinery, and more reliance on natural systems, but also changing the relationships between farmers and the rest of society. It means creating food networks that aren't dominated by seed and chemicals companies, grain barons, or supermarkets, but are independent and self-organized.[69, 70] In other words, it seeks food sovereignty. Farmers use their fine-grained knowledge of the land to find subtler ways of producing food,[71, 72] and their fine-grained knowledge of the market to find better ways of selling it.

Most of the farmland in Britain was carved from the wildwood at least a thousand years ago.[73] But on the part of the Cotswold plateau where Ian farms, in the center of England, the great trees of Wychwood Forest—hunting grounds of Saxon kings—were not felled until the 1860s. Only six generations have farmed this land, but already they have ripped it open.

The soil is thin, and has a stone content similar to Tolly's, but as the name of this fissile Jurassic rock—cornbrash—attests, it's reasonable land for grain. The hardest parts of the rock are the fossils it contains: in the fields they are often all that remain. So as I walked around Ian Wilkinson's rotation, everywhere I saw fragments of oysters, scallops, and brachiopods. It felt like trespassing on a dinosaur's beach.

Ian is a slim, lithe man in his late fifties, with a finely sculpted face and a gentle but determined manner. He has spent thirty years working at a private company called Cotswold Seeds. Soon after he and his wife, Celene, bought out the owner, the business started taking off.

"As the profits built, Celene and I couldn't think of anything we would rather do with the money than something like this."

Ian says "something like this," but there is nothing like this. His project, FarmED,[74] is a farm business that seeks to make no money: in fact he draws £100,000 a year from Cotswold Seeds to support its running costs. It's a controlled experiment, comparing conventional cereal-growing with agroecological methods. It's a community project, an education center, and a food economy in microcosm, all packed into 107 acres (43 hectares). It cost him a fortune.

"It was almost impossible to find a farm to buy. And land costs £10,000 an acre just about everywhere now. We needed a massive mortgage."

The sums sometimes keep him awake at night. But he has built something beautiful. For my purposes, it is also very useful, as his experiment permits direct comparisons between different ways of growing arable crops.

Though they are very different characters, Ian has a quality I also appreciate in Tolly, Fran, and Tim: he is honest about himself. This, in my view, is the definition of maturity. It has nothing to do with age: I've met mature thirteen-year-olds and immature eighty-year-olds. I believe what these four people tell me, because they tell me the downside. In every case, their approach is open and empirical, receptive to the lessons of failure. I cannot think of a more attractive quality.

When Ian bought the farm, it was growing one thing: barley for the Heineken brewing company. "No one had work except the farmer. You couldn't eat anything that came off the land. And the business wasn't economically viable. So whatever we do has to be an improvement on that." During the first year, he ran the old business model, to see how it worked. The barley cost him £11,020 to grow. He sold it to Heineken for £10,960. Result, misery.

He then started to build a system of his own. Like Tolly and Tim, Ian began with the soil. When he took on the farm, the soil was being treated as something for plants to stand up in. No effort had been made to build its fertility: this was supplied instead by lashings of chemical fertilizer. Its volume of organic matter,* a good indicator of soil health, stood at just 3 percent. Ian's intention is to build that up to 9 or 10 percent.

---

* This means molecules with carbon-hydrogen bonds.

He has kept one hectare in the old system, growing cereals with commercial seed and chemicals. This is the control against which he can compare his new, experimental techniques. He maintains 18 hectares of clay land as pasture, on which three cows graze. In one field he has planted heritage fruit trees. Two hectares are set aside for a community farming program, growing vegetables. The rest of the farmland—24 hectares—is devoted to building the soil, then using it to grow grain in a system that could scarcely differ more from raising barley to make bad beer.

On a wet, cold day in mid-May, when the valley of the River Evenlode beneath us and the wide lands that rose from it were blotted out by rain, Ian walked me through his system. The control crop, a modern milling wheat, was growing strongly: a dense lurid green tufting from the gingery soil. But he explained that achieving this state had required five passes of the tractor: three to spread nitrogen, one to apply weedkiller to slaughter the blackgrass that sprang up when the field was drilled, and one for fungicide (to kill the rusts that might otherwise harm the crop). Like Tim, he carried a sampling spade, but digging a spit was harder here, as the blade crunched painfully on stone. The soil he extracted from this chemical desert immediately fell apart. The roots of the developing crop were tiny and shallow.

Though it was hard to imagine on a day like this, Ian told me that the main limiting factor for crops on the plateau is soil moisture. When the carbon in the light soils here is depleted, they retain little water. Carbon, in the form of the polymer scaffolding extruded by microbes and soil animals, builds a healthy soil. A healthy soil holds water. So water needs carbon.

During the seven years in which Ian had owned the farm, the control crop had failed twice: once as a result of drought, once because of the thunderstorms that had also trashed Tim's crops in Shropshire.

The next field had also been planted with wheat the previous October, but you would scarcely recognize it as the same species. Though the plants were a little sparser, they were more than twice the height of the crop in the control field. This was heritage grain, saved and multiplied by an old friend of mine, the paleobotanist and plant breeder John Letts. Modern wheat has been bred for maximum grain and minimum straw: it tops out at around 50 centimeters. The old varieties rise to a meter or more.

Ian had first planted a mix of forty old breeds, an astonishing number to grow in one field. This, he told me, would increase the crop's resilience. Different varieties might use slightly different resources, which could reduce competition between the plants.[75] As he multiplied and resowed the grain in subsequent years, the breeds best suited to the land would start to dominate the seed mix.

Again, he shoved the spade into soil, metal scraping on stone. This spit held together slightly better than the first, bound by the longer roots of the wheat.[76] I could smell the difference. In the control plot, the soil scarcely smelled at all. Here it was rich and carroty. Though the test results were inconsistent (carbon levels vary greatly across a single field, so results tend to be unreliable),[77] Ian reckoned that in seven years the carbon content had roughly doubled. He told me that during the two years in which the control crop had failed, his heritage wheat had survived. Though the plants were taller, they had not been toppled* by the heavy weather, as their roots and stems were more robust, and the heads were lighter. This wheat was being grown without pesticides or fertilizer. Unlike the short, modern varieties, the ancient breeds can prosper without the use of herbicides, as taller plants tend to shoot past the weeds and shade them out.[78]

Like Tolly, Ian uses a long and complex rotation to raise the fertility of his soil. He has divided the 24 hectares of light land on top of the plateau into eight fields, through which his system cycles. For the first four years of the rotation, he grows a herbal ley, which means a mixture of grasses and wild flowers, several of which harbor nitrogen-fixing bacteria in the nodules on their roots. Ian explained that herbal leys had once been an essential part of mixed farming systems. But, thanks to the wide use of artificial nitrogen, they had vanished for a century. As progressive farmers sought to restore their soil, however, they were coming back into fashion. At Cotswold Seeds, 2,000 customers now buy his ley mix.

After the ley, there would be a year of wheat, undersown with clover, then a year of oats, then one of wild bird seed, which the

---

* The proper term for a crop falling over is lodging.

government pays him a handsome subsidy to grow. In the eighth year he would reseed the herbal ley.

Unlike Tolly, he uses animals. A neighboring farmer brings part of his flying flock of sheep to Ian's farm. A flying flock is less exciting than it sounds. It means that the sheep are moved from farm to farm as grazing becomes available. Ian shifts them quickly through his fields. He pens off a square of ley with electric fencing, and leaves the sheep on it for just twenty-four hours. This ensures that the animals eat only the best parts of the grass and wild flowers, so the plants can bounce back quickly when they've passed. Then he moves them into the next block.

He explained that by eating the tops of the plants, the sheep ensure the ground cover thickens and keeps growing. During their brief visit to each paddock, they eat roughly a third of the vegetation, trample a third and leave a third. This appears to help return carbon to the soil.

Like Tolly, Ian intends to plant rows of trees between his fields, partly to produce woodchip for compost and animal bedding, partly to provide shelter for the sheep. There's a cost, however. In the tropics, shade protects some crops from the scorching sun. In countries like the UK, seldom afflicted by excessive sunshine, it tends to hamper them. One experiment shows that small trees grown for woodchip roughly halve the yield of the cereal crops grown close to them.[79] Ian told me that he would try to coppice the trees (cut them to the ground without killing them), taking their branches for woodchip, just before he plants wheat. But given the complexity of his rotation, and the different lengths of the crop and wood cycles, I find it hard to see how he could make them synchronize.

After four years, he plows the ley. He describes the process of creating a seedbed from soil that's thick with the roots of herbs and grasses as "pretty brutal": first he runs a plow over it, then a spring tine,* then a power harrow, then a drill, then a roller. But because the soil remains unplowed for four years in every eight, this harsh treatment doesn't undo all the good work. At least some of the carbon and some of the improvements in soil quality that accumulated during the fallow period persist.

---

* A kind of giant rake.

I followed Ian around the rest of the farm. The vegetable growers were working in the field he leases to them. They run a Community Supported Agriculture (CSA) program. Its subscribers, who live in the nearby villages, each pay £35 a month for vegetables and a small amount of fruit. Some of them help the growers. This, Ian told me, is what he's trying to encourage: building a local food economy in which the land supports employment, and the employment builds a network of trust: people can see what they're buying, and need neither marque nor marketing to assure them it's good.

"Every parish should have a CSA.

"It works for us as well as them. We get some rental income. But most importantly, we get people. It ensures that more people pass through the farm and engage with the other things we do."

He showed me the three cows with their calves that another micro-business—a dairy—was running on his land. They were, to put it mildly, lightly stocked on the 18 hectares they roamed. Ian intended to let them follow the sheep around his rotation at certain times of year, expanding their range even further. He had explored the idea of bringing in chickens and pigs to forage in his cropping fields. But he had decided that, while their scratching and rootling and defecating might (if their numbers are very low) enhance the quality of the soil, it would mean breaking his rule of bringing no inputs onto his farm except for the control plot. The great majority of their feed would need to be grown on ghost acres. A lot of nonsense has been written about chickens and pigs supporting themselves by foraging. In reality, unless their numbers or productivity are tiny, most of their food must be delivered to them.

Ian had let his hedges run wild, and the birds loved them. When I returned to his land in the winter, goldfinches, linnets, and yellowhammers flitted through the long ragged tops, drawn to the farm by the birdseed he grew. On my next visit, the following September, hundreds of starlings filled them with chatter, and a sparrowhawk stormed through the rising trees.

He had demolished the old outhouses and erected in their stead two large and stunning buildings. Though airy and modern, they seemed to nod to the size and shape of the great tithe barns that once stood in these parts. One held offices and a cavernous lecture theater, the other kitchens and a vast banqueting hall. In the middle of the hall stood a domed bread oven, built from brick and lime, in which logs now glowed. We sat at the circular bar surrounding it and ate a delicious lunch: fava bean dip, herby potato salad, a pickled squash, and green salad with a pumpkin seed dressing. Matt, the baker who works here, walked in with a sack of flour on his shoulder, and started mixing dough.

He explained that he takes his own grain to the local mill, has it ground, then picks up the flour. Ian doesn't yet grow enough to supply him, so Matt must buy heritage wheat from other farms. At the moment he bakes here only once a week. He stays up all night, feeding the oven with dough. Like the vegetable growers, he delivers it to his subscribers. While he could easily charge £5 for one of his sourdough loaves, he deliberately keeps the price low, at £3, to make it accessible to as many people as possible. This, of course, is a lot dearer than the Chorleywood bread whose bakers Tim Ashton supplies, but the simple, slow process and smaller number of ingredients ensure that his bread is likely to be healthier.

A great deal has been written about the differences between heritage and modern wheat, and not all of it is true. In reality, there are no significant differences in important components such as fiber, B vitamins, and plant chemicals like phenols and terpenes, or in gluten content, starch composition, or glycemic load.[80, 81] In one respect (the level of an amino acid called asparagine, which, when baked, forms acrylamides: neurotoxins that are potentially carcinogenic), modern wheat appears to be healthier than the old varieties.[82] However, it also tends to contain fewer micronutrients, especially iron and zinc. The bigger issue is the way in which the wheat is processed and the bread is made: modern industrial methods tend to remove or destroy its healthy components.[83, 84, 85]

Matt's bakery brought the number of small businesses working on the farm to five: dairy, sheep, honey, vegetables, and bread. Ian is trying to build what he calls "an artisanal bioregional economy." The idea is to shorten the commercial chain, ensuring that farmers and the small businesses they supply earn more money from their products, and to connect local people to the land.

"If you had a county-sized trading area," Ian told me, "there would be enough consumption to allow a mixed farming system like ours to gain a foothold."

His system is much more profitable than the one it replaced.

"We don't use fertilizer or pesticides, so no money is spent on them. Cultivation costs £12 or £15 an acre, and drilling about the same. Seed hardly costs anything, as we now grow most of our own. The total costs are maybe £60 an acre. Some bakeries are paying £600 a ton for heritage wheat flour. That compares to roughly £150 a ton for conventional wheat. So farming like this should be very profitable."

His vision has inspired thousands. Roughly every other day, coachloads arrive to study Ian's techniques, to learn about the soil, listen to the dawn chorus, join workshops on creating a food business, to cook and eat, to discuss the future of farming.

Listening to him, sitting in the magnificent hall he'd built, I was struck by the different strategies he and Tim Ashton have used to navigate the price constraints imposed by cheap bread. Tim is trying to match his production to the market, while farming as well as he can within the limits it sets. Ian is trying to match the market to his production, enabling him to farm as well as he wants.

Ian's system is neat and appealing. But there's a problem. He told me that his heritage grain produces roughly half the yield per hectare of the conventional wheat he grows in the control plot. This would be troubling enough. But what makes it much worse is that, for six years out of eight, his fields produce no food at all, unless you count the small amount of lamb produced while the flock moves through them (at other times it flies to different farms). By contrast, Tolly crops his fields for five years out of seven.

Even in our changing climate, the two years in which the wheat in Ian's control crop was destroyed were exceptional. But, being generous to his figures, let's assume that conventional wheat fails twice in the course of every eight years. If, as Ian told me, the wheat he farms intensively here produces a little over 7 tons of grain per hectare, this would mean that over eight years his control plot would provide roughly 44 tons, albeit at a high environmental cost. By contrast, his heritage wheat in the same period would produce just over 7 tons. I think we can accept small differentials—10 or even 20 per-

cent—if other environmental issues are resolved. But a sixfold difference is surely a deal-breaker. In terms of land use, which I see as the
most important of all ecological metrics in farming, his wheat no longer
looks like grain, but more like meat. A kilogram of his heritage wheat
has a land footprint very close to that of a kilogram of chicken.[86]

Ian hopes that, as soil fertility builds, he might be able to introduce
another cash crop into the rotation. But this would still leave five
years out of eight in which no food leaves the field (except a few lamb
chops). To shorten the rotation any further and produce more food
would, he told me, defeat the point of his system, as he could no
longer build soil carbon and fertility.

Looking only at his farm, everything he does makes sense. He has
created a profitable circular economy that produces good food, albeit
at prices that not everyone can afford. But because his yields are so
low, the production required to feed us migrates to other places, where
ancient habitats might be destroyed to make way for crops, or land
that could otherwise be rewilded—returned to nature—continues to
be farmed. What is beautiful is not always right. What is right is not
always beautiful.

None of this should be seen as a general criticism of agroecology. It
can be transformative. In Karnataka, southern India, a strategy called
Zero Budget Natural Farming has released small farmers from the
grip of moneylenders and banks: by replacing the chemicals and seed
they bought with natural processes and seed they save themselves,
they have almost eliminated their expenses.[87] In Mzimba district in
Malawi, research teams led by farmers have reduced the need for
fertilizer, while raising their yields, protecting their soil, enhancing
women's power, and improving children's nutrition.[88]

To make it work, farmers need clear and secure land rights. Otherwise, they have no incentive to protect their soil, plant trees, and farm
as if tomorrow will come.[89] While such rights are a necessary condition for good farming, they are no guarantee, as examples in this
book attest. In some countries, I believe, farmers have too many rights
and freedoms: the right to build giant chicken barns without environmental permits or use vast tracts of land to produce tiny amounts of
food; the freedom to trash the soil, pollute the rivers, and intimidate

neighbors who object. But in other countries, they have too few, and can easily be evicted by land-grabbers. While the rights of big companies are guaranteed by international treaties, local people often have no protection, and governments and businesses sometimes collude to throw them off their land.[90]

One of the principles of agroecology is equality. This is hard to achieve when land costs £10,000 an acre (£25,000 per hectare). It's worth noting that all the farmers I have followed so far rely on what could be seen as an historical subsidy. Tolly has been able to develop his system as a result of the generosity of Sir Julian Rose, who charges him a land rent well below the market rate. Tim survives because he and his father, like Julian, inherited their estate, so they can farm without a mortgage on the land. Ian purchased his farm with money made elsewhere. Those with the money have the land, and those with the land have the power. Some will use it, as Julian, Tim, and Ian do, with magnanimity, working on behalf of other people or high ideals. Most will not.

Sometimes the yields from agroecology are higher than from conventional farming,[91] especially when all the produce small farmers take from their land is counted:[92] the leafy vegetables that might grow beneath the cereal crop, or the beans that might grow through it,[93] the nuts and fruit on the field margins, and the fish, crabs, and snails that might populate rice paddies and irrigation channels.[94] In some cases, growing a mixture of crops—such as grain and nuts and vegetables—produces more food than growing grain alone.[95, 96]

But too often, those who promote agroecology are yield-blind. One article, by an author who usually takes more care, enthused that an organic farm in California grows 50,000 pounds of food (mostly heritage wheat) on 130 acres.[97] This means a yield of less than half a ton per acre: one-fifth of the average in the U.S. (where wheat yields are low by international standards),[98] and one-sixteenth of what you would expect in the UK. Just as it's impossible to feed the world on pasture-fed meat, it's impossible to feed the world on low-yield agroecology. In every farming system, we should seek two properties: high yields and low impacts.

Both Tim's system and Ian's have great virtues and major drawbacks. Both, I feel, help us to understand what we should be looking for. We need healthy food that's cheap enough to let everyone eat well. We need high yields, to ensure that farming feeds the world without sprawling across the planet. We need healthy soils, whose fertility can be raised without either dousing the land with fertilizer or removing it from production for long periods. We need, as far as possible, to stop using herbicides and pesticides and to reduce the need for irrigation. We need farm landscapes that provide habitats and corridors for wildlife. The technologies we choose should be simple and cheap enough for small farmers to use, so that capital does not overwhelm labor. They should be varied enough to reverse the dangerous homogenization that creates the Global Standard Farm.

It sounds like a tall order. But I think it can be done.

While I was researching this chapter, the doorbell rang. Deliveries are usually an annoying distraction. But when I saw what the man was holding, I bounced down the stairs and flung open the door. I took receipt of a large cardboard box that contained a possible future. I asked my youngest daughter to open it, as it mostly belonged to her.

I'm sure she would have preferred a mobile phone. The contents were not the most thrilling present for a nine-year-old. There were five bags. Four of them contained flour, the fifth, rice. But when I showed her the remarkable photograph that accompanied them, and explained what it meant, she contracted a little of my excitement.

The photo was printed on a long thin strip of paper, folded in three. It opened to reveal two plants side by side, dug from the soil with their roots intact, shot against a black background. Both were members of the grass family. I recognized the one on the left. It was a modern, short-stemmed variety of wheat. The pale roots formed a thin goatee, a little longer than the plant was tall.

The grass on the right was a species I hadn't seen before. It was taller and bushier than the wheat. Its seedheads were long and thin. But the most interesting feature lay below ground: a ginger beard of roots that ZZ Top would have worn with pride, woolly and matted, blotting out the background and twisting down to the bottom of the third fold.[99] Using the wheat for scale, I reckoned it must have been over three meters long.

This was a specimen of intermediate wheatgrass.* The name is slightly misleading, as it has no close relationship to wheat. It's a wild plant, but the specimen in the photo had passed through several generations of selective breeding by a nonprofit group called the Land Institute,[100] which had sent me the box from its base in Salina, Kansas. Because intermediate wheatgrass is unlikely to lodge in the mind, the institute had given the plant a new name: Kernza. Four of the five bags contained flour made from its seeds.

There's nothing inherently amazing about flour made from grass seeds. After all, most of the grains we eat come from this family: wheat, rice, corn, oats, sorghum, millet, rye, and barley are all grass. Several businesses have brought new (or old) grass seeds onto the market, such as the heritage wheats that John Letts and others have revived, or spelt and einkorn. But what enthuses me about this flour is that it comes from a grass that differs fundamentally from all those mentioned in this paragraph. While those crops are annuals, which means that they must be replanted every year, Kernza is a perennial. It persists over several years, averting the need to clear and sow the ground for every harvest. Of all the possible agricultural technologies and techniques I've explored while researching this book, I find this the most exciting.

Large areas dominated by annual plants are rare in nature.[101] They tend to colonize ground in the wake of catastrophe: a fire, a flood, a landslide, or a volcanic eruption that exposes bare rock or soil. They survive, in these circumstances, only until perennial plants return and start to mend the broken land, whereupon they are usually overwhelmed.

But it's not hard to see why our ancestors selected them. Plants that colonize bare ground have evolved to grow fast and invest much of their energy in seeds, rather than in deep roots or dense foliage,[102] so that they can spread as far as possible before the new land closes up. Their seeds tend to be large and to germinate freely.[103]

The problem is that in cultivating annuals, we must keep the land in the catastrophic state they prefer. Every year, we must clear the soil of competing plants, puncture or turn it, and plaster it with the

---

* *Thinopyrum intermedium.*

nutrients required to raise a crop from seed to maturity in a few months. However sensitively it is conducted, annual grain production relies on sustaining an ecological disaster for its success.[104] But if instead we grew perennial grain crops, we would not depend on smashing living systems apart to produce our food.

For the past forty years, the Land Institute has been scouring the world for perennial species that could replace the annuals we grow. With other research groups, it has assessed thousands, looking for particular features:[105] high yield, synchronous development (so all the grain can be harvested at the same time), seed retention (which means that the grain stays on the plant until it's harvested),* and ease of growing and reaping (especially by small farmers).

Once they've found likely species, the researchers start selective breeding to raise their yields and improve their food and farming qualities. At first, they simply selected, crossed, and multiplied the individual plants with the highest yields. Now they use new technologies[†] to speed up the process.[106, 107] Breeding of the kind that once took centuries can be compressed into a couple of decades.

Sometimes a promising plant goes nowhere. Sometimes an unpromising plant takes off. For example, the Land Institute gave scarcely a thought to a perennial sunflower called *Silphium*, because in the wild it produces only fifteen or twenty seeds on every flowerhead. But by casually selecting the plants that had a few more seeds than average, within eight years its researchers found themselves harvesting heads with 150 seeds.[108]

Though Kernza—intermediate wheatgrass—is still being developed, it's one of the first crops that the Institute has tentatively brought to market.[109] So far, the plants have been bred to carry as many seeds as wheat does, but each one is only a quarter of the weight. The breeders hope to match wheat yields within thirty years.

Because the land remains covered for years at a time, bound by their massive roots and showered with their abundant litter, perennial grains like Kernza are likely to reduce erosion[110] and raise the amount of carbon in the soil.[111] One estimate suggests that, if the switch were

---

* The desirable trait is known as shatter resistance.
† Such as gene sequencing, genomic prediction, and marker-assisted selection.

made worldwide, agricultural soils would regain between a quarter and two-thirds of the carbon they've lost since they were first plowed.[112] Ecosystems dominated by perennials, such as forests and natural grasslands, support richer and more abundant soil life than fields growing annual crops.[113, 114, 115]

As their long roots draw nutrients from deep in the ground, perennial crops become their own green manures, their own herbal leys. So the land need not be taken out of production. The longer the plants stay in the ground, the stronger their relationships with bacteria that fix nitrogen, and microbes and fungi that seek out other nutrients. This means that they should, in principle, need less fertilizer.[116] One estimate suggests that perennial systems hold five times as much of the water that falls on the ground as annual crops do.[117]

When farmers switch from annuals to perennials, weeds are likely to be a major problem during the first couple of years,[118] as the land can't be plowed, and general herbicides would kill the crop. But once a perennial crop is established, its shade and litter tend to stifle weeds. Experiments with perennial energy crops suggest that no further weed control is needed.[119] Kernza, according to some of the farmers who have experimented with it, is "highly weed-suppressive."[120] Given that it forms great fibrous clumps covering the soil surface, it's not hard to see why.

Because weeding is likely to be needed only at the beginning, it might make economic sense for farmers to do it mechanically: hoeing or mowing between the rows with machinery (or, on very small farms, with hand tools) or, when the technology becomes widely available, briefly hiring a robot. The less the soil is disturbed, the fewer opportunities weeds have to take root: all else being equal, they should be easier to control in a perennial system.[121] Of course, these deep-rooted perennials could themselves become weeds, invading other fields or persisting when farmers want to change the crop.[122] This is one of many issues that needs to be investigated, for every new perennial plant that's bred.

Some papers argue that plant diseases could become a bigger problem in perennial grain farming, as they might proliferate from one year to another,[123] others that they could become a smaller problem, as stronger relationships with soil bacteria allow plants to ramp up their immune responses.[124]* Encouragingly, the wild plants from which Kernza and other grain crops are being bred seem highly resistant to the viruses that attack annual crops.[125]

The deep roots and tough structures of perennial plants are likely to make them better adapted to climate chaos. The perennial sunflowers I mentioned have sailed through two severe droughts, one of which entirely destroyed the annual sunflowers grown alongside them.[126] Farmers can greatly reduce their costs: once these crops are established, they should require little in the way of fertilizer, less irrigation, perhaps (as I'll discuss below) no pesticides, and less use of machinery. Because the poorest farmers have generally been pushed into the poorest lands, these crops, which tend to be less demanding than annual plants, could suit them better.[127]

As I opened the first bag of flour I felt a little anxious. What if it was disgusting? Because the Land Institute's milling had produced quite a heavy wholemeal, I mixed it, half and half, with a strong white wheat flour. It kneaded well, quickly becoming elastic, though it had the strange property of suddenly turning sticky after about five minutes. It rose nicely.

As it cooked, I found myself pacing up and down like an expectant father. After thirty-five minutes, I tapped the base of the loaf, and it answered with a nice hollow thud. So far so good. When it cooled, I cut it open, and was delighted to find a light, soft crumb. The flavor was fantastic: rich and slightly nutty, one of the best loaves I've ever tasted.

Over the following days, I made wraps, salt crackers, and digestive biscuits from the Kernza meal. The wraps stretched beautifully, then bubbled and puffed in the pan until they became almost spherical, the dough staying soft while it cooked all the way through. They had a slight beeriness, which subtly offset the spicy refried beans and tomato salsa with which I served them. The crackers, by contrast, were crisp and crunchy, the flour's nuttiness emphasized by chips of seasalt, and enhancing, in turn, the taste of the pumpkin seeds and crushed walnuts I had mixed into the dough. The digestive biscuits crumbled

---

*The process known as Induced Systemic Resistance.

nicely, with a trace of sugar cutting through the flour's tawny flavor. I found that the less I mixed the Kernza with other cereals, the better it tasted.

Even if we were to judge this crop only on taste, and ignore its ecological advantages, it would commend itself. I felt an unwarranted pride in being among the first people in my country to use it. The Land Institute had attempted something that had foiled other people for almost a century. The Soviet Union, for example, tried to breed a perennial cereal in the 1930s, which must have been a challenge after the inheritance of genetic traits was officially condemned as a capitalist delusion. Kernza is at least halfway to success.

The fifth bag goes all the way. The rice it contained came from a breeding program started by the Institute and continued by the brilliant plant scientist Fengyi Hu, at Yunnan University in China. They had crossed an annual Chinese rice variety with a wild African plant in the same genus,* to produce a perennial crop. Already its yields match, and in some cases exceed, those of modern annual rice breeds.[128] This cultivar, which goes by the unromantic name of PR23, has now been planted across 7,000 hectares of China, as a fully fledged commercial crop.[129]

Farmers can't get enough of it: all the available seed is immediately snapped up. Annual rice grown on slopes often causes devastating soil erosion. Sometimes entire terraces collapse as a result of excessive cultivation. But the long roots of PR23 help bind them and protect the soil. Perhaps the more immediate reason for its popularity is that rice farmers in many parts of China are desperately short of labor: many young people who might have done the arduous manual planting have left the countryside. Some fields of PR23 have now been harvested twelve times without resowing.[130]

I cooked the rice the Institute sent me, and found it, in taste and texture, indistinguishable from the short-grain annual rice on sale in my local Chinese supermarket. In other words, the perennial crop is essentially the same product. Now Professor Hu is crossing PR23 with other varieties, for different uses and flavors. The Greener Revolution has begun.

*

Oryza longistaminata.

The Land Institute and its partners are working on many other species, including beans,[131] sorghum, and oil seeds. They have hybridized ordinary wheat with Kernza, to create a new perennial crop that has already achieved between 50 and 70 percent of the yields of annual wheat.[132]

So here is one vision of what a new farming system could look like.

Cereals such as Kernza, perennial wheat, or rice could be planted in strips the width of two or three combine harvesters (or otherwise matched to the reaping technologies used by small farmers). Between every two strips of cereal grain is a strip of perennial beans. Beans house nitrogen-fixing bacteria, which fertilize the plant and add nitrates to the soil. Without annual plowing, nitrogen is likely to accumulate in the soil, helping to feed the neighboring cereals.[133] These crops, with their deeper roots and longer establishment, should find most of the other minerals they require. In some circumstances, it might be possible to eliminate the need for fertilizer.

Around each block of cereals and beans (or, for that matter, oilseeds and beans) is a band of native perennial wildflowers, about a meter wide, of the kind that Tolly uses. Both wild pollinators[134] and the predatory insects that eat crop pests[135] flourish better among perennial flowers than in annual flowerbeds. Farmers are often hostile to the idea of maintaining perennial flowerbeds in the midst of their crops,[136] as the tractor must work round them, and, according to one study, it takes about four years for their full benefits to be felt.[137] But in perennial fields, in which the tractor does less work and the crops remain in the ground, they might be more popular. The flowering banks—and possibly the crops themselves—become corridors and stepping stones for wild animals passing through the fields.

In annual systems, only the crops closest to the flowering banks are visited by the beneficial insects they harbor.[138, 139] In perennial systems, helpful insects and spiders should be able to travel further, as the denser plants provide them with refuge.[140, 141] If Tolly's results are anything to go by, natural predators could eliminate the need for pesticides. One study shows that, when flowering crops (such as beans or most kinds of oilseed) are grown, the net impact of pesticides on farmers' yields and profits is negative, as the harm inflicted by killing pollinators outweighs the benefits of killing pests.[142] Big Farmer has been conned by Big Pharma.

Perennial crops are not a panacea—nothing is. They cannot stay in the soil for ever: as they age, their yields are likely to decline, so after a few years they must be grubbed up and replaced with another species.[143] But a rotation of perennials breaks the ground less often than a rotation of annuals. In common with every system, a switch to perennial grains could, without good policy, trigger perverse and disastrous consequences: for example, because they can grow on marginal land,[144, 145] they could be used to produce biofuels in places unsuitable for other crops. Then they could become a powerful instrument of habitat destruction. There's a limit to how much we can eat. There's no limit to how much we can burn.

However successful the breeding programs might be, perennial crops will not be grown everywhere: farmers are often reluctant to abandon familiar strategies. But this isn't necessarily a bad thing: we need a variety of systems and technologies. They will introduce diversity and resilience to food production, challenging the dominance of the Global Standard Farm. By conserving the soil and reducing the use of machinery, water, and agrochemicals, they could, in principle, greatly ameliorate farming's environmental impacts. It is extraordinary that this crucial shift is being led not by governments or multilateral agencies, but by a small nonprofit working from a farmhouse in Salina.

A transition from short-term to long-term crops, and the short-term to long-term thinking that should accompany them, is one crucial step toward a better world. But there is further to go. We cannot complete the journey until those of us who have a choice of diets stop eating animals and find new sources of protein and fat.

7

Farmfree

Whether I was woken by a sound it made, or just happened to emerge from sleep at that moment, I don't know. But when I opened my eyes, it almost filled my field of vision. As everything was sideways, it took me a while to see what I was seeing. I was staring into a gilded eye. At length, I realized that it belonged to a hare, grazing on the grassy bank where I had been sleeping, a meter from my face. It sat up then slowly lolloped up the bank, across the little suburban road, along the verge on the other side then into a garden. It felt, in this Helsinki suburb, like an omen. But of what I cannot say.

Exhausted by the interview I'd just conducted, I had stepped outside for a power nap while the rest of the crew stayed in the laboratory, filming cutaways.[1] Pasi Vainikka, the man I'd been interrogating, is a brilliant scientist and a visionary entrepreneur, but I suspect he would fail the Turing Test. Several times he answered my enthusiastic questions with a mechanical "That is correct." Making conversation between takes was like rattling the handle of a locked door.

But not every insurgent makes stirring speeches or wears a bicorn hat. Pasi is building what might be the most important environmental technology ever developed. It requires no major breakthrough, no new instruments or materials, merely refinements and efficiencies. While his laboratory is a maze of pipes and wires, valves and gauges, pumps and propellers, he is simply—or not so simply—brewing. Everything he uses is standard equipment. His process was developed in the 1960s, by scientists working with NASA.[2,3] But he's fermenting a revolution.

Pasi's task is to multiply, as cheaply and efficiently as possible, one of those creatures I discussed in the first chapter: a soil bacterium.

Among the survival strategies of microbes in the soil is the ability to harvest energy in interesting ways. The species Pasi uses* draws its energy neither from photosynthesis nor from the products created by other organisms, but from hydrogen.[4] It is a hydrogen-oxygenating bacterium.

The view through a porthole in the fermentation tank, where Pasi's brew churned as if in a washing machine, was unpromising: a thin yellow sludge slapping against the glass. But when he extracted some of this primordial soup and piped it onto a heated drum, it began to seem more appetizing: it turned into a golden flour that smelled like scrambled egg. Pasi explained that this species produces beta carotenoids: the molecules that give carrots their color. More importantly, roughly 60 percent of the flour is protein,[5] and all of it is edible.

Out of sheer vanity, I asked his team to make me a pancake: I wanted to be the first person outside the laboratory to eat one made from these bacteria. Instead of mixing the dried microbes with eggs to raise their protein content, the lab staff had to mix them with wheat flour to reduce it. Otherwise they would have made an omelet. They added some oat milk (though water would have done as well) to make a light batter.

It's a long time since I've eaten a conventional pancake, made with eggs. I've tried gram flour substitutes, dosas composed of rice and lentil flour, and Korean versions made of wheat and corn flour, but none were as satisfying as the pancakes I stopped eating when I switched to a plant-based diet. What was missing was the protein—it's what makes a pancake succulent. The richest of these alternatives—gram or chickpea flour—has a protein content of around 20 percent, lower than the traditional Western European batter of wheat flour, egg, and milk. Had I not watched it being made, I could scarcely have believed that eggs had not been added to Pasi's pancake. I would have found it still harder to accept that the main ingredient was the desiccated bodies of bacteria. It tasted rich and mellow and filling: just like the pancakes I used to eat.

That's one small pancake for man, one giant flip for humankind. It represents, I believe, the beginning of the end of most agriculture.

* *Cupriavidus necator.*

In that thin crêpe is wrapped our best hope of restoring the living planet.

The reason is that this method of food production shrinks, to an astonishing degree, the most important environmental impact of all: our use of land. It's here that the potential lies for a radical transformation of our relationship with the living world and the restoration of planetary health.

Tomas Linder, an associate professor of agricultural sciences, has compared the land area needed to grow protein by a process similar to Pasi Vainikka's with the most efficient agricultural method: U.S. soybean farming.[6] In a typical year, soybeans occupy 36.5 million hectares of the U.S., an area greater than Italy. The land required to produce the same amount of protein by growing bacteria is 21,000 hectares: the size of the city of Cleveland, Ohio. In other words, you'd need 1,700 times less land to grow it.

That's not quite the end of the story. We would still require processing facilities, to kill and dry the bacteria, separate their components, and recycle the wastewater from the fermentation vats. But the same applies to processing soybeans (and, to a much greater extent, the carcasses of animals). More importantly, Pasi's brewing needs electricity, mostly to split water into hydrogen and oxygen through electrolysis. As the name of his company, Solar Foods, suggests, he has chosen to generate it from sunlight.[7] This is a land-hungry form of power generation. Even so, according to a paper in *Engineering Biology*, producing bacterial protein with solar panels needs between thirty and sixty times less land than soy protein.[8] If wind power were used instead, the paper maintains, the ratio would rise to between 150 and 400 times. If the wind turbines were built offshore, where much larger machines tend to be used, even less land (or seabed) would be needed. If the hydrogen were produced by fourth-generation nuclear reactors, the space required would be even smaller.

While crop plants take months to grow, the bacteria in these tanks double every three hours. So if you maintain good growth conditions, you can harvest half of them eight times a day, every day of the year.[9] This technology could release almost all the land currently needed to produce protein, whether it comes in the form of plants or animals. Much of our food supply could be farmfree.

As grazing occupies two-thirds of agricultural land, and grains grown to feed animals or protein crops for humans account for much of the rest, this could permit land-sparing on an otherwise unimaginable scale. We could withdraw our dire impacts from great tracts of the planet that we have plowed and fenced and grazed and doused with toxins. Indigenous people could reclaim and restore their lands; ecosystems could rebound. This transition could be our best hope of stopping the sixth great extinction.

A study in *Nature* proposes that if destructive activity ceased on just 15 percent of land in some parts of the world, 60 percent of the extinctions that would otherwise happen could be averted and 30 percent of all the carbon dioxide released since the Industrial Revolution could be extracted from the atmosphere.[10] But the promise of farmfree food is much greater than this: we could rewild most of the land now used for farming, while protecting the remaining wild places.

As well as land, Pasi's technology makes parsimonious use of other resources. He's designing his process to enable the water and carbon dioxide that the bacteria need to be drawn directly from the atmosphere outside the brewery: his ambition is to conjure food from air. In principle, the nitrogen the bacteria require could also be extracted from the sky,[11] as some hydrogen-oxygenating species can make direct use of it,[12] but Solar Foods uses the same source as farmers do: manufactured ammonia. Unlike the fertilizer applied in farming, however, almost none of the minerals the bacteria need are lost. The system is enclosed, and the water and its effluents can be recycled.[13] Plenty of steel is required to build the plant, but as the process is so efficient and the throughput so high, the impact will be relatively small: probably lower than the steel requirement for the world's slaughterhouses and meat-packing plants.

My calculation suggests that, if everything else remained the same, and we used Pasi's method to produce all the protein needed for full human nutrition, it would raise the world's electricity demand by 11 percent.* As his process wastes much less reactive nitrogen than

* The Reference Nutrient Intake (RNI) for protein, which means the amount required for full nutrition, is estimated at 55 grams/day for men and 45 for women (https://assets.publishing.service.gov.uk/government/uploads/system/uploads/attachment_data/

farming does, part of this 11 percent would be offset by reductions in the energy needed to make ammonia and urea. Even so, the extra power carries a significant materials cost: the steel, copper, lithium, cobalt, rare earths, and other minerals required to build generators and transport and store electricity.[14, 15] But against this we should weigh the savings in equipment and fuel used for plowing, drilling, spraying, and harvesting, housing and moving livestock, slaughtering them, and processing their meat.[16] As the geographic spread, scale of infrastructure and loads that must be shifted to grow farm protein are much greater than those required to grow microbes,[17, 18, 19] the overall use of materials will surely be smaller.

The electricity requirement is unlikely to remain as high as 11 percent of current supply, as more efficient ways of making hydrogen without fossil fuels are being fomented: solid oxide electrolysers, high-temperature steam electrolysis,[20] and the thermochemical cracking of water[21] (using concentrated solar plants,[22, 23] or small nuclear reactors).*[24] More importantly, everything else is unlikely to remain the same. The size of new energy systems based on clean electricity will be matched to the highest levels of demand: during midwinter in colder nations, for example, or at the hottest times of year (when most air conditioning is used) at lower latitudes. During the rest of the year, and in quiet hours in the peak months, clean energy systems will produce more electricity than their customers need. Precision fermentation is one of the technologies that might take advantage of this surplus, as hydrogen could be produced during the periods when unwanted electricity is most abundant, and therefore cheapest. If so,

---

file/384775/familyfood-method-rni-11dec14.pdf). For simplicity I've assumed that the entire global population is adult, though in reality, children's RNI is lower. On the other hand, some people argue that the Reference Nutrient Intake is too low, as not all proteins are equally digestible. So this might roughly balance out. 50 × 8 billion × 365 gives us a total global protein requirement of 146 billion kg/year. Solar Foods' current rate of electricity consumption is 10kWh/kg bacterial mass. A protein content of 60 percent means 16.7kWh/kg protein. So total electricity requirement is 2,438 TWh. The International Energy Agency estimates global electricity consumption at 22,315 TWh (https://www.iea.org/reports/electricity-information-overview). So producing the world's protein by these means would require 10.9 percent of the world's current electricity supply.

* In particular, High-Temperature Gas-Cooled Reactors.

making the new foods would scarcely contribute to the need for new generating capacity, and therefore to the need for materials.

As hydrogen production becomes more efficient, bacterial protein also becomes cheaper. At the moment, by far the greatest resource cost is electricity.[25] As the price of solar power falls, Pasi sees the cost of his produce falling within a few years to that of the cheapest available protein on earth (soy). If bacterial protein is widely accepted, for the first time in human history we will have a staple food that did not arise from photosynthesis.

Before examining the potential of microbial fermentation to feed humanity, it's worth reminding ourselves that people go hungry not because the world lacks food, but because they can't afford to buy it. Globally, we produce far more protein than we need: 81 grams per person per day is available for human beings,[26] though we require, on average, only 50 or 60 grams.[27] A greater quantity of plant protein is produced, but as much of it is passed through livestock before it reaches us, a lot is lost. Protein consumption is poorly distributed.[28] The rich tend to eat far more than they need, while the poor eat less.[29, 30]

Not only do the poor lack access to animal protein; they are disproportionately affected by the massive environmental cost of producing it, through climate breakdown, habitat destruction, river, sea, and air pollution, water loss, and soil erosion. As usual, they get most of the pain and least of the gain. I'm not proposing that microbial protein be eaten only by the world's poor. I believe it could form the basis of sustainable diets everywhere. But it could be instrumental in protecting the poor from hunger.

While the international cost of bacterial protein is unlikely to fall far below that of the cheapest soy protein in the near future, it could be more accessible. In many poor nations, people pay a premium for protein, which, as most food cannot be produced locally,* often needs to be bought with weak currencies on commodity markets denominated in dollars. But microbial flour can be generated anywhere with an electricity supply, at roughly the same cost. Many of the world's poorest nations have a massive potential for renewable power.[31]

* For the reasons I explained in chapter 5.

With fair technology transfer and distributed manufacturing, which are crucial, this protein could become widely and cheaply available, regardless of the vicissitudes of climate, soil, currencies, terms of trade, bottlenecks on shipping routes, and other hazards. Because the breweries can operate independently and most of the necessary ingredients can be obtained locally, regardless of where the plant is sited, making food this way could impede the transmission of economic and political shocks. In other words, it could reintroduce a crucial property to the food system: modularity.

Technologies like Pasi's also offer solutions to the other wicked dilemmas explained in chapter 2. They could reintroduce the back-ups, redundancy, and buffering that our agricultural systems often lack.[32, 33] In extreme cases, such as an asteroid strike, massive volcanic eruptions, or nuclear winter, they would allow us to keep producing food when farming becomes impossible.[34, 35] But my main concern is to avert the slower but more probable catastrophes of climate and ecological breakdown.

So what kind of food will these technologies make? Well, once their full potential begins to be realized, the possibilities are limited only by our imaginations. We can replace many of the protein- and fat-rich foods we eat today, but we need not stop there. We could also develop new cuisines, creating foods tailored to both our appetites and our health, with textures and tastes we have never encountered before. I can picture inventive chefs working with scientists to create, for example, a morsel that tastes like seared steak but with the texture of scallops. Or they might develop a mousse that breaks on the tongue like panna cotta but has the flavor of *jamón ibérico*. Who knows? Like the invention of farming, farmfree food could catalyze entirely new diets, impossible to picture in prospect, which could one day become as familiar as such agricultural innovations as bread and cheese.

Pasi's bugs produce all nine essential amino acids.[36, 37] Though more assessments are needed,[38] so far their digestibility and nutritional quality are reckoned to be roughly halfway between plant proteins and animal proteins.[39] The level of nucleic acids (which hold the cell's genetic information) is higher than in other foods. It needs to be reduced when the flour is processed, otherwise it could cause conditions like gout and kidney stones.[40, 41] While this flour contains no gluten or dairy proteins, it needs to be thoroughly tested for other possible allergens,[42] and any potential impacts on the microbes in our guts.

But the raw material that Pasi uses is just the beginning. For bacteria, by comparison to more complex lifeforms, are fantastically malleable. Even in nature, they can radically change their genetic composition, through horizontal gene transfer.

I oppose genetic engineering when its main purpose is to turn seeds into corporate property or to make plants resistant to proprietary herbicides: these uses enhance the power of big business while delivering scant environmental benefits. But the genetic engineering of certain microbes has so far been uncontroversial. Insulin has been produced this way since 1978, and engineered bacteria and yeast now account for at least 99 percent of the production of the drug.[43] The rennet needed to make hard cheese used to be extracted from the lining of the fourth stomach of an unweaned calf: it sounds like the missing ingredient from the brew in *Macbeth*. Calf stomachs must be sliced and mashed and the rennet extracted through chemical refinement.[44] Some is still produced this way, often to meet organic standards, which forbid the use of newer sources.[45] But most is now made by genetically engineered yeast and bacteria.[46] Scarcely anyone seems to mind.

Just as we have selected crop plants for specific characteristics, bacteria of the kind that Pasi grows can be bred to meet a great variety of tastes and needs. For example, their genomes could be edited or engineered to produce vitamin B12: several species of soil bacteria do so already. Many people are deficient in B12; vegans and vegetarians can be at particular risk, unless they deliberately eat foods that contain this vitamin.[47]*

Bacteria could be engineered to incorporate high levels of micronutrients, to help address the deficiencies afflicting 2 billion people.[48] They could be engineered to make long-chain omega-3 fatty acids, for which fish are mopped from the seas with devastating ecological consequences.

In the first instance, if it's deemed acceptable by regulators and

* Good sources include purple laver seaweed, yeast flakes, and some varieties of yeast extract, some brands of plant-based milk, and, of course, supplement pills.

customers, Pasi expects his product to be bought as a generic ingredient: high-protein flours are already used in bread, pasta, shakes, and many kinds of fast food and ready meals. But they can also be employed, perhaps more effectively than plant proteins, to supply the booming market for meat substitutes. The environmental impacts of plant-based meats are already much lower than those of the animal flesh they seek to replace: one review found that their average greenhouse gas emissions are 34 percent smaller than those of chicken meat and 93 percent smaller than beef's.[49] Because we eat the soy and other grains they contain directly, rather than churning them through animals, plant-based meats use 40 percent less land than chickens, and 98 percent less than beef cattle, while greatly reducing water consumption, pollution,[50] and the use of fertilizers and pesticides.[51]

Microbial proteins, with a land footprint and chemical use much smaller than soy's, would shrink these impacts even further. As bacteria could be tailored to make the specific proteins and fats that might ensure meat replacements became more realistic, the long list of ingredients and the ultra-processing for which plant-based meats have rightly been criticized[52, 53] could be sharply reduced.

When I began my research for this book, I imagined that new proteins and fats would be used to create cultured (or cell-based) meat: flesh that's biologically identical to meat from animals, but is reared in a bioreactor, not a farm. But the more I've read about cultured meat and fish, and the more I've come to appreciate the phenomenal complexities involved in growing cells on a scaffold to make something that looks and feels and tastes like steak or tuna,[54, 55, 56, 57, 58] the more I doubt this vision will come to pass. As the anticipated cost curves[59] fail to materialize, the initial enthusiasms of venture capital will give way to frustration and disillusionment. I doubt the money invested so far is patient enough to march through the hundreds of steps required to bring good products to market.

Instead, I think we will see the development of a hybrid system.[60] Already a cultured meat ingredient—soy leghemoglobin—produced by brewing a genetically modified microbe, is being used by the company Impossible Foods to badge its plant-based burgers with blood.[61] Other companies are using advances in 3D printing to produce plant extracts with fibrous textures, which, according to some of those who have tasted them, are remarkably similar to the muscle texture of cuts of meat,[62] without the immense complexity and cost of trying to grow them on scaffolds in bioreactors. Ingredients such as pea protein and coconut fat can steadily be replaced by proteins and fats made with precision fermentation. The result might be revolutionary, but every step is simple and logical.

In a Zoom interview, the two young scientists who have founded a company called Hoxton Farms[63]—Ed Steele and Max Jamilly—explained the Fat Problem to me. In lumps of animal, fat is cloistered in cells. This is what makes meat juicy and accounts for much of its taste. Animal fat, sealed in membranes, holds its structure when it's cooked. But the plant fats added to meat substitutes are loose, which is why these products are often greasy rather than juicy. Most of the oil melts out in the pan, or even on the shelf.

This is the main reason for the long list of ingredients in plant-based meats: their manufacturers seek to compensate for the deficiencies of the oils they use, and to hide their strong tastes. Ed and Max, using a fermentation process similar to Pasi's, are attempting something much easier than the construction of full-cultured meat: the encapsulation of fats. They hope to bring cellular fats to market within five years, as an ingredient for plant-based meat. When they do so, we can expect meat substitutes to become healthier (as there will be no need for the additives used to address the problems with plant oils), simpler, and more realistic. Attacking the plant or microbial meats of the future on the grounds of their poor quality today is like warning people not to fly on a jetliner because of the Hindenburg disaster.

In her excellent book *The Way We Eat Now*, Bee Wilson discusses some of the reasons for the massive global expansion of chicken production.[64] She points out that it's partly driven by the fact that industrial chicken meat is so bland that it scarcely seems like meat at all. It's a generic white protein to which flavorings, crusts, or sauces can be added to create an endless array of fast foods, ready meals, stews, curries, and stir fries. This is one of those remarks that seems obvious in retrospect, but has the effect of pulling back a curtain.

I immediately thought of the 66 billion chickens[65]—almost eight for every person on the planet—slaughtered each year to produce this uniform protein, and the abominable cruelty the great majority suffer:

the sensory deprivation; the crowding and the attacks and killing this causes; the extreme weight for which modern varieties have been bred, which might buckle their legs or make some birds keel forward and become stranded on the ground; the diseases and parasites and the drugs they necessitate; the reduction of animals whose wild ancestors lived varied and complex lives to little more than machines. It is one of the world's dirty secrets. We don't want to know what happens inside those neat steel factories. We don't want to consider the scale of the suffering our diet demands.

Generic white protein is not a difficult product to make. Quorn, for example, manufactures a version from fungal mycelium, which involves no animal suffering. It will become easier still, and much cheaper, through new kinds of precision fermentation.

There might still be a remote possibility that better plant or bacterial substitutes will pave the way for fully cultured meat or fish,[66] but it seems much more likely that they'll render the need to produce it redundant. New microbial ingredients and 3D printing could make them almost indistinguishable from chicken nuggets, burgers, sausages, and other processed meats. This will land the meat industry with a massive problem.

Because margins are tight, making a profit from animal meat depends on finding a market for as many parts of the carcass as possible: an issue known to insiders as "carcass balancing." North American beef production exemplifies the problem. As 62 percent of beef demand in the U.S. is for ground meat (the kind that goes into hamburgers and most ready beef meals), the proportion of the carcass that gets minced has risen to match it.[67] The value of the rest of the corpse—the prime cuts—has fallen as demand has dropped, but cattle still need to be fed to a standard that ensures these cuts are worth buying. This further grinds the margins, making the industry highly vulnerable to disruption. If even part of the market for ground beef is captured by plant or microbial substitutes, it leaves beef producers with even less balanced sales, and little option but to raise the price of prime cuts to compensate. As the think tank RethinkX points out, it would take only a small bump to push the industry into a spiral of collapse.[68]

Chicken and pork producers, who also rely for much of their profit on ground portions (for nuggets, hot dogs, and the rest), face a similar problem, though it will take longer to materialize. Dairy might be even more vulnerable than beef. This is because milk is mostly water. The most valuable fractions—casein and whey proteins—account for just 3.3 percent of its volume.[69] Roughly one-third of these proteins are currently separated from milk and used as ingredients for infant formula, desserts, and other processed food, or sold to bodybuilders. Using microbes to produce them is simple, and several companies are already in the market. So far, the fermentation firms laying siege to the industry have focused on valuable products like ice cream and cheese.[70, 71] But their costs will fall. By contrast, most of the possible efficiencies in producing milk from cows have already been extracted, which helps to explain the desperate and destructive behavior of many dairy farms. If I were in that business, I would get out now.

At first sight, eggs look harder to crack. But some 30 percent are sold to businesses as ingredients for cakes and other processed foods. In many cases, manufacturers want their eggs separated, sometimes into isolated proteins.* It won't be long before they're cheaper to grow in vats.

Such transitions will be slow at first, then fast. They are likely to reach critical thresholds, beyond which existing systems will tip. In other words, just as the internal dynamics of complex systems can detonate sudden and dangerous regime shifts of the kind I described in chapter 2, they can also catalyze sudden but beneficial changes of state.[72] Issues such as the problems meat substitutes create for carcass balancing, or the dangers that manufactured milk proteins present to dairy production, could quickly erode the margins of existing industries, which already tend to survive only with the help of public subsidies. As businesses learn from each other and synergies develop between their novel processes, the diffusion of new technologies is likely to be self-accelerating.[73]

Pasi's version of precision fermentation is one of dozens of options,[74, 75, 76] and the bacterial species he grows one of thousands of candidates. The proteins and fats they generate could be used directly, or could feed other microbes making specialist products. Already, Solar Foods has several commercial competitors.

---

* Ovalbumin, ovotransferrin, and ovomucoid.

These technologies, as they become cheaper, could one day threaten not only animal products but also some of the most damaging plant extracts, such as palm oil, olive oil, and coconut oil. You may be surprised to learn that I've listed these oils in escalating order of destructive impact. Some farmers now barbarically vacuum olives from their trees at night, killing millions of roosting birds.[77] Coconut production is sweeping across tropical islands, threatening or exterminating species that live nowhere else.[78] Advanced agricultural logic has a tendency to turn everything good into something bad.

Innovators in the food industry live in fear of something they call "the chasm." This is the gap between those who are prepared to experiment with a new product—an adventurous minority—and the rest of the market that the product needs to reach to stay on the shelves.[79] Between 70 and 80 percent of new food lines fail, partly as a result of wariness.[80] But there are also countervailing forces: scientific convergence, economies of scale, and the self-reinforcing adoption of new technologies. I think it's fair to say that, for several plant-based meats and milks, the chasm has already been crossed. Others will flop, but the general trend is in one direction.

We are likely to witness a techno-ethical shift, similar to the impact of the contraceptive pill. The pill (and other scientific methods of family planning) accelerated the liberation of women.[81] It intensified impatience with the status quo, hastening a transition that was already beginning to happen. It helped to drive a virtuous spiral of social change, making what was once scarcely imaginable quickly seem inevitable.

As meat is challenged by plant proteins, then plant proteins are challenged by microbial proteins, and as farmfree products become cheaper, better, and healthier than the foods with which they compete, the existence of good alternatives will sharpen our growing disquiet with the treatment of livestock, the destruction of our life-support systems, and the pandemics caused by animal farming. Why should we eat the products of this cruel, dangerous, ecocidal system when we no longer need to? Only when something becomes amendable does it become intolerable. Governments would then find it easier to cut subsidies and tighten the rules on environmental impacts and animal welfare, making livestock farming still less viable and accelerating the transition. Arguably, this techno-ethical shift could happen more swiftly than the liberation of women (still a very long way from completion), as it needs to occur in just one sphere—food—rather than in every social domain.

There are social as well as technological tipping points: for better or worse, most people side with the status quo, whatever it may be. When enough people change their habits or views, others sense that the wind has changed, and tack round to catch it. There are plenty of rapid changes of state in recent history, such as the remarkably swift reduction in smoking; the shift, in nations like the UK and Ireland, away from homophobia; and the #MeToo movement, which, in a matter of weeks, significantly reduced the social tolerance of sexual abuse and everyday sexism. Experiments suggest that a critical threshold is likely to be passed when the size of a committed minority reaches roughly 25 percent of the population.[82] At this point, social conventions might suddenly flip, and the great majority alters its behavior.

To avert systemic environmental collapse, we need systemic economic and social change. Our lives might depend on triggering what scientists call a cascading regime shift: a flip from one equilibrium state to another, followed by a beneficial hysteresis: in other words, permanent system change. We have sought to address our existential crisis by attending to minutiae, such as changing the gut microbes of cows, so that they produce slightly less methane,[83] or tweaking farm subsidies to allow a few small corners to be planted with trees. These are the agricultural equivalent of trying to prevent catastrophic climate breakdown by changing the cotton buds we buy, an approach I call Micro-Consumerist Bollocks.[84] In most respects, our response to the greatest crises that humanity has ever faced has been narrow and timid. Where our thinking needs to be bold, complex, and holistic, it has been siloed, blinkered, and incremental. We wonder how we might modify the industries that are driving us toward disaster, while the scale of our crisis demands that we replace them. Our challenge is not to tinker with existing models, but to discover the feedback loops that push them past their tipping points.

As we begin to explore the vast possibilities offered by precision fermentation, I suspect that mimicking animal products will become ever less important. The new foods we invent, as unimaginable today as Camembert was to the first people to catch a live aurochs, might be healthier, cheaper, and tastier than either meat or meat substitutes.

Are you horrified by the idea of eating bacteria? If so, I have bad news: you do it with every meal. In fact, your digestion, immune system, and general health depend on consuming them, which helps to explain (if not justify) the existence of a billion-dollar industry selling microbial supplements.

For thousands of years, we've eaten foods that rely for their quality and character on bacterial contamination: cheese, yogurt, fermented fish, and vegetables. Live yogurts are sold at a premium: the adjective means they contain active bacteria, which should surely revolt you more than the dead cells used to make the pancake I ate. Worse still, every human cell is packed with miniature components, some of which shift and stretch and wriggle, that were once free-living germs. Your enemy's not just within the gate. It's you.

There is a strange tension in foodie culture between modernist experimentation and the quest for an "authentic" folk cuisine. The search for authenticity (which by definition is self-defeating) is encapsulated in a maxim by the wonderful food writer Michael Pollan, a man for whom I have great respect, though I find his rule hard to comprehend: "Don't eat anything your great-great-great grandmother wouldn't recognize as food."[85]

I have no idea what my great-great-great-grandmother recognized as food. But my grandma, who was born in 1911, would be a great- or great-great-grandmother to most of those alive today. She was everything a nostalgic foodie would celebrate: a tough, skilled, and knowledgeable countrywoman, connected to the land, who caught or collected some of her food, and made her meals from scratch.

She was strict and unbending, intolerant of vulnerability or distress. When my sisters and I visited, she insisted that we were up by six and outdoors after breakfast, regardless of the weather. But staying with her was the highlight of my school holidays.

She taught me to make tiny imitations of insects from scraps of fur and feathers, and use them to deceive fish in the river behind her house. We caught trout between the beds of crowfoot, chub under the trees, dace in the shallows where the cattle lumbered into the river to drink and cool their flanks. We brought the trout home and filleted and fried them for our dinner. In August and September we gathered mushrooms, which some years were so abundant they turned the pastures white. She taught me to watch, to listen, to name the birds and flowers. I studied to be quiet.

We visited her friends, a fantastic collection of crazy old ladies, living in genteel poverty on narrowboats and smallholdings. One woman had wild gray hair and a skirt held up with baler twine. She wore her wellington boots indoors and served us tea in old jam jars while her chickens clucked in and out of the kitchen and a pig plowed her garden. We came away with a couple of hens' eggs or a gigantic goose egg, which we boiled and ate together the following morning. We bought unpasteurized milk from the farm up the lane.

My grandmother recognized the following items as food:

Bacon

Eggs

Tea

Porridge (which in times of need could double as Ordinary Portland Cement)

Beef stew with suet dumplings

Shepherd's pie

Red lentil and pearl barley stew

Split peas

Potatoes

Homemade bread (even the ducks couldn't eat it, because it sank too fast)

Butter

Marmalade

Vegetables (boiled until they fused with the saucepan)

Preserved tongue

Canned ham (with a salt content somewhere north of seawater)

Cheddar cheese

Trout (caught from the river)

Cod (caught a week ago and sold "fresh")

Lardy cake (there's a clue in the name)

Rock cakes (likewise)

Tapioca pudding

Rhubarb and custard

Apple crumble

Canned fruit

Gingernut biscuits

Wild mushrooms (for three weeks of the year)

Blackberries (for two weeks of the year)

Homemade wine (handy for stripping the paint off doors and
    window frames)

And not much else.

Her diet was probably slightly healthier than the Global Standard Diet is today. It contained plenty of fiber and not much sugar. But it was awash with salt, saturated fat, and preservatives, including sodium nitrite. It was deficient in fresh fruit and vegetables. And much of it was frankly disgusting. What a miserable world it would be if we ate only what she recognized as food!

Almost everything I eat would have been unfamiliar to her. For breakfast, I have muesli (I work for *The Guardian*—it's in the contract), to which I add oat milk, pumpkin seeds, and hazelnuts. The last item is the only one of these foods she would have recognized.

For lunch I might eat the salt crackers I now bake from Kernza, with homemade hummus, frozen peas, and my own chili sauce. I make the hummus from canned chickpeas, garlic, tahini, olive oil, and lemon juice. The only one of these ingredients my grandmother would have recognized as food was lemon juice, but on the rare occasions on which she used it, it came from a yellow plastic bottle in the shape of a lemon. Olive oil would have been familiar, but not as food: it was sold in pharmacies for softening earwax. She knew what garlic was, but defined herself against it: a foreign abomination.

Frozen peas were alien to her. The local grocery stocked canned peas. Its stinking freezer, a dark and intimidating chest whose contents were replaced from the top in stratigraphic layers, contained fish, beef, and lamb but no vegetables. Much of it was occupied by a substance that would astonish anyone younger than me.

Whale meat.

It was sold in blocks of marbled, gray-mauve flesh in bloody wrappers. It was not for human consumption, but for cats and dogs. It's a reminder not to romanticize the primary industries of yesteryear.

Like my grandmother, I usually cook dinner for myself and my partner from scratch. One of my favorite meals is a recipe by the cookery writer Anna Jones: green peppercorn and lemongrass coconut broth.[86] I've adapted it a little. My version contains fresh ginger, garlic, spring onions, coriander leaves, ground turmeric, green peppercorns, coconut milk, lime juice, tamari, lemongrass, kuri squash, tofu, mustard greens, Asian purple laver (for the B12),[87] and brown rice noodles. The only ingredient my grandmother would have acknowledged is the spring onion.

In other words, I eat almost nothing she would have recognized as food. Yet my diet is healthier than hers. Perhaps surprisingly, in view of the wide-ranging ingredients, it also has a lower environmental impact (though I've recently discovered I need to drop the coconut milk and change the source of the olive oil).

I wonder how many people who recite Michael Pollan's mantra—and there are plenty—abide by it. I don't know any foodie who eats lardy cake or preserved tongue, but I know plenty who sample cuisines from around the world, to which their great-great-great-grandmothers would probably have reacted with suspicion and disgust. Their self-proclaimed conservatism bears no relationship to how they live.

Let's imagine it was the other way round. Imagine that the world was currently producing most of its protein and much of its fat from microbes in breweries, occupying, in total, the land area of a small European province, and fed and powered by clean electricity. Imagine that my evil, anagrammatic twin, Tom Go-Bioregen, wrote a book with the following argument.

"I've got this great idea. Let's shut down the food factories. Let's replace the food they make by catching some wild animals—aurochs, wild boar, jungle fowl, and a woolly ruminant from Mesopotamia would do—modifying them drastically and breeding them in stupendous numbers. Let's separate the young from their mothers, castrate them, dock their tails, clip their beaks, teeth, and horns without anesthesia, herd them into barns and cages, subject them to extreme boredom and sensory deprivation for their short, distressing lives,[88] then corral them into giant factories where we stun them, cut their throats, skin, pluck, and hack their bloody flesh into chunks that you, the lucky customer, will want to eat (oh yes you will!). I've done the sums—we'd need to slaughter only 75 billion animals a year.[89]

"Let's kill the baby aurochs, extract a chemical from the lining of their fourth stomachs and mix it with milk from lactating mothers of the same species, to create a wobbly mass of fat and protein. We'll stir in some live bacteria to digest this mass, then let their excrements sit till they go hard and yellow and start to stink. You're really going to want this!

"Let's fell the forests, drain the wetlands, seize the wild grasslands, expel the indigenous people, kill the large predators, exclude the wild herbivores, trigger the global collapse of wildlife, climate breakdown, and the destruction of the habitable planet. Let's fence most of this land for our captive animals to graze, and plant the rest with crops to make them fat. Let's spray the crops with biocidal toxins and minerals that'll leach into the soil and water. Let's divert the rivers and drain the aquifers. Let's pour billions of tons of shit into the sea. Let's trigger repeated plagues, transmitted to humans by the animals we've captured, and destroy the efficacy of our most important medicines.

"Sure, it will trash everything after a while, but think of the fun we'll have. Come on, you know you want this."

I hope you would run this scoundrel out of town.

I'm not naive about the challenge. I know that proposals of the kind this book makes will be met with bitter resistance. Niccolò Machiavelli explained:

> It ought to be remembered that there is nothing more difficult to take in hand, more perilous to conduct, or more uncertain in its success, than to take the lead in the introduction of a new order of things. Because the innovator has for enemies all those who have done well under the old conditions, and lukewarm defenders in those who may do well under the new. This coolness arises partly from fear of the opponents, who have the laws on their side, and partly from the incredulity of men, who

do not readily believe in new things until they have had a long experience of them.[90]

In the European Union and several U.S. states, legislators, lobbied and funded by the meat industry, have sought to suffocate the nascent plant-based milk and meat industries, partly by banning any recognizable names for them. They've tried to prohibit terms like burger and sausage for foods that aren't made from animals:[91, 92, 93] vegan and vegetarian products would have to be sold as "disks" or "pucks" or "tubes."[94] They've insisted that the words milk, cream, butter, cheese, and yogurt may be applied only to the products of lactation.[95] Some have even tried to forbid certain packaging styles (such as butter blocks and milk cartons)[96] for anything other than traditional foods. They sweetly insist they're seeking to protect consumers from "confusion." But given that the selling point of plant-based products is that they're, ahem, plant-based, and manufacturers want their customers to know it, it's unsurprising that there's no evidence for this confusion.[97]

The lobbyists know that words are a powerful weapon. The way we name and frame things can determine the way we see them.[98] If Moses had promised the Israelites a land of mammary secretions and insect vomit, I doubt many would have followed him into Canaan, though these are accurate descriptions of milk and honey.

If legislators are to insist on food literalism, they should at least be consistent. There is no dinosaur in dinosaur chicken nuggets (and not much chicken). If a vegetarian hot dog is ruled out on the grounds that it contains no meat, the meat version should be ruled out on the grounds that it contains no dog. Minced meat is meat, but mincemeat isn't. Nor are sweetmeats, though sweetbreads are. Buffalo wings are an obvious attempt to mess with our heads. Shepherd's pie, or so the printed ingredients claim, contains no shepherd. And don't get me started on jelly babies.

The meat industry's attempts at dissuasion are reinforced by gastronomes and some environmental campaigners, railing against "fake food."[99, 100] In what sense is the meat produced in the filthy factory we call a barn by one cultured organism (the chicken or the pig) more real than the cells produced by a different cultured organism—a bacterium—in a much cleaner one?

Too often, it seems to me, food writers and campaigners take the missionary position: seeking to convert people to their own diets, without apparent recognition that these might be impossible to universalize. I would love to imagine that everyone might subsist on leafy greens, wild herbs, fruit, nuts, pulses, and raw cereals, which they could either grow or gather themselves or buy from local suppliers and cook at home. But I recognize that for many, this proposal falls somewhere on the spectrum between impossible dream and atavistic nightmare.

Apart from anything else, the great majority of the world's people simply can't afford to eat like this: remember that a healthy diet currently costs five times as much as one that is merely adequate in terms of calories.[101] If food writers who promote their own diets also called for a radical distribution of wealth from the rich to the poor, I might support their prescriptions. But to propose this menu without an economic transformation is to tantalize and mock.

Traditionally, rich and diverse diets prepared from fresh ingredients often depended (and still depend in some parts of the world) on the near-servitude of the women who cooked them. Today, when large numbers of people are simultaneously poor and time-poor, many couldn't cook like this even if they wished. Responding to a survey by the market-research company Euromonitor, 25 percent of people in the U.S. and UK, 30 percent of French people, and almost 40 percent of Germans reported that they don't have time to cook.[102] Only around a third of respondents in Germany, the UK, and the U.S. say that they cook "almost every day." In the UK, 18 percent stated that they cook either less than once a month or never. Some apartments are now sold or rented without kitchens.[103, 104, 105]

In poorer nations, the proportion of those unable to prepare their own food is likely to be much higher. I worked briefly in the Nairobi suburb of Kibera, which at the time was the biggest slum in Africa. Here, a great number of people, living in tiny huts without even adequate space to sleep, let alone cook, rely on street vendors for their subsistence. The cheapest meal on sale is *githeri*: kidney beans and coarse white corn (the kind that in rich nations is used as cattle feed) boiled together on a charcoal stove. This stew should be cooked for around two and a half hours to release the nutrients it contains, but fuel is expensive, and the vendors tend to skimp. The result is that Kibera's staple food does not lack fiber, and contains some vitamins and minerals, but tends to be deficient in every other property required to make it nutritious or even digestible. One of the common causes of hospital admissions among the children of Kibera is a prolapsed rectum.

Above all, there is a general failure to investigate whether, even in terms of physical space, let alone ecological impact, everyone could be fed on the diets that food writers and celebrity chefs prescribe. The classic case is the promotion of pasture-fed meat. In this field, as in others, we should apply Immanuel Kant's Categorical Imperative:

> Act only according to that maxim whereby you can, at the same time, will that it should become a universal law.[106]

In other words we should ask ourselves: "If everyone were to eat like this, what would the outcome be?"

Professor Robert Paarlberg, who studies food and agricultural policy, compares the tacit alliance of foodies and the meat industry to the inadvertent coalition between Baptists and bootleggers in the early twentieth century.[107] By lobbying successfully for the prohibition of alcohol, the Southern Baptists accidentally created a new economy dominated by armed gangsters, trading in stronger and more dangerous drink, corrupting police and legislators, embedding organized crime in public life.

By attacking plant or microbial meats and milks, the food Baptists don't advance the cause of healthy or sustainable food one iota. Instead, they extend the meat corporations' license to keep selling the cruel, destructive, and unhealthy products the new foods seek to replace, such as burgers, nuggets, and sausages.

Our aim should be to start where people are, to recognize the constraints they face, and develop healthier, cheaper, and less damaging versions of familiar and accepted foods.[108] This means—among other changes—fast food and ready meals that are less processed and contain less salt, sugar, and hard fat, and more fiber, vitamins, and minerals. Like Fran Gardner at the Rose Hill Community Centre, I believe governments should also subsidize the price of fruit and vegetables: this would be a much more effective way of reducing the cost of good food than the public money they shower on farmers. It could also help to ensure that growers like Tolly are paid fair prices for their produce and can earn a decent living. I suspect these measures would make a greater contribution to human and planetary health than a million foodies exhorting other people to be more like them.

None of this is to suggest that genuine issues be dismissed. Producing new foods through microbial fermentation is not a perfect system: nothing is or could be. Ensuring that the equipment and growth media remain sterile requires, in some cases, the use of antibiotics. This will be less profligate and better contained than their deployment in the livestock industry; even so, it needs to be closely regulated and continually checked.

There's a real danger that this revolution could be captured by big business, creating a system that replicates some of the faults and frailties of the Global Standard Farm.[109] At the moment, corporations producing protein are consolidating and growing.[110] Those of us who care about human nutrition and the living world should engage with these issues now: not to shut down the new foods, but to ensure they are used fairly and openly.[111]

There are two useful instruments. The first is strong antitrust laws. They should be used in every sector, to prevent a few large corporations from dominating the industry. Thanks to corporate lobbying, in recent years such laws have seldom been invoked. The result is extreme concentration, especially in digital, retail, and conventional food and farming: anticompetitive but seldom challenged by governments.

We mustn't let the same thing happen to farmfree foods. Already, large meat companies are muscling into the plant-based sector,[112, 113] and it's not always clear whether they seek to expand or contain it. Their products tend to be of lower quality than those produced by smaller start-ups: after all, their aim is not to replace processed meats, but to create a parallel market.

The second instrument is the restraint of intellectual property rights. Developing new foods from microbes and other sources is generally easier and cheaper than creating new drugs or new machines.

So the property rights that surround them should be weaker. There should be provision for "compulsory licensing": granting poorer nations use of these technologies without demanding they pay prohibitive fees to the originators. There's strong evidence to suggest that compulsory licensing, far from smothering innovation, stimulates it.[114]

While there might be arguments for patenting certain production methods, there are good reasons not to patent the molecules, genes, and living organisms with which food technologists work. Life on Earth belongs to everyone and no one, and its enclosure, as we've seen in agriculture,[115] favors only the biggest players.

As in all scientific fields, advances could be obstructed by patent trolls: companies using the property rights they amass to demand ransoms from the innovators who want to work with them.[116] The new food revolution could be snarled up by rights asserted over entire classes of technology. Such interference is already happening in the case of CRISPR gene editing.[117] Any patents granted in this field should be narrow and ephemeral. If they exist at all, they should be attached to processes, not to biology (in other words, not to genes, proteins, cells, seeds, or whole organisms).

To the greatest extent possible, farmfree food should be open source.[118] It's a gift to the world, which arrives just as we need it most. In many cases (Solar Foods is one) it is being developed with the help of public universities and public money.[119, 120] Let's not squander it.

The wide diffusion of these technologies introduces an intriguing opportunity. While it is impossible to feed the world with local agriculture, for the reasons I explained in chapter 5, it is possible to meet the world's need for protein and fat with local farmfree foods. If we prevent their capture by large corporations, the new fermentation technologies could be used by local businesses to serve local markets.[121] As some of the world's poorest nations are rich in ambient energy—the sunlight that strikes them—they could produce the new foods cheaply. Microbial protein horrifies some of those who demand food sovereignty and food justice. But it could deliver both more effectively than farming does.

These are all political choices. These are systems created by humans, which can be improved by humans. We should insist that the ownership and benefits of these emergent technologies are widely distributed, that their prices go low and their quality goes high, that they are properly tested, regulated and labeled, that they're accessible to those who need them most.

The Counter-Agricultural Revolution will be extremely disruptive. Not only livestock farmers, but workers employed in slaughterhouses and packing plants will lose their jobs. We should not mourn a sector with a long and disgraceful record of industrial injuries, starvation wages, and the exploitation of migrant workers.[122, 123] But, by contrast to the brutal way in which coal mining was shut down in my country, leaving the miners without employment and allowing their communities to collapse, we should demand effective government support for those who will need to find work elsewhere. RethinkX proposes what should be a universal rule: "Protect people, not companies or legacy industries."[124] The vast government subsidies spent on livestock farming should be repurposed: from helping people to stay in the industry to helping them leave it.

The transition is likely to happen, however fiercely the defenders of the old dispensation resist it: it appears to possess an inexorable economic logic. Our task is to ensure that the process is both swift and just. The new industries that will take the place of livestock farming and meatpacking are likely to employ large numbers. We should demand that they do so on better terms.

It might allow us to realize the wild vision I entertained in a previous book, *Feral*:[125] of reintroducing some of the Earth's megafaunas. A megafauna—which means a community of enormous animals—is the default state of all ecosystems. Elephants, rhinos, and lions once lived throughout Europe, Asia, Africa, and the Americas, alongside many other great beasts which have since become everywhere extinct. Whales, great sharks, and giant tuna were common sights in many of our coastal seas. The vast and magnificent creatures that once dominated the planet helped to anchor and sustain its living systems.[126, 127, 128] Everywhere, the great beasts have been either erased or severely reduced by people, at first through hunting, then through farming and fishing. Until recently, it seemed as if this process could travel in only one direction. But changing our sources of protein and fat unlocks the possibility of rewilding on such a tremendous scale that we could create Serengetis on every continent.[129] Unlike the original Serengeti, these ecosystems could and should continue to be richly habited by the people who live there today, who could flourish in a new, nature-based economy.[130]

As forests, steppes, savannas, wetlands, mangroves, kelp forests, and sea floors recover, they will draw down carbon dioxide on a massive scale. While rewilding, even of this magnitude, cannot counteract our industrial emissions,[131] and needs to happen alongside the decarbonization of our economies, it could absorb enough carbon from the atmosphere to help prevent planetary catastrophe.[132, 133, 134] In fact, stopping climate breakdown might be impossible without a mass restoration of the living world.[135]

For the first time since the Neolithic, thanks to the possibilities created by microbial protein and fat, we have the opportunity to transform not only our food system but our entire relationship to the living world. Vast tracts of land could be released from both intensive and extensive farming. The age of Extinction could be succeeded by an age of Regenesis.

It is both ironic and appropriate that our liberation from farming—the most destructive human activity ever to have blighted the Earth—should begin with a microbe discovered in the soil.

8

Pastures New

What stops us from seeing that something needs to change? Why do we ignore or tolerate or even justify levels of environmental destruction and social exclusion that—if they were inflicted by any other industry—we would furiously oppose? Why are those who call for a shift in the way our food is produced so fiercely denounced? Well, these are, by definition, visceral issues, and food and identity are closely entwined. But over many years as an environmental campaigner, I've slowly reached an outrageous conclusion. One of the greatest threats to life on Earth is poetry.

In the seventh century BC, the Greek poet Hesiod wrote of a Golden Age long gone, in which humans "lived like gods." They "dwelled in ease and peace upon their lands," healthy and strong, "free from toil" and "merry with feasting," as the Earth "bore them fruit abundantly." Similar ancestral stories are told in other parts of the world.

The Golden Age myth is generally treated as a ridiculous fantasy. But I wonder. We now have data showing that our hunter-gatherer ancestors were taller, stronger, and healthier than the farmers who succeeded them.[1,2,3] We know from contemporary studies that hunter-gatherers work shorter hours than peasant farmers,[4] and their work bears so little relation to the quotidian drudgery suffered by other people that it scarcely qualifies as toil.[5] They tend to devote more of their time to making merry with feasting, or, to be precise, the talking, singing, and dancing that help to strengthen their social networks.[6]

As an inhabitant of a maritime state—Boeotia—Hesiod may have heard travelers' tales of the hunter-gatherers still living on the fringes of the known world. If so, he might have learned of people who were, by comparison to the toilers in his own land, huge, hale, and leisured.

There was no prospect of returning to this Golden Age, as hunting and gathering in his day could not have supported the Boeotian population. They would be still less capable of supporting us today: one analysis suggests that between 10 and 50 square kilometers of land is needed to support one hunter-gatherer, while 10 square kilometers of modern, productive farming can feed 4,000.[7] Even so, it seems to me that the myth had some foundation in truth. It was not until the mid-twentieth century that people in the rich world regained the height (if not the strength) of our distant ancestors. We live much longer today, though our teeth are worse.[8]

In the third century BC, the poet Theocritus retold the Golden Age myth in a form that was entirely fanciful. His leisured people, who live like gods, are shepherds and cowherds. Unlike real herders, who often labor from dawn till dusk, they appear to do no work, but spend their languid hours singing, playing pipes, and dying exquisitely of unrequited love. Looking back to his native Sicily from the busy cauldron of Alexandria, his idylls, while richly homoerotic,[9] cast the shepherd's life as innocent and pure. He established a tradition that came to be known as pastoral poetry.

The theme was taken up by other Greek poets and then, to vivid effect, by Virgil in the first century BC. His eclogues were set in a different utopia, which was, like Theocritus' Sicily, both real and imagined: the rocky core of the Peloponnese, Arcadia. Here, in Pan's dissolute but ethereal realm, the poet and his friends become shepherds, recovering their lost innocence in a dreamworld of harmony, repose, and sexual promise. His poems tend to be more allegorical and political than the Greek pastorals, but they reinforce the notion that the shepherd's life exemplifies virtue and simplicity, in contrast to the seething corruption of the city. The good shepherd in Arcadia becomes the ideal ruler, governing an ideal land.

A remarkable conjunction knits Virgil's pastoral poems into another tradition, developed independently by the authors of the Old Testament. They were, on the whole, the literate, urban descendants of nomadic herders. From Genesis onward, they cast shepherds as the beloved of God: "Abel was a keeper of sheep, but Cain was a tiller of the ground."[10] God told Moses to set Joshua over the people of Israel so that "the congregation of the Lord be not as sheep which have no shepherd."[11] The Israelites belonged to God's pasture: they were "the sheep of his hand."[12] The prophets denounced the complex, compromising urban lives they led: "Woe to the bloody city! It is all full of lies and robbery; the prey departeth not."[13]

Virgil's fourth eclogue, written around 40 BC, prefigures to an extraordinary degree the Christian story.[14] It speaks of a "boy's birth in whom / The iron shall cease, the golden race arise." In his "glorious age," we will be released from "our old wickedness." The boy will "reign over a world at peace," in which the sheep need not fear the lion. Some Christians, influenced by Constantine the Great in the fourth century CE, see Virgil as a prophet.

Jesus told his people, "I am the good shepherd: the good shepherd giveth his life for the sheep."[15] He was also, according to John, "the Lamb of God, which taketh away the sin of the world."[16] He sent forth his disciples to "feed my sheep."[17] They and their successors became his "pastors," or shepherds, whose bishops still carry a pastoral staff, shaped in the Western Church like a shepherd's crook.

The sacred and secular traditions fused spectacularly in the Renaissance, at first through the work of Dante, Petrarch, and Giovanni Boccaccio, then through such Italian and Spanish poets and playwrights as Jacopo Sannazaro, Battista Guarini, Isabella Andreini, Garcilaso de la Vega, and Jorge de Montemayor. They developed a rich, often allegorical or satirical literature, which tended to reinforce powerful contemporary values. Toward the end of the sixteenth century, the form began to proliferate in England, after the publication of Edmund Spenser's *The Shepheardes Calender*.[18] The craze was taken up by Sir Philip Sidney, his sister Mary Herbert, Thomas Lodge, Christopher Marlowe, and dozens of others. They glued Arcadia, the Garden of Eden, and the Golden Age together to create an ideal world in which, paradoxically, existing power structures were often valorized and celebrated. While the good leader became the good shepherd, the good subject became a good sheep. As Keith Thomas notes in his book *Man and the Natural World*, "Loyal, docile animals obeying a considerate master were an example to all employees."[19]

But the treatment of these ancient themes could also be ironic and self-reflexive. Shakespeare's *As You Like It* approaches the pastoral with characteristic ambivalence, lampooning the traditional trope of fleeting the time carelessly in the golden world. Having fled to the forest of Arden, the Duke claims that the countryside is "More free from peril than the envious court." Here, in a life "exempt from public haunt," he "Finds tongues in trees, books in the running brooks, / Sermons in stones, and good in everything." This fantasy takes human form, in the shape of the old shepherd Corin (whose name surely nods to Corydon, a mythical shepherd celebrated by both Theocritus and Virgil): when we meet him, he is engaged in a vintage pastoral eclogue about unrequited love with the young herder, Silvius.

But the poetic fate of young shepherds is demolished by Rosalind with a laconic flick: "men have died from time to time and worms have eaten them, but not for love." The verbal duel between Corin and Touchstone, the court jester, suggests that insight is to be found neither in the tortuous wordplay of the court nor in rustic banalities. When the Duke's position is restored, he and his men troop merrily back from Arden/Arcadia to the intrigues and power plays of the court.

The tradition lingered for a couple more centuries, especially in elegies that compared dead poets to Arcadian shepherds.[20, 21] But in 1783 it was dealt a mortal blow by George Crabbe's lithic plea for rural realism: *The Village*.[22] He argued that while poets on their downy couches "in smooth alternate verse, / Their country's beauty or their nymphs' rehearse," the reality of rural life, for genuine shepherds and farm laborers, was grinding misery. He evokes a kind of poetic Resource Curse: just as wealth makes the enslaved people who extract it "doubly poor," so the extravagances of pastoral poetry belittle and therefore sharpen the poverty of rural workers. He also had a message for nostalgic gourmets, as pertinent now as it was then. What your great-great-great-grandmother recognized as food is likely to have been:

> Homely not wholesome, plain not plenteous, such
> As you who envy would disdain to touch.

But by then the pastoral myth had become what the cognitive historian Jeremy Lent—whom I see as one of the greatest thinkers of our age—calls a "root metaphor."[23] A root metaphor is an idea so deeply embedded in our minds that it shapes our understanding and affects our preferences without our conscious knowledge. It lays a cognitive trail, which, he suggests, we follow like a path already trodden through a field of tall grass.

When we encounter something that aligns with a root metaphor, it can create a sense of comfort: "God's in his heaven—All's right with the world!" When we encounter something that conflicts with a root metaphor, it can trigger confusion, anger, and cognitive dissonance. The pastoral story is one that urban civilization tells against itself without a flicker of disquiet: the shepherds and their sheep are good and pure, while the city is base and venal. Why do we see sheep farmers tending to their flocks in a blizzard as more romantic figures than office workers trudging through the same storm? Perhaps because the poets' herds have trampled such a trail through the grass that departing from it requires a deliberate cognitive effort.

In the twentieth and twenty-first centuries, the old myth was revived in the two media that count most: television and children's books.

A remarkable number of books for infants tell an almost identical story: of a farmyard, with or without a rosy-cheeked farmer, in which the animals talk to each other or to the reader.[24] In most cases, there is one cow, one pig, one horse, one chicken, one dog, and one cat, living together as if they were a family. There is of course no hint of why animals might be kept on a farm, what happens to them in life, or how and why they die. At the very dawning of consciousness, we learn that the livestock farm is a place of comfort and safety, a harmonious world removed from stress and conflict.

Petting farms and play farms reify this tale, ensuring that the only livestock enterprises most people encounter appear to validate the infant pastoral. What we learn during the first stirrings of consciousness can become embedded more deeply than anything we learn afterward, and is harder and more painful to unlearn: even grim news and images of industrial farming cannot displace the original impression from our minds. As adults, we subconsciously seek the places of comfort and safety we encountered when we were very young, and associate them with what is good and right. When someone challenges our mythic conception of farming, we can quickly become angry and upset.

I've often heard farmers complain that people would be more

sympathetic toward their industry if they knew more about it. I suspect the opposite is true. I think our benign perception of animal farming is sustained only by a remarkable ignorance of what it involves. This ignorance might help to explain why, when over 90 percent of U.S. citizens eat meat,[25] according to one survey, 47 percent want to ban slaughterhouses.[26] W. H. Auden's anti-pastoral poem *Et in Arcadia Ego* portrays us as creeping past the realities of rural life, not daring to look.[27]

The inane pastoral vision we imbibe as children is reinforced in adulthood by primetime television programs. On British TV, at least once a week, often at peak time on a Sunday evening, weary urban people lose themselves in a bucolic fantasy of livestock farming. For decades, programs about lambing, sheepdog trials, and rugged shepherds rescuing their strays have colonized the schedules: if the BBC were any keener on sheep it would be illegal. As urban life becomes more complex and frenetic, these films appear to become more popular.

So many programs about the countryside are presented by sheep farmers that it seems like a qualification. It's as if you had to be an oil driller to talk about climate breakdown. In this artless revival, all the allegory, allusion, and irony of the pastoral tradition has been stripped out. The idyll is presented as reality. One reader wrote to me about a sheep farmer who has often been idolized on television, alleging a long record of slurry dumping, waste burning, and intimidation of his neighbors. But a sheep farmer needs to know just one thing to be fêted as a hero: how to gaze meaningfully into the distance while standing on a hill.

What these programs never tell you is how the farmers make their living. It's not through keeping sheep. In economic terms, the sheep are ornamental, or worse. In Wales, where many of these idylls are filmed, farmers lose 33 pence for every kilogram of lamb they produce (roughly £13 per beast).[28] In England, lowland farms on which sheep or cattle graze earn minus £16,300 a year from farming, and upland farms, minus £16,600.[29] The real business of the farm takes place on the computer, filling in subsidy forms: livestock graziers in the UK are entirely dependent on the taxpayer.[30, 31] But no one wants to watch that.

There's probably more money being made in this country by talking about sheep farming than by doing it. But these programs are popular and lucrative for the very reason that they trade in escapist fantasy.

A similar story is told in the USA, though in this case the heroes herd cattle rather than sheep. I see the cowboy story as a species of the pastoral idyll, in which we escape our complex lives into a fantasy of independence and simplicity of thought and action. As the great historian Eric Hobsbawm argued, the myth of the Wild West creates a confrontation between nature and freedom on one side, and civilization and social constraint on the other.[32] The cowboy is a fugitive from urban life, seeking refuge in the wild, with only his horse, his gun, and his cattle for company.

Sometimes he will meet another wanderer and sit beside the fire under the stars, where, like Theocritus' shepherds, they might tell tall tales and play simple instruments (though until *Brokeback Mountain*, that's as far as they were allowed to go). The literature and cinematography of the Wild West creates a myth of endless vistas and opportunities, an imaginary frontier that rolls toward the Pacific but never arrives. The West, once the indigenous people who lived there had been extirpated, became white America's Arcadia.

This myth also dies hard. The "Oath Keepers" of Nevada, and the "Three Percenters" of Idaho, two of the militias that helped lead the attack on the U.S. Capitol building in January 2021, established themselves as a political force by defending the pastoral ideal against state and federal authorities.[33] After the rancher Cliven Bundy was ordered to remove the cattle he had illegally herded onto public lands near Bunkerville, Nevada, harming the brittle desert ecosystem, these militias, equipped with semiautomatic weapons, set up what they called their "Liberty Camp" to defend him. In an armed confrontation on the freeway, they forced federal agents to back down. Then they stalked, harassed, and threatened to kidnap officials: several had to flee the region and hide in safe houses. Though they committed crimes that in any other circumstances would have been treated as terrorism, these paramilitaries got away with it: few were prosecuted or even arrested.

So strong were the Western myths, and their embodiment in Bundy's act of defiance, that no one in power dared move against them. Their impunity in Nevada is likely to have encouraged their attack on the Capitol.

This is an extreme example of the way in which the pastoral fantasy bleeds into the real world. Those of us with an interest in politics tend to emphasize the economic power of lobby groups. But over the years, I've come to see that, in nations like mine, in a contest between economic and cultural power, cultural power wins. Governments are sometimes prepared to move against oil and mining companies, pharmaceutical giants, and even banks.[34] But they seldom dare to confront livestock farmers, even when they inflict great environmental harm. Instead, they shower them with money.

Every year the world's governments spend between $500 billion and $600 billion on farm subsidies.[35, 36] By comparison, the rich world's long-standing promise to spend $100 billion a year helping poorer nations to curtail and survive climate chaos has yet to be fulfilled.[37] There are several excuses for this lavish use of our money. One of the commonest is that farm subsidies reduce the price of food. If this were their aim, it's hard to think of a less efficient way of doing it. Roughly half the world's subsidies, which take the form of "market price support,"[38] raise food prices. Part of the other half could, in theory, reduce them, but only in convoluted and ineffective ways. Alternatively, when subsidy payments increase the price of land, as they do in some cases,[39] even the money not allocated to price support might raise the price of food. If governments wanted food—ideally healthy food—to be cheaper, they would subsidize it at the point of sale.

Another excuse is that farm subsidies reduce rural hardship. But the great majority of the money tends to be mopped up by the biggest and wealthiest farmers. In the European Union, for example, money is paid by the hectare: the more land you control, the more cash you are given:[40] these payments could be the most regressive form of major public spending on Earth today. Taxpayers of all stations kindly donate their hard-earned income to the dukes, oil sheikhs, Russian oligarchs, corrupt politicians, and other aristocrats and tycoons who own great tracts of land.[41] An investigation by the European Court of Auditors discovered that the EU has no useful data on farm incomes, and therefore no knowledge of whether its subsidies serve any social purpose.[42]

In the U.S., 10 percent of farmers—generally the biggest and richest— harvest 77 percent of subsidies.[43] Some of the recipients are absentee owners, who may never have set foot on their land. For many years the federal government systematically discriminated against Black farmers, denying them farm payments and forcing many out of agriculture,[44] as it sought to encourage "the right kind of farmer."[45]

There may well be an argument for a rural hardship fund: poverty in the countryside is often more severe than in the cities. But it's hard to see why it should be reserved for farmers. There is no more reason to favor their profession with an exclusive source of public charity than there is to provide a fund for distressed lawyers or plumbers. Farm laborers, many of whom are paid very badly, some not at all,[46] are surely more appropriate recipients of public welfare than the owners who exploit them. Money should be disbursed according to need, not occupation.

A third excuse is that subsidies help raise farm production. Many of the payments have no such aim, and those that do are often perverse. In the Indian states of Punjab and Haryana, for example, the only funds available are for growing wheat in the winter and rice in the summer.[47] While these once helped greatly to increase food production, today they discourage farmers from bringing other crops into their rotation, which would improve the soil, reduce their use of water, and enhance their livelihoods. There's a similar effect in the U.S., where federal support encourages farmers to concentrate on just a few crops. The government's underwriting of insurance also gives farmers an incentive to ignore environmental risks,[48] which can cause catastrophic harvest failure.[49]

But perhaps the most misleading claim of all is that these payments protect the living world. In most cases they do the opposite. In the European Union, you don't receive your money unless your land is in "agricultural condition." This doesn't mean it must produce food: you can take the full payment in some nations without delivering a single ear of wheat or liter of milk. It means it must be almost bare. If it harbors what the rules call "ineligible features," and the rest of us call wildlife habitats—such as regenerating woodland, ungrazed marshes, ponds, and reedbeds—it is disqualified from the main source of subsidies: the EU's basic payments program.[50, 51] Destruction is not an accidental outcome of the subsidy regime; it is a contractual requirement.

In 2016, I spent a few weeks in Transylvania, exploring some of the richest wood pastures on the planet: a mosaic of flowering meadows, marshes, and trees, which had sprung back to life as livestock grazing had declined or ended altogether. I watched golden orioles, hoopoes, honey buzzards, red-backed and great gray shrikes, lesser spotted eagles, black storks, roe deer, wild boar, and bears. Cuckoos were so common that they flew around in flocks. All nine species of European woodpecker live in one small valley in which we stayed; so do bee-eaters, goshawks, corncrakes, quails, nightjars, tortoises, tree frogs, pine martens, wildcats, lynx, and wolves. But the farmers there had begun to realize that they would be paid for cleansing the land. I saw the heartbreaking results: stunning places cleared and burned simply to meet the European rules.[52] Though the European Commission hasn't bothered to collect the data, across the EU, hundreds of thousands of hectares of land that might otherwise have been left for nature are likely to have been trashed for the sole purpose of harvesting subsidies. This perverse incentive must rank among the world's most powerful drivers of environmental destruction.

A high proportion of subsidies are directed at livestock farmers: in the EU, these payments amount to more than half the farm budget, around €30 billion.[53] It's hard to see how the majority of livestock operations—arguably the most damaging of all industrial sectors—could survive without this money. Extensive livestock farming is everywhere an economic fantasy, sustained either by lashings of public money or public tolerance of massive environmental destruction, or both. Intensive livestock farms, which sometimes claim to take no farm subsidies, might receive major grants of other kinds, such as the wood-pellet payments that keep many of the big chicken factories solvent, while trashing both forests and rivers. If governments wanted to reduce the harm caused by producing meat, as they sometimes claim, they wouldn't need to tax it. They could simply stop paying for it.

An analysis by the World Bank found that only 5 percent of farm subsidies in the world's richest nations have any environmental component.[54] Even this money often does more harm than good. In the EU, while "Pillar One" of the subsidy system—the basic payment—prompts farmers to destroy the wildlife on their land, "Pillar Two" pays them to put some of it back. But only a little.[55] If they restore it too well, they can forfeit their Pillar One payments. In my country, where the subsidy system, at the time of writing, still shadows the European program, Pillar Two payments provide 30 percent of the cash that farmers receive for keeping their animals on unproductive land.[56] Without these "green" funds, the greatest cause of habitat and wildlife destruction in countries like mine—livestock grazing on infertile land—would in many cases cease, and forests and other rich habitats would return. The money that farmers are given first to destroy and then minimally to restore upland ecosystems—£37,000 a year on average in the UK[57]— could be used instead to help them rewild their land.

These gifts of free money resist all attempts at reform. Every few years, politicians in the U.S., the EU, Japan, South Korea, India, and other nations announce their intention radically to reduce or repurpose the public money they give away, but by the time the new policies are passed into law, the only changes are either cosmetic or replace one perversity with another.[58] Though we provide the money, the lay public appears to have no influence over how it is used. Only the farm lobby is heard. It's taxation without representation.

Brexit provided an excellent opportunity to test the cultural power of farming. Leave campaigners raged ceaselessly against the European Union's misuse of taxpayers' money. But while they denounced it for tiny items of expenditure, some of which turned out to be imaginary,[59] somehow they managed to overlook the obvious issue. At the time, 40 percent of the European Union's entire budget was spent on farm subsidies, most of which consisted of payments for owning land. Yet the Leavers wouldn't touch it. On the rare occasions when they mentioned these payments, they did so only to assure UK farmers that they wouldn't be reduced. In fact, two of the most prominent Leave campaigners in Parliament suggested raising them.[60, 61]

As well as throwing money at livestock farming, the European Union has a separate budget for promoting its products to consumers. Over three years, it has spent €71 million on encouraging us to eat more meat.[62] In print advertisements showing cool-looking people grinning over a plate of meat, it encourages us to "Become a Beefetarian." A TV ad claims that we "support sustainable farming by choosing European beef."[63] Following deep thought by the EU Sheep Meat Reflection Group (I picture them sitting cross-legged on the floor, contemplating a leg of mutton), the European Commission decided that "it is vital to appeal to and convert younger consumers" to eating lamb as their "everyday protein choice."[64]

Even after we voted to leave the European Union, some of this money was spent in the UK. Following some frankly mendacious claims about sheep farming protecting wildlife and sequestering carbon,[65] the EU's promotional material in this country issued the following dire warning:

> Without sheep breeding, these abandoned meadows would evolve into unproductive forests for human consumption. It would also mean that the land is solely being used for breeding, thus preventing the usage of this land for other activities: such as tourism.[66]

I have no idea what this means, unless it's arguing that tourists will be deterred by people having sex in the woods. But I'm sure it was public money well spent.

The countryside is neither innocent nor pure. In some places, it is more corrupt than the city, its politics dominated by landed elites, hereditary power, and a culture of deference. Pollution, now caused primarily by agriculture in many nations, is the physical manifestation of corruption. In rivers around the world, you can smell the political stench. If we judged farming by the standards of any other industry, we would be incensed by their transformation into open sewers, alongside the destruction of much of the rest of life on Earth. But the cultural power of the industry insulates it from both criticism and regulation.

We grant farming an uncontested political space offered to no other profession. When farm lobbyists erect a "No Trespassing" sign in front of their sector, insisting it's no business of ours, we humbly accept that their cows are sacred and their lambs are holy, and walk away, even though these issues are arguably the most important of all. Bucolic nostalgia shuts down our moral imagination, unstrings our critical faculties, stops us from asking urgent and difficult questions. But at a time of global ecological catastrophe, we cannot afford this indulgence.

While we immerse ourselves in ancient myths, we neglect the wonderful and thrilling stories that science has to tell us, especially the story of the soil on which we subsist. There are more things in the earth than are dreamed of in our philosophy. We should replace the tired and groundless fables that chain us to destruction with the fresh and grounded revelations that could unlock our minds and allow the world to flourish.

In his influential essay *The Quants & The Poets*, my friend and antagonist Paul Kingsnorth argues that "the green movement has torpedoed itself with numbers."[67] He uses the terms sometimes taught on MBA courses—"quants" do the numbers, "poets" do the words—and complains that "the green movement is being taken over by quants":

> We live in a remarkably literal-minded and reductionistic culture . . .
> the kind of culture which produces an environmental movement made
> up of frustrated, passionate people who feel obliged to act like speak-
> your-weight machines just to be heard.

It's time, he contends, for the poets to dominate.

He makes some good points, but where food is concerned, the contrary problem prevails. A remarkable feature of this debate, at least in the media and on social media, is that it proceeds largely in the absence of numbers: it is blinded by poetry. To a greater extent than in almost any other field, our loyalties are to the aesthetics, not the evidence. We are seduced by the way things look, and overlook the way they function. But beauty is seldom truth, and truth is seldom beauty.

On the rare occasions when numbers are cited, they are often wrong, and seldom contextualized. On one side sits brute corporate power; on the other, Arcadian fantasy.

Who stands between them? Where are those who care about food, care about people, care about the living world, yet also care about the math? They exist, but they are few. It's time we became obsessed by numbers. We need to compare yields, compare land uses, compare the diversity and abundance of wildlife, compare emissions, erosion, pollution, costs, inputs, nutrition, across every aspect of food production. Visceral as these issues are, we cannot resolve the questions they raise through gut instinct. They require a massive effort of research and quantification, much of which has yet to be attempted.

I can see why people balk at this. We cling to the warm and comforting beliefs with which we were raised, seek to burrow back into our own Arcadia. And numbers mean work. Before I could begin this book, I had to learn enough about soil ecology to have taken a degree. Since then, I've read over 5,000 scientific papers and a shelf of books, but I feel I've barely scratched the surface. I still have everything to learn.

I don't have all the answers, and nor does anyone. But we should at least try to build the new ethics that our times demand on facts, not fairy tales. This book could be seen as The Revenge of the Quants.

George Eliot, my favorite English novelist, remarked that "skepticism, as we know, can never be thoroughly applied, else life would come to a standstill: something we must believe in and do."[68] I think she's half right. We must believe in something, and act on that belief, but we can build our beliefs through the thorough application of skeptical inquiry.

I agree, however, that numbers on their own are not enough. Paul Kingsnorth is right on this point: we also need stories. We need stories that tell us where we are, how we got here, and where we need to go. We need, in particular, stories that follow the plot structure that works most effectively in public life: the Restoration Narrative.[69] Yet I see no reason why such stories cannot be informed by numbers.

So what would a new restoration story about food, which could carry us through this century and into those that follow, sound like? Perhaps something like this.

The world has been thrown into chaos by powerful forces: the rise of the Global Standard Farm, global corporations, agricultural sprawl, cultural myths, plows, poisons, and pollutants. This chaos threatens our life-support systems, drives other species to extinction, and harms human health. The powerful forces are jeopardized in turn by both their internal dynamics—which threaten collapse and hysteresis—and the pressures that bear upon them, above all climate breakdown and the loss of irrigation water.

But ranged against these forces are the heroes our crisis demands. They are pioneering a new science of soil ecology, discovering novel ways of working with the life of the soil, developing crops that can deliver large yields with small impacts, igniting a farmfree revolution. With their help, we can create a transition as profound as the Neo-lithic shift. We can avert our looming environmental catastrophe and reverse much of the damage we have inflicted on the living world, while ensuring that healthy diets are available to everyone. We can make peace with the planet.

There are crucial subplots that need to be resolved. We should strive to protect the livelihoods of small farmers and farm workers. This means, above all, developing a new agronomy, based on an Earth Rover Program: the precise mapping of soils and an advanced knowledge of their biology, leading to a new understanding of fertility of the kind that Tolly is developing, and of how to enhance it. Using this knowledge, farmers and scientists working together can evolve exact and minimal organic treatments, tailored to local conditions. We need an army of public consultants trained not to sell pesticides and fertilizers, but to do the opposite: to help release farmers from dependency on corporate seeds and chemicals, reduce their costs, and ensure they receive a greater portion of the value of the food they produce.

We must find ways to resolve the conundrum I mentioned in Chapter 5: if most of our food has to be grown, for simple mathematical reasons, far from where we live, how do we prevent transnational companies from exerting an ever more powerful grip on the food chain? The answer might lie in a revitalized fair-trade movement, in which companies buying bulk commodities are pressed to source them from small, productive farmers.

We need to help livestock farmers leave the industry through a just transition,[70] paying them to shift to new sources of income or new forms of employment. Given the dire economics of livestock farming, generating better incomes and more jobs might not be a high bar to jump. A study of a rewilding project in the Netherlands found that it employs six times as many workers as the dairy farms it replaced, as small businesses have proliferated to serve the people coming to watch the abundant wildlife now living there.[71] An analysis of twenty rewilding projects in England by Rewilding Britain discovered that they created, on average, a 47 percent increase in full-time equivalent jobs.[72]

Effective rewilding, in its early stages, requires plenty of work.[73, 74] Fences need to be removed, drains un-dug, rivers un-straightened and wetlands restored, missing species reestablished, trees seeded or planted where they cannot regenerate spontaneously. Excessive nutrients in the soil, introduced with fertilizer and manure, might need to be reduced, for example by cutting and removing hay for several years. Paying farmers to restore the living world is surely a better use of public money than paying them to harm it.

Subsistence farmers working outside the cash economy could be less affected by the new food revolution than those who depend on selling their produce. Even so, foreign aid for agriculture should be redeployed, away from its almost exclusive focus on promoting the Global Standard Farm, and, if new research leads to the breakthroughs I hope to see, toward helping small farmers practice high-yielding agroecology.

Governments should finance the rapid development of perennial grain crops, to reduce damage to the soil and the need for water and fertilizer: it's ridiculous that this crucial technology has been left to a small nonprofit with limited funds. In the meantime, they should help farmers like Tim Ashton to find new ways of protecting the soil, while minimizing or eliminating their use of herbicides. They should also help farmers to apply the burgeoning science of biological control: using predators to manage pests. They should help ensure that any land we continue to farm is hospitable to wildlife, creating stepping stones and corridors between protected places. Crops should be grown for food, not to produce animal feed, fuel, or bioplastic. In other respects, governments should get out of the way, and stop imposing pointless restrictions on the development of microbial proteins, plant-based, and cultured foods. They should work instead to prevent these crucial new technologies from being monopolized by a few corporations or billionaires.

Debates about the future of food tend to divide people on technological lines. Some believe that the answer to feeding the world lies in extending the Green Revolution, which, from the 1950s onward, has greatly increased the yields of major cereal crops, through a combination of plant breeding and farm chemicals. Others believe that the answer lies in rejecting this highly technological approach, and turning instead to "natural" processes: integrating livestock and crop plants, replacing intensive animal farming with pastured livestock, reviving a rotational system that incorporates long fallow periods.

But technology, I believe, is the wrong battleground. The systems we should favor are those that deliver high yields with low environmental impacts. The systems we should reject are those that deliver high yields but with high environmental impacts, or low yields. Low yields necessarily mean high impacts, because of the area of land they need to produce a given volume of food. As I argued in chapter 3, land use should be seen as perhaps the most important of all environmental issues.

The movements that—often for good reasons—have challenged the Green Revolution, its heavy use of fertilizers, pesticides, and irrigation water, and its tendency to favor multinational corporations above the needs of small farmers and local markets, have themselves often suffered from a disastrous flaw. Because they have tended to be yield-blind, they accidentally promote agricultural sprawl: the use of large areas of land to produce small amounts of food. Sprawl, above almost all other forces, threatens the survival of our life-support systems.

The crucial question we should ask about technology is not "How sophisticated is it?," but "Who owns it?" If a productive technology has a low environmental impact and its ownership is either distributed or public, we should be ready to embrace it. ("Public," incidentally, needn't mean state-owned: it can also mean community-owned.) If its ownership is both concentrated and private, we should challenge this pattern, as concentration contributes to the advance of the Global Standard Diet and the Global Standard Farm, both of which we should resist. Whether or not a system is technologically sophisticated is, in its own right, irrelevant.

We need almost to sweep away the debates that have raged until now, and start over. They are ill-matched to the environmental, social, and agricultural challenges we face. We need to build a new food and environment movement, ready to embrace high-yield, low-impact production. The practices it favors might range all the way from using ramial woodchip to multiplying bacteria by precision fermentation.

The new movement should begin by acknowledging an uncomfortable but well-established reality, a reality that has all too often been swept under the carpet: farming, whether intensive or extensive, is the world's major cause of ecological destruction. This movement should ask itself three fundamental questions when considering any new system: "Does this deliver more food with less farming?," "Who owns and controls it?," and "Is the food it produces healthy, cheap, and accessible?"

A new movement, informed by these questions, needs a manifesto. It might look something like this:

> To allow human beings and the rest of life on Earth to flourish, we should:
>
> Become food-numerate
> Change the stories we tell ourselves
> Limit the land area we use to feed the world
> Minimize our use of water and farm chemicals
> Launch an Earth Rover Program to finely map the world's soils
> Enhance fertility with the smallest possible organic interventions
> Research and develop a high-yield agroecology
> Stop farming animals
> Replace the protein and fat from animals with precision fermentation
> Break global corporations' grip on the food chain
> Diversify the global food system
> Use our understanding of complex systems to trigger cascading change
> Rewild the land released from farming

The COVID-19 pandemic blew away two political ideas that had defined the previous forty years: that governments should not govern, and that people are not prepared to put the public interest ahead of individual interest. The notion that governments should be passive—the can't-do culture—assiduously promoted by the self-hating state, collapsed. So did the idea that citizens will not respond to the signals they send. When governments want to govern, they can, in some countries (Taiwan, New Zealand, Kenya, South Korea, and Vietnam, for example) to great effect. When money is needed, it is found. When people are mobilized, they respond.

The changes we were asked to make to contain the virus were far more extreme than those required to stop environmental collapse. We were asked to stop working, shopping, playing sports, visiting bars and restaurants, throwing parties, going to games and concerts, taking holidays: activities we considered essential to our lives and identity. Children were kept from school, travel ground to a halt, we were asked to cover our faces in public, sterilize our hands, keep our distance, to live, for long periods, as if we were under house arrest.

No reasoned environmental demand comes anywhere near the severe measures the pandemic required. And yet, when we were asked to make these drastic changes, most of us did so willingly. We recognized our public duty and acted on it. Changing the sources of some of our foods, slightly altering our diets, reducing the land area occupied by farming, ending its most damaging practices, expanding protected places: these changes are tiny by comparison. It's not human capacity we lack; it's the political will to invoke it.

If governments communicated our environmental crisis—which presents a threat to humanity far greater even than COVID-19—with the urgency it requires, changed the rules, and explained our public duty to respond, we would do so. Political systems were made by people. They can be changed by people.

It is time to take back control of the global food system, to overthrow the corporate lobbyists and special-interest groups that dominate it. It is time to create a new, rich, productive, and, ideally, organic agriculture, no longer dependent on livestock, growing food that is cheap, healthy, and available to everyone. It is time to develop a new and revolutionary cuisine, based on farmfree food. It is time to release large parts of the planet from our devastating impacts, reverse its dysbiosis, restore its living systems, enhance our own prosperity and prospects of survival.

We can now contemplate the end of most farming, the most destructive force ever to have been unleashed by humans. We can envisage the beginning of a new era, in which we no longer need to sacrifice the living world on the altar of our appetites. We can resolve the greatest dilemma with which we have ever been confronted, and feed the world without devouring the planet.

9

The Ice Saints

It was the best fruit set I had ever seen. Across the entire pollination season—the second half of April and the beginning of May—the days were balmy and the nights were warm. For the first time in my experience, there wasn't even a ground frost. Bees and hoverflies swarmed the blossom, and almost every flower, as the petals fell to the ground, began swelling into fruit. I warned the other families in our orchard group that we would need to set aside a couple of days in June for fruit-thinning, otherwise we would find ourselves with thousands of apples the size of walnuts. It was a good problem to have. I blocked out a weekend in October for the mother of all cider pressings.

When I saw a frost forecast for the night of May 12, it didn't worry me. The moment of vulnerability—when the blossom has opened—had passed. I visited the orchard the next day, to discover whether it had scorched the last few blooms.

It took me a while to register what I was seeing. Every fruit on every tree had begun to shrivel, and the stems to yellow and weaken. I wandered round and round the orchard, checking that my eyes weren't deceiving me. When I could fool myself no longer, I almost cried. The entire crop had been destroyed. Later, I discovered that the frost had been much harder than the forecast had advised: it was a freak freeze, deep enough to break the usual rule that once the blossom is over, the fruit is safe.

I trudged toward the gate, shaking my head in disbelief. On the path I met Stewart, our allotment neighbor, who had given us his row of trees. He and another veteran digger, Mike, were contemplating their ruined potatoes, engaged in the perennial conference of hoary gardeners the world over, on the vicissitudes of the weather.

"I can't believe it," I told them, "we've lost the whole lot."

"Well," said Stewart, "we've only got ourselves to blame. We forgot to pray to the Ice Saints."

"The what?"

"Look them up."

I did, and discovered that May 11, 12, and 13 are, respectively, the days of Saints Mamertus, Pancras, and Servatius. These three days, in several parts of Europe, are lamented as the "blackthorn winter," because, allegedly, when the blackthorn is in flower, winter returns to strike a final blow. According to an unusual meteo-agri-theological paper in the journal *Weather*, this period is associated in Britain with northerly winds caused by high pressure over Greenland and weak cyclonic activity in the Atlantic.[1] It told me that "he who shears before St. Servatius Day loves his wool more than his sheep," presumably because they are left to shiver in the late frosts. In several countries, especially in Eastern Europe, the three saints (whose identity shifts from one nation to another) are widely petitioned to stay the hand of the weather.

Frustratingly I've not been able to find, anywhere in the scientific literature, a controlled trial of the prayers or libations most efficacious for enhancing the strength of Atlantic cyclones. Nor is it clear why, when the switch from the Julian to the Gregorian calendar in the sixteenth century knocked the saints' days back by ten days, the weather shifted with them. But frankly, given my despair at yet another failed harvest, I would try anything.

Yet somehow, the following month, I found myself in the allotments again, scything the meadow, picking up the old shoes that a fox was collecting—who knows from where—and leaving in our grass, summer-pruning the trees. I realized I was preparing the orchard for the following year, in the expectation that our luck would change. Hope had once again triumphed over experience.

The experience of environmental campaigners has so far been bitter. We appeal to humanity's fabled survival instinct, and find it missing. We assemble the evidence, explain the problem, propose a solution, and are received like Dr. Stockmann in Henrik Ibsen's play *An Enemy of the People*: with anger, denial, and obloquy.

But the experience of all effective movements also shows that success is a function of their preparedness for a moment of transformation. It may come unexpectedly, caused by entirely unrelated forces. Sometimes the role of entire generations is simply to make ready, develop their arguments, tell their stories, build their campaigns, for an opportunity their successors can seize.

I think we are beginning to see an alignment of technological change, systemic fragility, and public disquiet sufficient to trigger a techno-ethical shift, of the kind catalyzed by the printing press and the contraceptive pill, that could allow us to recast our relationship with the living planet. No prayers are required to provoke this change of state, just the hard work of a small number of committed people, and the willingness of others to support them. We are, I think, soon to encounter a moment when conditions change.

Notes

CHAPTER I

1. The Orchard Project, 2013. *Winter Wassail 2013*. https://www.the orchardproject.org.uk/blog/winter-wassail-2013.
2. Our World in Data, 2018. Calorie Supply by Food Group, 2017. https://ourworldindata.org/grapher/calorie-supply-by-food-group?cou ntry=GBR~CHN~SWE~USA~BRA~IND~BGD.
3. Tiehang Wu, Edward Ayres, Richard D. Bardgett et al., 2011. "Molecular study of worldwide distribution and diversity of soil animals." *Proceedings of the National Academy of Sciences*, vol. 108: 43, pp. 17720–5. https://doi.org/10.1073/pnas.1103824108.
4. David C. Coleman, Mac A. Callaham Jr., and D. A. Crossley Jr, 2018. *Fundamentals of Soil Ecology*. Academic Press, Cambridge, MA. https://doi.org/10.1016/C2015-0-04083-7.
5. Tiehang Wu, Ayres, Bardgett et al., Ibid.
6. Coleman, Callaham, and Crossley, Ibid.
7. Radnorshire Wildlife Trust. *Yellow Meadow Ant*. https://www.rwt wales.org/wildlife-explorer/invertebrates/ants/yellow-meadow-ant.
8. Nick Baker, May 2020. Hidden Britain: Ant Woodlouse. BBC Wildlife. https://www.yumpu.com/news/en/issue/7785-bbc-wildlife-issue-052020/read?page=21.
9. Coleman, Callaham, and Crossley, Ibid., p. 10.
10. Ibid.
11. A. Pascale et al., 2020. "Modulation of the root microbiome by plant molecules: The basis for targeted disease suppression and plant growth promotion." *Frontiers in Plant Science*, vol. 10, article 1741. https://doi.org/10.3389/fpls.2019.01741.
12. Coleman, Callaham, and Crossley, Ibid.

13. David R. Montgomery and Anne Biklé, 2016. *The Hidden Half of Nature: The Microbial Roots of Life and Health*. W. W. Norton and Company, New York.

14. Merlin Sheldrake, 2020. *Entangled Life: How Fungi Make Our Worlds, Change Our Minds and Shape Our Futures*. Bodley Head, London.

15. Patrick Lavelle et al., 2016. "Ecosystem Engineers in a Self-Organized Soil" *Soil Science*, March/April, vol. 181:3/4, pp. 91–109. https://doi.org/10.1097/SS.0000000000000155.

16. Hongwei Liu et al., 2020. "Microbiome-mediated stress resistance in plants." *Trends in Plant Science*, vol. 25:8, pp. 733–43. https://doi.org/10.1016/j.tplants.2020.03.014.

17. Lavelle et al., Ibid.

18. Coleman, Callaham, and Crossley, Ibid., p. 50.

19. Dilfuza Egamberdieva et al., 2008. "High incidence of plant growth-stimulating bacteria associated with the rhizosphere of wheat grown on salinated soil in Uzbekistan." *Environmental Microbiology*, vol. 10:1, pp. 1–9. https://doi.org/10.1111/j.1462-2920.2007.01424.x.

20. Andrew L. Neal et al., 2020. "Soil as an extended composite phenotype of the microbial metagenome." *Scientific Reports*, vol. 10, article 10649. https://doi.org/10.1038/s41598-020-67631-0.

21. Ioannis A. Stringlis et al., 2018. "MYB72-dependent coumarin exudation shapes root microbiome assembly to promote plant health." *Proceedings of the National Academy of Sciences*, vol. 115:22, article E5213–E5222. https://doi.org/10.1073/pnas.1722335115.

22. Pascale et al., Ibid.

23. Hongwei Liu et al., Ibid.

24. Shamayim T. Ramírez-Puebla et al., 2012. "Gut and root microbiota commonalities." *Applied and Environmental Microbiology*, vol. 79:1, pp. 2–9. https://doi.org/10.1128/AEM.02553-12.

25. Rodrigo Mendes and Jos M. Raaijmakers, 2015. "Cross-kingdom similarities in microbiome functions." *The ISME Journal*, 9, pp. 1,905–7. https://doi.org/10.1038/ismej.2015.7.

26. Kateryna Zhalnina et al., 2018. "Dynamic root exudate chemistry and microbial substrate preferences drive patterns in rhizosphere microbial community assembly." *Nature Microbiology*, vol. 3, pp. 470–80. https://doi.org/10.1038/s41564-018-0129-3.

27. Ed Yong, 2016. *I Contain Multitudes: The Microbes Within Us and a Grander View of Life*. Vintage, London.

28. Maureen Berg and Britt Koskella, 2018. "Nutrient- and dose-dependent microbiome-mediated protection against a plant pathogen." *Current*

Biology, vol. 28:15, pp. 487–2492, e2483. https://doi.org/10.1016 /j.cub.2018.05.085.

29. Paulo José P. L. Teixeira et al., 2019. "Beyond pathogens: Microbiota interactions with the plant immune system." *Current Opinion in Microbiology*, vol. 49, June, pp. 7–17. https://doi.org/10.1016 /j.mib.2019.08.003.

30. Mathias J. E. E. E. Voges, 2019. "Plant-derived coumarins shape the composition of an *Arabidopsis* synthetic root microbiome." *Proceedings of the National Academy of Sciences*, vol. 11:25, pp. 12558–65. https://doi.org/10.1073/pnas.1820691116.

31. Stringlis et al., Ibid.

32. Viviane Cordovez et al., 2019. "Ecology and evolution of plant microbiomes." *Annual Review of Microbiology*, vol. 73:1, pp. 69–88. https:// doi.org/10.1146/annurev-micro-090817-062524.

33. Hongwei Liu et al., Ibid.

34. Pascale et al., Ibid.

35. Stephen A. Rolfe, Joseph Griffiths, and Jurriaan Ton, 2019. "Crying out for help with root exudates: Adaptive mechanisms by which stressed plants assemble health-promoting soil microbiomes." *Current Opinion in Microbiology*, vol. 49, pp. 73–82. https://doi.org/10.1016/j.mib.2019.10.003.

36. Sergio Rasmann et al., 2005. "Recruitment of entomopathogenic nematodes by insect-damaged maize roots." *Nature*, 434, pp. 732–7. https:// doi.org/10.1038/nature03451.

37. D. R. Strong et al., 1996. "Entomopathogenic nematodes: Natural enemies of root-feeding caterpillars on bush lupine." *Oecologia*, vol. 108:1, pp. 167–73. https://doi.org/10.1007/BF00333228.

38. Pinar Avci et al., 2018. "In-vivo monitoring of infectious diseases in living animals using bioluminescence imaging." *Virulence*, vol. 9:1, pp. 28–63. https://doi.org/10.1080/21505594.2017.1371897.

39. Geraldine Mulley et al., 2015. "From insect to man: *Photorhabdus* sheds light on the emergence of human pathogenicity," *PLoSONE*, 10:12, e0144937. https://doi.org/10.1371/journal.pone.0144937.

40. Matt Soniak, 2012. "Why some Civil War soldiers glowed in the dark." April 5, 2012. Mental Floss.

41. Montgomery and Biklé, Ibid.

42. E. J. N. Helfrich et al., 2018. "Bipartite interactions, antibiotic production and biosynthetic potential of the *Arabidopsis* leaf microbiome." *Nature Microbiology*, vol. 3, pp. 909–19. https://doi.org/10.1038/s41564 -018-0200-0.

43. Coleman, Callaham, and Crossley, Ibid.

44. Pascale et al., Ibid.

45. Thimmaraju Rudrappa et al., 2008. "Root-secreted malic acid recruits beneficial soil bacteria." *Plant Physiology*, 148:3, pp. 1547-56. https://doi.org/10.1104/pp.108.127613.

46. Montgomery and Biklé, Ibid.

47. Gabriele Berg et al., 2017. "Plant microbial diversity is suggested as the key to future biocontrol and health trends." *FEMS Microbiology Ecology*, vol. 93:5. https://doi.org/10.1093/femsec/fix050.

48. Charisse Petersen and June L. Round, 2014. "Defining dysbiosis and its influence on host immunity and disease." *Cellular Microbiology*, vol. 16:7, pp. 1024-33. https://doi.org/10.1111/cmi.12308.

49. Cordovez et al., Ibid.

50. Rodrigo Mendes, Paolina Garbeva, and Jos M. Raaijmakers, 2013. "The rhizosphere microbiome: Significance of plant beneficial, plant pathogenic, and human pathogenic microorganisms." *FEMS Microbiology Reviews*, vol. 37:5, pp. 634-63, https://doi.org/10.1111/1574-6976.12028.

51. Rodrigo Mendes and Jos M. Raaijmakers, 2015. "Cross-kingdom similarities in microbiome functions." *The ISME Journal*, 9, pp. 1905-7. https://doi.org/10.1038/ismej.2015.7.

52. Rodrigo Mendes et al., 2011. "Deciphering the rhizosphere microbiome for disease-suppressive bacteria." *Science*, vol. 332:6033, pp. 1097-100. DOI: 10.1126/science.1203980.

53. Niki Grigoropoulou, Kevin R. Butt, and Christopher N. Lowe, 2008. "Effects of adult *Lumbricus terrestris* on cocoons and hatchlings in Evans' boxes." *Pedobiologia*, 51, pp. 343-9.

54. Susanne Wurst, Ilja Sonnemann, and Johann G. Zaller, 2018. "Soil Macro-Invertebrates: Their Impact on Plants and Associated Aboveground Communities in Temperate Regions." *Aboveground-Belowground Community Ecology, Ecological Studies (Analysis and Synthesis)*, vol. 234, pp. 175-200. https://doi.org/10.1007/978-3-319-91614-9_8.

55. M. Blouin et al., 2013. "A Review of earthworm impact on soil function and ecosystem services." *European Journal of Soil Science*, vol. 64:2, pp. 161-82. https://doi.org/10.1111/ejss.12025.

56. Blouin et al., Ibid.

57. Christian Feller et al., 2003. "Charles Darwin, earthworms and the natural sciences: Various lessons from past to future." *Agriculture, Ecosystems & Environment*, vol. 99:1-3, pp. 29-49. https://doi.org/10.1016/S0167-8809(03)00143-9.

58. Coleman, Callaham, and Crossley, Ibid.

59. Jan Willem van Groenigen et al., 2014. "Earthworms increase plant

production: A meta-analysis." *Scientific Reports*, vol. 4:6365. https://doi.org/10.1038/srep06365.

60. Ruben Puga-Freitas et al., 2012. "Signal molecules mediate the impact of the earthworm *Aporrectodea caliginosa* on growth, development and defence of the plant *Arabidopsis thaliana*." *PLOS One*, vol. 7:12. e49504. https://doi.org/10.1371/journal.pone.0049504.

61. Manuel Blouin et al., 2005. "Belowground organism activities affect plant aboveground phenotype, inducing plant tolerance to parasites." *Ecology Letters*, vol. 8:2, pp. 202–8. https://doi.org/10.1111/j.1461-0248.2004.00711.x.

62. Zhenggao Xiao et al., 2018. "Earthworms affect plant growth and resistance against herbivores: A meta-analysis." *Functional Ecology*, vol. 32:1, pp. 150–60. https://doi.org/10.1111/1365-2435.12969.

63. Maria J. I. Briones, 2018. "The serendipitous value of soil fauna in ecosystem functioning: The unexplained explained." *Frontiers in Environmental Science*, vol. 6, article 149. https://doi.org/10.3389/fenvs.2018.00149.

64. Andrew L. Neal et al., 2020. "Soil as an extended composite phenotype of the microbial metagenome." *Scientific Reports*, vol. 10, article 10649. https://doi.org/10.1038/s41598-020-67631-0.

65. Ruben Puga-Freitas and Manuel Blouin, 2015. "A review of the effects of soil organisms on plant hormone signalling pathways." *Environmental and Experimental Botany*, vol. 114, pp. 104–16. https://doi.org/10.1016/j.envexpbot.2014.07.006.

66. Neal et al., Ibid.

67. Ibid.

68. Yakov Kuzyakov and Evgenia Blagodatskaya, 2015. "Microbial hotspots and hot moments in soil: Concept & review." *Soil Biology and Biochemistry*, vol. 83, pp. 184–99. https://doi.org/10.1016/j.soilbio.2015.01.025.

69. G. E. Hutchinson, 1957. "Concluding remarks." *Cold Spring Harbor Symposia on Quantitative Biology*, vol. 22, pp. 415–27. https://www2.unil.ch/biomapper/Download/Hutchinson-CSHSymQunBio-1957.pdf.

70. Robert K. Colwell and Thiago F. Rangel, 2009. "Hutchinson's duality: The once and future niche." *Proceedings of the National Academy of Sciences*, November, vol. 106, supplement 2, pp. 19651–8. https://doi.org/10.1073/pnas.0901650106.

71. Samuel Pironon et al., 2017. "The 'Hutchinsonian niche' as an assemblage of demographic niches: Implications for species geographic ranges." *Ecography*, vol. 41:7, pp. 1103–13. https//doi: 10.1111/ecog.03414.

72. Kuzyakov and Blagodatskaya, Ibid.

CHAPTER 2

1. Philipp de Vrese, Stefan Hagemann, and Martin Claussen, 2016. "Asian irrigation, African rain: Remote impacts of irrigation." *Geophysical Research Letters*, vol. 43:8, pp. 3737–45. https://doi.org/10.1002/2016GL068146.

2. Dirk Helbing, 2013. "Globally networked risks and how to respond." *Nature*, vol. 497, pp. 51–9. https://doi.org/10.1038/nature12047.

3. Robert K. Merton, 1936. "The unanticipated consequences of purposive social action." *American Sociological Review*, vol. 1:6, pp. 894–904. https://doi.org/10.2307/2084615.

4. Andrew G. Haldane, April 28, 2009. "Rethinking the financial network." Bank of England at the Financial Student Association, Amsterdam. https://www.bankofengland.co.uk/speech/2009/rethinking-the-financial-network.

5. Tim G. Benton et al., 2017. *Environmental Tipping Points and Food System Dynamics: Main Report.* The Global Food Security programme, UK. https://dspace.stir.ac.uk/bitstream/1893/24796/1/GFS_Tipping%20Points_Main%20Report.pdf.

6. Ibid.

7. Timothy M. Lenton et al., 2019. "Climate tipping points—too risky to bet against." *Nature*, vol. 575, pp. 592–5. doi: https://doi.org/10.1038/d41586-019-03595-0.

8. Timothy M. Lenton, 2020. "Tipping positive change." *Philosophical Transactions of the Royal Society B, Biological Sciences*, vol. 375:1794. https://doi.org/10.1098/rstb.2019.0123.

9. Benton et al., Ibid.

10. Zeynep K. Hansen and Gary D. Libecap, 2004. "Small farms, externalities, and the Dust Bowl of the 1930s." *Journal of Political Economy*, vol. 112:3. https://doi.org/10.1086/383102.

11. Ibid.

12. Lenton, Ibid.

13. Stefano Battiston et al., 2016. "Complexity theory and financial regulation." *Science*, vol. 351:6275, pp. 818–19. https://doi.org/10.1126/science.aad0299.

14. Leonhard Horstmeyer et al., 2020. "Predicting collapse of adaptive networked systems without knowing the network." *Scientific Reports*, vol. 10, article 1223. https://doi.org/10.1038/s41598-020-57751-y.

15. Battiston et al., Ibid.

16. Haldane, Ibid.

17. Miguel A. Centeno et al., 2015. "The emergence of global systemic risk." *Annual Review of Sociology*, vol. 41, pp. 65–85. https://doi.org /10.1146/annurev-soc-073014-112317.

18. Flaviano Morone, Gino Del Ferraro, and Hernán A. Makse, 2019. "The k-core as a predictor of structural collapse in mutualistic ecosystems." *Nature Physics*, vol. 15, pp. 95–102. https://doi.org/10.1038 /s41567-018-0304-8.

19. Paolo D'Odorico et al., 2018. "The global food-energy-water nexus." *Reviews of Geophysics*, vol. 56:3, pp. 456–531. https://doi.org/10 .1029/2017RG000591.

20. Chengyi Tu, Samir Suweis, and Paolo D'Odorico, 2019. "Impact of globalization on the resilience and sustainability of natural resources." *Nature Sustainability*, vol. 2, pp. 283–9. https://doi.org/10.1038/s41893 -019-0260-z.

21. Dirk Helbing, 2013. "Globally networked risks and how to respond." *Nature*, vol. 497, pp. 51–9. https://doi.org/10.1038/nature12047.

22. Charles D. Brummitt, Raissa M. D'Souza, and E. A. Leicht, 2012. "Suppressing cascades of load in interdependent networks." *Proceedings of the National Academy of Sciences*, vol. 109:12, e680–e689. https://doi .org/10.1073/pnas.1110586109.

23. Sara Kammlade et al., 2017. *The Changing Global Diet*. International Center for Tropical Agriculture (CIAT). https://ciat.cgiar.org/the -changing-global-diet.

24. Ibid.

25. Colin K. Khoury et al., 2014. "Increasing homogeneity in global food supplies and the implications for food security." *Proceedings of the National Academy of Sciences*, vol. 111:11, pp. 4001–6. https://doi .org/10.1073/pnas.1313490111.

26. Bee Wilson, 2019. *The Way We Eat Now: How the Food Revolution Has Transformed Our Lives, Our Bodies, and Our World*. Basic Books, New York.

27. D'Odorico et al., Ibid.

28. Christopher Bren d'Amour and Weston Anderson, 2020. "International trade and the stability of food supplies in the global south." *Environmental Research Letters*, vol. 15:7. https://doi.org/10.1088/1748-9326/ab832f.

29. Ryan Walton, J. O. Miller, and Lance Champagne, 2019. *Simulating Maritime Chokepoint Disruption in the Global Food Supply*. Winter Simulation Conference, National Harbor, December 8–11, 2019, pp. 1708–18. https://doi.org/10.1109/WSC40007.2019.9004883.

30. Michael J. Puma et al., 2015. "Assessing the evolving fragility of the global food system." *Environmental Research Letters*, vol. 10:2. https://doi.org/10.1088/1748-9326/10/2/024007.

31. Walton, Miller, and Champagne, Ibid.

32. FAO, 2009. *How to Feed the World in 2050*. Food and Agriculture Organization of the United Nations, October 12, 2009. http://www.fao.org/fileadmin/templates/wsfs/docs/expert_paper/How_to_Feed_the_World_in_2050.pdf.

33. Puma et al., Ibid.

34. Christopher Bren d'Amour et al., 2016. "Teleconnected food supply shocks." *Environmental Research Letters*, vol. 11:3. https://doi.org/10.1088/1748-9326/11/3/035007.

35. David Seekell et al., 2017. "Resilience in the global food system." *Environmental Research Letters*, vol. 12:2. https://doi.org/10.1088/1748-9326/aa5730.

36. Puma et al., Ibid.

37. M. Nyström et al., 2019. "Anatomy and resilience of the global production ecosystem." *Nature*, vol. 575, pp. 98–108. https://doi.org/10.1038/s41586-019-1712-3.

38. Chengyi Tu, Suweis, and D'Odorico, Ibid.

39. Samir Suweis et al., 2015. "Resilience and reactivity of global food security." *Proceedings of the National Academy of Sciences*, vol. 112:22, pp. 6902–7. https://doi.org/10.1073/pnas.1507366112.

40. D'Odorico et al., Ibid.

41. Nyström et al., Ibid.

42. Adam Smith, 1759. *The Theory of Moral Sentiments*. Part IV, "Of the effect of utility upon the sentiment of approbation," p. 165. https://www.ibiblio.org/ml/libri/s/SmithA_MoralSentiments_p.pdf.

43. FAO, 2006. *Building on Gender, Agrobiodiversity and Local Knowledge—A Training Manual*. Food and Agriculture Organization of the United Nations, 2006. http://www.fao.org/3/y5956e/Y5956E03.htm.

44. Ravi P. Singh et al., 2011. "The emergence of Ug99 races of the stem rust fungus is a threat to world wheat production." *Annual Review of Phytopathology*, vol. 49, pp. 465–81. https://doi.org/10.1146/annurev-phyto-072910-095423.

45. Ian Heap and Stephen O. Duke, 2018. "Overview of glyphosate-resistant weeds worldwide." *Pest Management Science*, vol. 74:5, pp. 1040–9. https://doi.org/10.1002/ps.4760.

46. Patricio Grassini, Kent M. Eskridge, and Kenneth G. Cassman, 2013. "Distinguishing between yield advances and yield plateaus in historical

crop production trends." *Nature Communications*, vol. 4, article 2918. https://doi.org/10.1038/ncomms3918.

47. Ibid.

48. David Tilman et al., 2002. "Agricultural sustainability and intensive production practices." *Nature*, vol. 418, pp. 671–7. https://doi.org/10.1038/nature01014.

49. Ibid.

50. Kenneth G. Cassman et al., 2003. "Meeting cereal demand while protecting natural resources and improving environmental quality." *Annual Review of Environment and Resources*, vol. 28. pp. 315–58. https://doi.org/10.1146/annurev.energy.28.040202.122858.

51. Nyström et al., Ibid.

52. Patrick Woodall and Tyler L. Shannon, 2018. "Monopoly power corrodes choice and resiliency in the food system." *The Antitrust Bulletin*, vol. 63:2, pp. 198–221. https://doi.org/10.1177/0003603X18770063.

53. Sophia Murphy, David Burch, and Jennifer Clapp, 2012. *Cereal Secrets: The World's Largest Grain Traders and Global Agriculture*. Oxfam Research Reports, August 2012. https://www-cdn.oxfam.org/s3fs-public/file_attachments/rr-cereal-secrets-grain-traders-agriculture-30082012-en_4.pdf.

54. Adam Putz, 2018. "The ABCDs and M&A: Putting 90% of the global grain supply in fewer hands." *Pitchbook*, February 21. https://pitchbook.com/news/articles/the-abcds-and-ma-putting-90-of-the-global-food-supply-in-fewer-hands.

55. Philip Howard and Mary Hendrickson, 2020. "The state of concentration in global food and agriculture industries," in Hans Herren and Benedikt Haerlin, 2020. *Transformation of Our Food Systems: The Making of a Paradigm Shift*. IAASTD. https://philhoward.net/2020/09/27/the-state-of-concentration-in-global-food-and-agriculture-industries.

56. Jennifer Clapp and Joseph Purugganan, 2020. "Contextualizing corporate control in the agrifood and extractive sectors." *Globalizations*, vol. 17:7, pp. 1265–75, https://doi.org/10.1080/14747731.2020.1783814.

57. Jennifer Clapp, 2018. "Mega-mergers on the menu: Corporate concentration and the politics of sustainability in the global food system." *Global Environmental Politics*, vol. 18:2, pp. 12–33. https://doi.org/10.1162/glep_a_00454.

58. Pat Mooney et al., 2017. *Too Big to Feed: Exploring the Impacts of Mega-Mergers, Concentration of Power in the Agri-Food Sector*. International Panel of Experts on Sustainable Food Systems (IPES-Food),

October 2017. http://www.ipes-food.org/_img/upload/files/Concentra
tion_FullReport.pdf.

59. Susanne Gura and François Meienberg, 2013. *Agropoly—A Handful of Corporations Control World Food Production*. Berne Declaration (DB) & EcoNexus, Zurich. https://www.econexus.info/sites/econexus /files/Agropoly_Econexus_BerneDeclaration.pdf.

60. Mooney et al., Ibid.

61. Clapp and Purugganan, Ibid.

62. Woodall and Shannon, Ibid.

63. Michael L. Katz, 2019. "Multisided platforms, big data, and a little antitrust policy." *Review of Industrial Organization*, vol. 54, pp. 695–716. https://doi.org/10.1007/s11151-019-09683-9.

64. Laura Wellesley et al., 2017. "Chokepoints in global food trade: Assessing the risk." *Research in Transportation Business & Management*, vol. 25, pp. 15–28. https://doi.org/10.1016/j.rtbm.2017.07.007.

65. Bren d'Amour et al., Ibid.

66. Evan D. G. Fraser, Alexander Legwegoh, and Krishna KC, 2015. "Food stocks and grain reserves: Evaluating whether storing food creates resilient food systems." *Journal of Environmental Studies and Sciences*, vol. 5, pp. 445–58. https://doi.org/10.1007/s13412-015-0276-2.

67. Christophe Gouel, 2013. "Optimal food price stabilisation policy." *European Economic Review*, vol. 57, pp. 118–34. https://doi.org/10 .1016/j.euroecorev.2012.10.003.

68. Fraser, Legwegoh, and Krishna, Ibid.

69. Jennifer Clapp and S. Ryan Isakson, 2018. "Risky returns: The implications of financialization in the food system." *Development and Change*, vol. 49:2. pp. 437–60. https://doi.org/10.1111/dech.12376.

70. José Azar, Martin C. Schmalz, and Isabel Tecu, 2018. "Anticompetitive effects of common ownership." *The Journal of Finance*, vol. 73:4, pp. 1513–65. https://doi.org/10.1111/jofi.12698.

71. Dirk Helbing, 2013. "Globally networked risks and how to respond." *Nature*, vol. 497, pp. 51–9. https://doi.org/10.1038/nature12047.

72. *Chicago SRW Wheat—Volume, Futures and Options*. Daily Exchange Volume Chart. https://www.cmegroup.com/trading/agricultural/grain -and-oilseed/wheat_quotes_volume_voi.html#tradeDate=20191216.

73. M. Graziano Ceddia, 2020. "The super-rich and cropland expansion via direct investments in agriculture." *Nature Sustainability*, vol. 3, pp. 312–18. https://doi.org/10.1038/s41893-020-0480-2.

74. Land Matrix. https://landmatrix.org.

75. Land Matrix. https://landmatrix.org/region/africa.

76. Ward Anseeuw and Giulia Maria Baldinelli, 2020. *Uneven Ground: Land Inequality at the Heart of Unequal Societies.* International Land Coalition & Oxfam. https://oi-files-d8-prod.s3.eu-west-2.amazonaws .com/s3fs-public/2020-11/uneven-ground-land-inequality-unequal -societies.pdf.

77. Private Eye, 2015. *Selling England (and Wales) by the Pound.* https:// www.private-eye.co.uk/registry.

78. Mario Herrero, 2017. "Farming and the geography of nutrient production for human use: A transdisciplinary analysis." *The Lancet Planetary Health,* vol. 1:1, pp. e33–e42. https://doi.org/10.1016/S2542-5196(17) 30007-4.

79. Nyström et al., Ibid.

80. Rong Wang, 2012. "Flickering gives early warning signals of a critical transition to a eutrophic lake state." *Nature,* vol. 492, pp. 419–22. https://doi.org/10.1038/nature11655.

81. Jon Greenman, Tim Benton, and Joseph Travis, 2003. "The amplification of environmental noise in population models: Causes and consequences." *The American Naturalist,* vol. 161:2. https://doi.org/10.1086 /345784.

82. Benton et al., Ibid.

83. Richard S. Cottrell et al., 2019. "Food production shocks across land and sea." *Nature Sustainability,* vol. 2, pp. 130–7. https://doi.org/10 .1038/s41893-018-0210-1.

84. FAO, IFAD, UNICEF, WFP and WHO, 2020. *The State of Food Security and Nutrition in the World 2020: Transforming Food Systems for Affordable Healthy Diets.* Food and Agriculture Organization of the United Nations, Rome. https://doi.org/10.4060/ca9692en.

85. Jennifer Clapp, 2017. "Food self-sufficiency: Making sense of it, and when it makes sense." *Food Policy,* vol. 66, pp. 88–96. https://doi.org /10.1016/j.foodpol.2016.12.001.

86. Bren d'Amour et al., Ibid.

87. Tiziano Distefano, 2018. "Shock transmission in the international food trade network." *PLoS ONE,* vol. 13:8, e0200639. https://doi.org/10 .1371/journal.pone.0200639.

88. Frederick Kaufman, 2011. "How Goldman Sachs created the food crisis." https://foreignpolicy.com/2011/04/27/how-goldman-sachs-created -the-food-crisis.

89. Angelika Beck, Benedikt Haerlin, and Lea Richter, 2016. *Agriculture at a Crossroads: IAASTD Findings and Recommendations for Future Farming.* Foundation on Future Farming. https://www.globalagriculture

.org/fileadmin/files/weltagrarbericht/EnglishBrochure/Brochure
IAASTD_en_web_small.pdf.

90. Marianela Fader et al., 2013. "Spatial decoupling of agricultural production and consumption: Quantifying dependences of countries on food imports due to domestic land and water constraints." *Environmental Research Letters*, vol. 8:1. https://doi.org/10.1088/1748-9326/8/1/014046.

91. Therea Falkendal et al., 2021. "Grain export restrictions during COVID-19 risk food insecurity in many low- and middle-income countries." *Nature Food*, vol. 2, pp. 11–14. https://doi.org/10.1038/s43016-020-00211-7.

92. Franziska Gaupp, 2020. "Extreme events in a globalized food system." *One Earth*, vol. 2:6, pp. 518–21, https://doi.org/10.1016/j.oneear.2020.06.001.

93. Marcy Nicholson, August 26, 2021. "Canada sees supply of main crops dropping 26% on drought impact." Bloomberg News. https://www.bloomberg.com/news/articles/2021-08-26/canada-sees-supply-of-main-crops-dropping-26-on-drought-impact.

94. Baird Langenbrunner, 2021. "Water, water not everywhere." *Nature Climate Change*, vol. 11, p. 650. https://doi.org/10.1038/s41558-021-01111-9.

95. Kanat Shaku, Akin Nazli, and Will Conroy, August 19, 2021. "Kazakhstan set to lose quarter of grain crop as drought hammers Eurasia." Intellinews. https://www.intellinews.com/kazakhstan-set-to-lose-quarter-of-grain-crop-as-drought-hammers-eurasia-218502.

96. France 24, June 9, 2021. "Historic drought threatens Brazil's economy." https://www.france24.com/en/live-news/20210609-historic-drought-threatens-brazil-s-economy.

97. Zhao Yimeng, July 30, 2021. "Agriculture officials work to ensure food supply after floods Damage Crops." *China Daily*. https://global.chinadaily.com.cn/a/202107/30/WS6103a913a310efa1bd6658fb.html.

98. Agnieszka de Sousa and Megan Durisin, September 2, 2021. "Global food costs jump back near decade-high on harvest woes." https://www.bloomberg.com/news/articles/2021-09-02/global-food-prices-jump-back-near-decade-high-on-harvest-woes.

99. Beck, Haerlin, and Richter, Ibid.

100. Max Roser, Hannah Ritchie, and Esteban Ortiz-Ospina, 2013. *World Population Growth*. OurWorldInData.org. https://ourworldindata.org/world-population-growth.

101. Nikos Alexandratos and Jelle Bruinsma, 2012. *World Agriculture Towards 2030/2050: The 2012 Revision*. Food and Agriculture

Organization of the United Nations. https://www.researchgate.net /publication/270890453_World_Agriculture_Towards_20302050 _The_2012_Revision.

102. These calculations, updated to take account of further declines in the human population growth rate, are explained at: George Monbiot, November 19, 2015. *Pregnant Silence*. https://www.monbiot.com /2015/11/19/pregnant-silence.

103. David Tilman et al., 2011. "Global food demand and the sustainable intensification of agriculture." *Proceedings of the National Academy of Sciences*, vol. 108:50, 20260–4. https://doi.org/10.1073/pnas.11164 37108.

104. Beck, Haerlin. and Richter, Ibid.

105. Rob Cook, October 23, 2021. *World Cattle Inventory by Year*. https:// beef2live.com/story-world-cattle-inventory-1960-2014-130-111523.

106. Beck, Haerlin, and Richter, Ibid.

107. Nikos Alexandratos and Jelle Bruinsma, 2012. *World Agriculture Towards 2030/2050: The 2012 Revision*. Food and Agriculture Organization of the United Nations. https://www.researchgate.net /publication/270890453_World_Agriculture_Towards_20302050 _The_2012_Revision.

108. D'Odorico et al., Ibid.

109. Henri de Ruiter, et al., 2017. "Total global agricultural land footprint associated with UK food supply 1986–2011." *Global Environmental Change*, vol. 43, pp. 72–81. https://doi.org/10.1016/j.gloenvcha.2017 .01.007.

110. Department for Environment, Food and Rural Affairs, 2020, October 8, 2020. *Farming Statistics—Provisional Arable Crop Areas, Yields and Livestock Populations at 1 June 2020 United Kingdom*. https://assets .publishing.service.gov.uk/government/uploads/system/uploads/attach ment_data/file/931104/structure-jun2020prov-UK-08oct20i.pdf.

111. Emily S. Cassidy et al., 2013. "Redefining agricultural yields: From tonnes to people nourished per hectare." *Environmental Research Letters*, vol. 8:3, 034015. https://doi.org/10.1088/1748-9326/8/3/034015.

112. Maria Cristina Rulli et al., 2016. "The water-land-food nexus of first-generation biofuels." *Scientific Reports*, vol. 6, article 22521. https:// doi.org/10.1038/srep22521.

113. Mitchell C. Hunter et al., 2017. "Agriculture in 2050: Recalibrating targets for sustainable intensification." *BioScience*, vol. 67:4, pp. 386–91. https://doi.org/10.1093/biosci/bix010.

114. Deepak K. Ray et al., 2013. "Yield trends are insufficient to double

global crop production by 2050." *PLoS ONE*, vol. 8:6, e66428. https://doi.org/10.1371/journal.pone.0066428.

115. Colin Raymond, Tom Matthews, and Radley M. Horton, 2020. "The emergence of heat and humidity too severe for human tolerance." *Science Advances*, vol. 6:19. https://doi.org/10.1126/sciadv.aaw1838.

116. Ibid.

117. Nir Y. Krakauer, Benjamin I. Cook, and Michael J. Puma, 2020. "Effect of irrigation on humid heat extremes." *Environmental Research Letters*, vol. 15:9. https://doi.org/10.1088/1748-9326/ab9ecf.

118. Raymond, Matthews, and Horton, Ibid.

119. Luke J. Harrington and Friederike E. L. Otto, 2020. "Reconciling theory with the reality of African heatwaves." *Nature Climate Change*, vol. 10, pp. 796–8. https://doi.org/10.1038/s41558-020-0851-8.

120. Chi Xu et al., 2020. "Future of the human climate niche." *Proceedings of the National Academy of Sciences*, May, vol. 117:21, pp. 11350–5. https://doi.org/10.1073/pnas.1910114117.

121. UN Food and Agriculture Organization, 2018. "Nigeria: Small family farms country fact sheet." http://www.fao.org/3/i9930en/I9930EN.pdf.

122. Nigerian Price, 2021. "Tractor prices in Nigeria," September. https://nigerianprice.com/tractor-prices-in-nigeria.

123. Vincent Ricciardi, 2018. "How much of the world's food do smallholders produce?" *Global Food Security*, vol. 17, pp. 64–72. https://doi.org/10.1016/j.gfs.2018.05.002.

124. Deepak K. Ray et al., 2019. "Climate change has likely already affected global food production." *PLoS ONE*, vol. 14:5, e0217148. https://doi.org/10.1371/journal.pone.0217148.

125. Lindsey L. Sloat et al., 2020. "Climate adaptation by crop migration." *Nature Communications*, vol. 11, article 1243. https://doi.org/10.1038/s41467-020-15076-4.

126. David Makowski et al., 2020. "Quantitative synthesis of temperature, CO_2, rainfall, and adaptation effects on global crop fields." *European Journal of Agronomy*, vol. 115. https://doi.org/10.1016/j.eja.2020.126041.

127. Xuhui Wang et al., 2020. "Emergent constraint on crop yield response to warmer temperature from field experiments." *Nature Sustainability*, vol. 3, pp. 908–16. https://doi.org/10.1038/s41893-020-0569-7.

128. M. Zampieri et al., 2019. "When will current climate extremes affecting maize production become the norm?" *Earth's Future*, vol. 7, pp. 113–22. https://doi.org/10.1029/2018EF000995.

129. Chuang Zhao et al., 2017. "Temperature increase reduces global yields of major crops in four independent estimates." *Proceedings of the*

National Academy of Sciences, vol. 114:35, pp. 9326–31. https://doi.org/10.1073/pnas.1701762114.

130. Rory G. J. Fitzpatrick et al., 2020. "How a typical West African day in the future-climate compares with current-climate conditions in a convection-permitting and parameterised convection climate model." *Climatic Change*, vol. 163, pp. 267–96. https://doi.org/10.1007/s10584-020-02881-5.

131. Samuel S. Myers et al., 2014. "Increasing CO_2 threatens human nutrition." *Nature*, vol. 510, pp. 139–42. https://doi.org/10.1038/nature13179.

132. Robert H. Beach et al., 2019. "Combining the effects of increased atmospheric carbon dioxide on protein, iron, and zinc availability and projected climate change on global diets: A modelling study." *The Lancet Planetary Health*, vol. 3:7, e307–17. https://doi.org/10.1016/S2542-5196(19)30094-4.

133. J. I. Macdiarmid and S. Whybrow, 2019. "Nutrition from a climate change perspective." *Proceedings of the Nutrition Society*, vol. 78, pp. 380–7. https://doi.org/10.1017/S0029665118002896.

134. Ibid.

135. Tim Spector, 2020. *Spoon-Fed: Why Almost Everything We've Been Told About Food Is Wrong.* Jonathan Cape, London.

136. Elizabeth R. H. Moore et al., 2020. "The mismatch between anthropogenic CO_2 emissions and their consequences for human zinc and protein sufficiency highlights important environmental justice issues." *Challenges*, vol. 11:(1):4. https://doi.org/10.3390/challe11010004.

137. Matthew R. Smith and Samuel S. Myers, 2018. "Impact of anthropogenic CO_2 emissions on global human nutrition." *Nature Climate Change*, vol. 8, pp. 834–9. https://doi.org/10.1038/s41558-018-0253-3.

138. Danielle E. Medek, Joel Schwartz, and Samuel S. Myers, 2017. "Estimated effects of future atmospheric CO_2 concentrations on protein intake and the risk of protein deficiency by country and region." *Environmental Health Perspectives*, vol. 125:8. https://doi.org/10.1289/EHP41.

139. M. R. Smith, C. D. Golden, and S. S. Myers, 2017. "Potential rise in iron deficiency due to future anthropogenic carbon dioxide emissions." *GeoHealth*, vol. 1:6, pp. 248–57. https://doi.org/10.1002/2016GH000018.

140. Samuel S. Myers et al., 2017. "Climate change and global food systems: Potential impacts on food security and undernutrition." *Annual Review of Public Health*, vol. 38:1, pp. 259–77. https://doi.org/10.1146/annurev-publhealth-031816-044356.

141. M. R. Smith and S. S. Myers, 2019. "Global health implications of nutrient changes in rice under high atmospheric carbon dioxide." *GeoHealth*, vol. 3:7, pp. 190–200. https://doi.org/10.1029/2019GH000188.

142. Andrew S. Ross, 2019. "A shifting climate for grains and flour." Cereals & Grains Association, Cereal Foods World. https://www.cerealsgrains.org/publications/cfw/2019/September-October/Pages/CFW-64-5-0050.aspx.

143. Edward D. Perry, Jisang Yu, and Jesse Tack, 2020. "Using insurance data to quantify the multidimensional impacts of warming temperatures on yield risk." *Nature Communications*, vol. 11, article 4542. https://doi.org/10.1038/s41467-020-17707-2.

144. Noah S. Diffenbaugh, 2020. "Verification of extreme event attribution: Using out-of-sample observations to assess changes in probabilities of unprecedented events." *Science Advances*, vol. 6:12, e2368. https://doi.org/10.1126/sciadv.aay2368.

145. ReliefWeb, 2021. "2 years since Cyclone Idai and Mozambique has already faced an additional 3 cyclones." https://reliefweb.int/report/mozambique/2-years-cyclone-idai-and-mozambique-has-already-faced-additional-3-cyclones.

146. Xiaogang He and Justin Sheffield, May 13, 2020. "Lagged compound occurrence of droughts and pluvials globally over the past seven decades." *Geophysical Research Letters*, vol. 47:14. https://doi.org/10.1029/2020GL087924.

147. Xu Yue and Nadine Unger, 2018. "Fire air pollution reduces global terrestrial productivity." *Nature Communications*, vol. 9, article 5413. https://doi.org/10.1038/s41467-018-07921-4.

148. Walton, Miller, and Champagne, Ibid.

149. Laura Wellesley et al., 2017. "Chokepoints in global food trade: Assessing the risk." *Research in Transportation Business & Management*, vol. 25, pp. 15–28. https://doi.org/10.1016/j.rtbm.2017.07.007.

150. Jon Gambrell and Samy Magdy, March 24, 2021. "Massive cargo ship becomes wedged, blocks Egypt's Suez Canal." AP News. https://apnews.com/article/cargo-ship-blocks-egypt-suez-canal-5957543bb555ab31c14d56ad09f98810.

151. Wellesley et al., Ibid.

152. Walton, Miller, and Champagne, Ibid.

153. Kyle Frankel Davis et al., 2017. "Water limits to closing yield gaps." *Advances in Water Resources*, vol. 99, pp. 67–75. https://doi.org/10.1016/j.advwatres.2016.11.015.

154. Y. Wada et al., 2016. "Modeling global water use for the 21st century: The water futures and solutions (WFaS) initiative and its approaches."

Geoscientific Model Development, vol. 9:1, pp. 175–222. https://doi .org/10.5194/gmd-9-175-2016.

155. Zhongwei Huang et al., 2019. "Global agricultural green and blue water consumption under future climate and land use changes." *Journal of Hydrology*, vol. 574, pp. 242–56. https://doi.org/10.1016/j.jhydrol .2019.04.046.

156. D'Odorico et al., Ibid.

157. J. S. Famiglietti et al., 2011. "Satellites measure recent rates of ground-water depletion in California's Central Valley." *Geophysical Research Letters*, vol. 38:3. https://doi.org/10.1029/2010GL046442.

158. Justin Fox, April 13, 2015. "Cows suck up more water than almonds." Bloomberg. https://www.bloomberg.com/opinion/articles/2015-04-13 /cows-suck-up-more-of-california-s-water-than-almonds.

159. Mesfin Mekonnen and Arjen Hoekstra, 2011. "The green, blue and grey water footprint of crops and derived crop products." *Hydrology and Earth System Sciences Discussions*, vol. 15, pp. 1577–1600. https:// doi.org/10.5194/hess-15-1577-2011.

160. World Resources Institute, Aqueduct. *Aqueduct Tools*. https://www.wri .org/aqueduct.

161. Daniel Viviroli et al., 2020. "Increasing dependence of lowland popula-tions on mountain water resources." *Nature Sustainability*, vol. 3, pp. 917–28. https://doi.org/10.1038/s41893-020-0559-9.

162. W. W. Immerzeel et al., 2019. "Importance and vulnerability of the world's water towers." *Nature*, vol. 577, pp. 364–9. https://doi.org/10 .1038/s41586-019-1822-y.

163. Hamish D. Pritchard, 2019. "Asia's shrinking glaciers protect large populations from drought stress." *Nature*, vol. 569, pp. 649–54. https:// doi.org/10.1038/s41586-019-1240-1.

164. Ibid.

165. H. Biemans et al., 2019. "Importance of snow and glacier meltwater for agriculture on the Indo-Gangetic plain." *Nature Sustainability*, vol. 2, pp. 594–601. https://doi.org/10.1038/s41893-019-0305-3.

166. Pritchard, Ibid.

167. Immerzeel et al., Ibid.

168. Pritchard, Ibid.

169. Sadaf Taimur, October 15, 2020. "India, Pakistan, and the coming climate-induced scramble for water." Salzburg Global Seminar. https:// www.salzburgglobal.org/news/opinions/article/india-pakistan-and-the -coming-climate-induced-scramble-for-water.

170. Yue Qin et al., 2020. "Agricultural risks from changing snowmelt."

Nature Climate Change, vol. 10, pp. 459–65. https://doi.org/10.1038/s41558-020-0746-8.

171. C. A. Scott et al., 2014. "Irrigation efficiency and water-policy implications for river-basin resilience." *Hydrology and Earth System Sciences*, vol. 18, pp. 1339–48. https://doi.org/10.5194/hess-18-1339-2014.

172. Ibid.

173. Isaac M. Held and Brian J. Soden, 2006. "Robust responses of the hydrological cycle to global warming." *Journal of Climate*, vol. 19:21, pp. 5686–99. https://doi.org/10.1175/JCLI3990.1.

174. Chang-Eui Park et al., 2018. "Keeping global warming within 1.5°C constrains emergence of aridification." *Nature Climate Change*, vol. 8, pp. 70–4. https://doi.org/10.1038/s41558-017-0034-4.

175. Miroslav Trnka et al., 2019. "Mitigation efforts will not fully alleviate the increase in water scarcity occurrence probability in wheat-producing areas." *Science Advances*, vol. 5:9, e2406. https://doi.org/10.1126/sciadv.aau2406.

176. Matti Kummu et al., 2021. "Climate change risks pushing one-third of global food production outside the safe climatic space." *One Earth*, vol. 4:5, pp. 720–9. https://doi.org/10.1016/j.oneear.2021.04.017.

177. David B. Lobell, 2014. "Climate change adaptation in crop production: Beware of illusions." *Global Food Security*, vol. 3:2, pp. 72–6. https://doi.org/10.1016/j.gfs.2014.05.002.

178. Commission of the European Communities, 2006. "Proposal for a directive of the European Parliament and of the Council establishing a framework for the protection of soil and amending Directive 2004/35/EC." Access to European Union Law, Document 52006PC0232. https://eur-lex.europa.eu/legal-content/EN/TXT/?uri=CELEX:52006PC0232.

179. National Farmers' Union, May 22, 2014. *Withdrawal of Soil Framework Directive Welcomed*. https://www.nfuonline.com/withdrawal-of-soil-framework-directive-welcomed.

180. Defra Strategic Evidence and Partnership Project, 2011. https://issuu.com/westcountryriverstrust/docs/9850_theriverstrustdseppreport.

181. Ibid.

182. *Soil Protection Review*, 2013. The Farming Forum. https://thefarmingforum.co.uk/index.php?threads/soil-protection-review.15170/page-3.

183. Department for Environment, Food and Rural Affairs, 2020. "Farming Statistics—Land Use, Livestock Populations and Agricultural Workforce at 1 June 2020—England." https://assets.publishing.service.gov.uk/government/uploads/system/uploads/attachment_data/file/928397/structure-landuse-june20-eng-22oct20.pdf.

184. George Monbiot, November 23, 2004. "Fuel for nought." *The Guardian.* https://www.theguardian.com/politics/2004/nov/23/greenpolitics.uk.

185. The original online article has been deleted—https://www.fginsight .com/home/arable/taking-maize-for-energy-production-to-the-next -level/59704.article—but I cited it at the time, here: George Monbiot, March 14, 2014. "How a false solution to climate change is damaging the natural world." *The Guardian.* https://www.theguardian.com/envi ronment/georgemonbiot/2014/mar/14/uk-ban-maize-biogas.

186. Richard Gaughan, May 10, 2018. "How much land is needed for wind turbines?" *Sciencing.* https://sciencing.com/much-land-needed-wind -turbines-12304634.html.

187. R. C. Palmer and R. P. Smith, 2013. "Soil structural degradation in SW England and its impact on surface-water runoff generation." *Soil Use and Management*, vol. 29:4, pp. 567–75. https://doi.org/10.1111 /sum.12068.

188. Nils Klawitter, August 30, 2012. *Corn-Mania: Biogas Boom in Germany Leads to Modern-Day Land Grab.* Spiegel International. https:// www.spiegel.de/international/germany/biogas-subsidies-in-germany -lead-to-modern-day-land-grab-a-852575.html.

189. Pasquale Borrelli et al., 2017. "An assessment of the global impact of 21st-century land use change on soil erosion." *Nature Communications*, vol. 8, article 2013. https://doi.org/10.1038/s41467-017-02142-7.

190. Martina Sartori et al., 2019. "A linkage between the biophysical and the economic: Assessing the global market impacts of soil erosion." *Land Use Policy*, vol. 86, pp. 299–312. https://doi.org/10.1016/j.landusepol.2019 .05.014.

191. World Bank Group, 2017. *Republic of Malawi Poverty Assessment.* World Bank, Washington DC. https://openknowledge.worldbank.org /handle/10986/26488.

192. Solomon Asfawa, Giacomo Pallanteb, and Alessandro Palma, 2020. "Distributional impacts of soil erosion on agricultural productivity and welfare in Malawi." *Ecological Economics*, vol. 177. https://doi.org /10.1016/j.ecolecon.2020.106764.

193. Luca Montanarella, Robert Scholes, and Anastasia Brainich (eds.), 2018. *The IPBES Assessment Report on Land Degradation and Restoration.* Secretariat of the Intergovernmental Science-Policy Platform on Biodiversity and Ecosystem Services, Bonn, Germany. https://doi.org/10 .5281/zenodo.3237392.

194. Ibid.

195. Pasquale Borrelli et al., 2020. "Land use and climate change impacts

on global soil erosion by water." *Proceedings of the National Academy of Sciences*, vol. 117:36, pp. 21994–22001. https://doi.org/10.1073/pnas .2001403117.

196. Margaret R. Douglas, Jason R. Rohr, and John F. Tooker, 2014. "Neo-nicotinoid insecticide travels through a soil food chain, disrupting biological control of non-target pests and decreasing soya bean yield." *Journal of Applied Ecology*, vol. 52, pp. 250–60. https://doi.org/10 .1111/1365-2664.12372.

197. Cláudia de Lima e Silva et al., 2017. "Comparative toxicity of imidaclo-prid and thiacloprid to different species of soil invertebrates." *Ecotoxicol ogy*, vol. 26, pp. 555–64. https://doi.org/10.1007/s10646-017-1790-7.

198. Bo Yu et al., "Effects on soil microbial community after exposure to neonicotinoid insecticides thiamethoxam and dinotefuran." *Science of the Total Environment*, vol. 725. https://doi.org/10.1016/j.scitotenv .2020.138328.

199. Peng Zhang et al., 2018. "Sorption, desorption and degradation of neonicotinoids in four agricultural soils and their effects on soil micro-organisms." *Science of the Total Environment*, vol. 615, pp. 59–69. https://doi.org/10.1016/j.scitotenv.2017.09.097.

200. Montanarella, Scholes, and Brainich (eds.), Ibid.

CHAPTER 3

1. Tim Maugham, November 30, 2021. "The modern world has finally become too complex for any of us to understand. No one's driving." https://onezero.medium.com/the-modern-world-has-finally-become -too-complex-for-any-of-us-to-understand-1a0b46fbc292.

2. Wales Environment Link, December 18, 2020. *Statement in response to Natural Resources Wales' new advice on the River Wye Special Area of Conservation (SAC)*. https://www.waleslink.org/sites/default/files/wel _statement_on_nrw_phosphate_in_river_wye_sac_final.pdf.

3. United Kingdom Government Legislation, 2016. *The Environmental Permitting (England and Wales) Regulations 2016, Intensive Farming*. UK Statutory Instruments, 2016, no. 1154, sch. 1, pt. 2, ch. 6, section 6.9. https://www.legislation.gov.uk/uksi/2016/1154/schedule/1/part/2 /chapter/6/crossheading/intensive-farming/made.

4. E-mailed responses from the two county councils and the regulators to the Rivercide team, pers comm.

5. Nicola Cutcher, July 15, 2021. "Counting chickens." https://cutcher .co.uk/linklog/2021/07/15/counting-chickens.

6. Wil Crisp, February 12, 2021. "Revealed: no penalties issued under 'useless' English farm pollution laws." *The Guardian.* https://www .theguardian.com/environment/2021/feb/12/revealed-no-penalties -issued-under-useless-uk-farm-pollution-laws.

7. Gordon Green, 2021. "River phosphate aspects of poultry farming in Powys—A case study." https://www.wyesalmon.com/wp-content /uploads/2021/04/A-Study-of-Poultry-Farming-and-its -Impact-on-Water-Quality-in-the-Wye.pdf.

8. The Wye & Usk Foundation, June 8, 2020. *Nation's "Favourite" River Facing Ecological Disaster.* https://www.wyeuskfoundation.org/news /nations-favourite-river-facing-ecological-disaster.

9. Brecon & Radnor Branch of Campaign for the Protection of Rural Wales and, Herefordshire and Shropshire Branches of Campaign for the Protection of Rural, 2019. *Intensive Chicken Production Units: Here-fordshire, Shropshire & Powys.* http://www.brecon-and-radnor-cprw .wales/wp-content/uploads/2019/07/IPU-ALLdataV4-Master -20190707-3-Counties-FINAL-2.0-20190711.pdf.

10. The Wye & Usk Foundation, Ibid.

11. "Rivercide: The world's first live investigative documentary." https:// www.youtube.com/watch?v=5IDoVAUNANA.

12. CPRW Brecon & Radnor Branch Press Release, July 2, 2020. *5 Years this July: Remember, Remember the Disappearing Wildlife and Clean Rivers of Powys.* https://m.facebook.com/notes/river-wye-pollution -and-conservation/cprw-brecon-radnor-branch-press-release-2nd -july-2020/2633300743624910.

13. Kate Bull at Change.org, 2020. "Save the River Wye!" *Demand Mora-torium on all New Poultry Units in Powys.* https://www.change.org/p /powys-county-council-save-the-river-wye-demand-moratorium-on-all -new-poultry-units-in-powys?recruiter=66808012&utm_source=share _petition&utm_medium=facebook&utm_campaign=psf_combo_share _initial&utm_term=share_petition&recruited_by_id=4570935e-5370 -4756-9804-1bdb777cb9b4&utm_content=fht-23011731-en-gb %3Av12&fbclid=IWAR0WEWKRbef-EZ2ZzroS9Yz_516 GNEKYMG8nNylcr4z8mw7w46PJzg3Schc.

14. Natural Resources Wales, December 17, 2020. "NRW issues new advice to safeguard the River Wye special area of conservation." https://natu ralresources.wales/about-us/news-and-events/news/nrw-issues-new -advice-to-safeguard-the-river-wye-special-area-of-conservation /?lang=en.

15. Salmon & Trout Conservation, February 9, 2021. "NRW's planning

advice on Wye pollution ineffective, say conservation organisations *Fish Legal* and *Salmon & Trout Conservation.*" https://salmon-trout .org/2021/02/09/nrws-planning-advice-on-wye-pollution-ineffective -say-conservation-organisations.

16. Elgan Hearn, February 4, 2021. "Plans for 150,000-bird chicken farm near Welshpool rejected." *Powys County Times.* https://www.county times.co.uk/news/19066507.plans-150-000-bird-chicken-farm -near-welshpool-rejected.

17. George Monbiot, October 5, 2015. "Think dairy farming is benign? Our rivers tell a different story." *The Guardian.* https://www.theguard ian.com/environment/2015/oct/05/think-dairy-farming-is-benign-our -rivers-tell-a-different-story.

18. Ibid.

19. Crisp, Ibid.

20. Madeleine Cuff, September 24, 2020. "Farmers on average receive a pollution inspection from the Environment Agency every 263 years." *iNews.* https://inews.co.uk/news/environment/farmers-pollution-inspection -environment-agency-chemicals-pollutants-659701.

21. John Cossens, 2019. *River Axe N2K Catchment Regulatory Project Report.* Salmon & Trout Conservation. https://www.salmon-trout.org /wp-content/uploads/2020/03/Final-Axe-Regulatory-Report.pdf.

22. Andrew Wasley et al., 2017. *Dirty Business: The Livestock Farms Polluting the UK.* The Bureau of Investigative Journalism, August 21, 2017. https://www.thebureauinvestigates.com/stories/2017-08-21/farming -pollution-fish-uk.

23. *Red Tractor.* https://redtractor.org.uk.

24. United Kingdom Parliamentary Business, November 22, 2018. *UK Progress on Reducing Nitrate Pollution.* Commons Select Committees, Environmental Audit. https://publications.parliament.uk/pa/cm201719 /cmselect/cmenvaud/656/65605.htm.

25. Cossens, Ibid.

26. UK Environment Agency, December 1, 2017. *Corporate Report: Environment Agency Statistics on Serious Pollution Incidents, their Impacts on Air and Water and the Sectors Responsible.* https://www.gov.uk /government/publications/pollution-incidents-evidence-summaries.

27. Natural Resources Wales, April 29, 2019. *Sea Trout Stock Performance in Wales 2018.* https://cdn.cyfoethnaturiol.cymru/media/688881/sea -trout-stock-performance-in-wales-2018_1.pdf?mode=pad&rnd =132013665370000000.

28. Meat Promotion Wales, 2018. *Little Book of Meat Facts: Compendium of Welsh Red Meat and Livestock Industry Statistics 2018*. https:// meatpromotion.wales/images/resources/Little_Book_of_Meat_Facts _2018_web.pdf.

29. Ella McSweeney, September 28, 2020. "'We've crossed a threshold': Has industrial farming contributed to Ireland's water crisis?" *The Guardian*, Animals farmed, Ireland. https://www.theguardian.com /environment/2020/sep/28/weve-crossed-a-threshold-has-industrial -farming-contributed-to-irelands-water-crisis.

30. Ministry for the Environment & Stats NZ, 2019. *Embargoed until 10.45am 18 April 2019: New Zealand's Environmental Reporting Series: Environment Aotearoa 2019*. https://www.documentcloud.org /documents/5954379-Environment-Aotearoa-2019-Embargoed.html.

31. Yaara Bou Melhem, March 15, 2021. "New Zealand's troubled waters." ABC. https://www.abc.net.au/news/2021-03-16/new-zealand-rivers -pollution-100-per-cent-pure/13236174.

32. Fiona Proffitt, July 22, 2010. *How Clean Are Our Rivers?* National Institute of Water and Atmospheric Research (NIWA). https://niwa .co.nz/publications/wa/water-atmosphere-1-july-2010/how-clean-are -our-rivers.

33. Maryna Strokal et al., 2016. "Alarming nutrient pollution of Chinese rivers as a result of agricultural transitions." *Environmental Research Letters*, vol. 11:2. https://doi.org/10.1088/1748-9326/11/2/024014.

34. Franziska Schulz et al., 2013. *Unterschiede der Fütterung ökologischer und konventioneller Betriebe und deren Einfluss auf die Methan- Emission aus der Verdauung von Milchkühen*. Braunschweig, Johann Heinrich von Thünen-Institut, pp. 189–205. http://www.pilotbetriebe .de/download/Abschlussbericht%202013/5-8_Schulz%20et%20al %202013.pdf.

35. Rhian Price, February 12, 2018. "How dairy farmers are making four cuts of 11.3ME silage." *Farmers Weekly*. https://www.fwi.co.uk/live stock/how-dairy-farmers-are-making-four-cuts-of-11-3me-silage.

36. Volac International Limited, 2021. *Grow More Milk With Ecosyl*. https://uk.ecosyl.com/grow-more-milk-with-ecosyl#.

37. Paul Cawood, pers comm.

38. UK Department for Environment, Food & Rural Affairs and Environ- ment Agency, April 2, 2018. *Rules for Farmers and Land Managers to Prevent Water Pollution*. https://www.gov.uk/guidance/rules-for-farmers -and-land-managers-to-prevent-water-pollution.

39. Fly Fishing & Fly Tying. *Welsh Slurry Contractors Say Regulation is Required on Spreading.* https://flyfishing-and-flytying.co.uk/blog/view/welsh_slurry_contractors_say_regulation_is_required_on_spreading.

40. Alexandra Leclerc and Alexis Laurent, 2017. "Framework for estimating toxic releases from the application of manure on agricultural soil: National release inventories for heavy metals in 2000–2014." *Science of The Total Environment*, vols 590–1, pp. 452–60. https://doi.org/10.1016/j.scitotenv.2017.01.117.

41. Xiaoyong Qian et al., 2018. "Heavy metals accumulation in soil after 4 years of continuous land application of swine manure: A field-scale monitoring and modeling estimation." *Chemosphere*, vol. 210, pp. 1029–34. https://doi.org/10.1016/j.chemosphere.2018.07.107.

42. Organization for Economic Cooperation and Development (OECD), November 10, 2016. *Policy Insights—Antimicrobial Resistance.* https://www.oecd.org/health/health-systems/AMR-Policy-Insights-November2016.pdf.

43. U.S. Food and Drug Administration (FDA), 2013. *Guidance for Industry: New Animal Drugs and New Animal Drug Combination Products Administered in or on Medicated Feed or Drinking Water of Food-Producing Animals: Recommendations for Drug Sponsors for Voluntarily Aligning Product Use Conditions with GFI #209.* December, no. 213. https://www.fda.gov/media/83488/download.

44. W.-Y. Xie, Q. Shen, and F. J. Zhao, 2018. "Antibiotics and antibiotic resistance from animal manures to soil: A review." *European Journal of Soil Science*, vol. 69:1, pp. 181–95. https://doi.org/10.1111/ejss.12494.

45. Evelyn Walters, Kristin McClellan, and Rolf U. Halden, 2010. "Occurrence and loss over three years of 72 pharmaceuticals and personal care products from biosolids-soil mixtures in outdoor mesocosms." *Water Research*, vol. 44:20, pp. 6011–20. https://doi.org/10.1016/j.watres.2010.07.051.

46. Ya He et al., 2020. "Antibiotic resistance genes from livestock waste: Occurrence, dissemination, and treatment." *Nature Partner Journal (NPJ)*, Clean Water, 3, article 4. https://doi.org/10.1038/s41545-020-0051-0.

47. Yong-Guan Zhu et al., 2019. "Soil biota, antimicrobial resistance and planetary health." *Environment International*, vol. 131. https://doi.org/10.1016/j.envint.2019.105059.

48. Paulo Durão, Roberto Balbontín, and Isabel Gordo, 2018. "Evolutionary mechanisms shaping the maintenance of antibiotic resistance." *Trends in Microbiology*, vol. 26:8, pp. 677–91. https://doi.org/10.1016/j.tim.2018.01.005.

49. Wan-Ying Xie et al., 2018. "Long-term effects of manure and chemical fertilizers on soil antibiotic resistome." *Soil Biology and Biochemistry*, vol. 122, pp. 111–19. https://doi.org/10.1016/j.soilbio.2018.04.009.

50. Fenghua Wang et al., 2020. "Fifteen-year application of manure and chemical fertilizers differently impacts soil ARGs and microbial community structure." *Frontiers in Microbiology*, vol. 11. https://doi.org/10.3389/fmicb.2020.00062.

51. Heather Storteboom et al., 2010. "Tracking antibiotic resistance genes in the South Platte River Basin using molecular signatures of urban, agricultural, and pristine sources." *Environmental Science & Technology*, vol. 44:19, pp. 7397–404. https://doi.org/10.1021/es101657s.

52. S. Koike et al., 2007. "Monitoring and source tracking of tetracycline resistance genes in lagoons and groundwater adjacent to swine production facilities over a 3-year period." *Applied and Environmental Microbiology*, vol. 73:15, pp. 4813–23. https://doi.org/10.1128/AEM.00665-07.

53. Zhishu Liang et al., 2020. "Pollution profiles of antibiotic resistance genes associated with airborne opportunistic pathogens from typical area, Pearl River estuary, and their exposure risk to humans." *Environment International*, vol. 143. https://doi.org/10.1016/j.envint.2020.105934.

54. Fangkai Zhao et al., 2019. "Bioaccumulation of antibiotics in crops under long-term manure application: Occurrence, biomass response and human exposure." *Chemosphere*, vol. 219, pp. 882–95. https://doi.org/10.1016/j.chemosphere.2018.12.076.

55. Yu-Jing Zhang et al., 2019. "Transfer of antibiotic resistance from manure-amended soils to vegetable microbiomes." *Environment International*, vol. 130, article 104912. https://doi.org/10.1016/j.envint.2019.104912.

56. OECD, Ibid.

57. Holger Heuer, Heike Schmitt, and Kornelia Smalla, 2011. "Antibiotic resistance gene spread due to manure application on agricultural fields." *Current Opinion in Microbiology*, vol. 14:3, pp. 236–43. https://doi.org/10.1016/j.mib.2011.04.009.

58. World Health Organization (WHO Europe), 2011. *Tackling Antibiotic Resistance from a Food Safety Perspective in Europe*. https://www.euro.who.int/__data/assets/pdf_file/0005/136454/e94889.pdf.

59. Jim O'Neill, 2014. *Review on Antimicrobial Resistance—Antimicrobial Resistance: Tackling a Crisis for the Health and Wealth of Nations*. https://wellcomecollection.org/works/rdpck35v.

60. Aude Teillant et al., 2015. "Potential burden of antibiotic resistance on surgery and cancer chemotherapy antibiotic prophylaxis in the USA: A literature review and modelling study." *The Lancet Infectious Diseases*, vol.15:12,pp.1429–37.https://doi.org/10.1016/S1473-3099(15)00270-4.

61. Thomas P. Van Boeckel et al., 2015. "Global trends in antimicrobial use in food animals." *Proceedings of the National Academy of Sciences*, vol. 112:18, pp. 5649–54. https://doi.org/10.1073/pnas.1503141112.

62. Biosolids Assurance Scheme (BAS). *About Biosolids.* https://assured biosolids.co.uk/about-biosolids.

63. Crispin Dowler and Zach Boren, 2020. *Revealed: Salmonella, Toxic Chemicals and Plastic Found in Sewage Spread on Farmland.* Unearthed, Greenpeace UK. https://unearthed.greenpeace.org/2020/02/04/sewage -sludge-landspreading-environment-agency-report.

64. Environment Agency, October 2019. *Perfluorooctane Sulfonate (PFOS) and Related Substances: Sources, Pathways and Environmental Data.* Department for Environment, Food & Rural Affairs. https:// consult.environment-agency.gov.uk/environment-and-business /challenges-and-choices/user_uploads/perfluorooctane-sulfonate-and- related-substances-pressure-rbmp-2021.pdf.

65. Gareth Simkins, 2020. "EA 'sat on sewage sludge risks research,' report says." ENDS Report. https://www.endsreport.com/article/1672915/ea -sat-sewage-sludge-risks-research-report-says.

66. Environment Agency, December 1, 2015. *News Story: Farm Suppliers Warned over Waste Materials Rules; New Enforcement Programme Targets Inappropriate Landspreading.* UK Government. https://www .gov.uk/government/news/farm-suppliers-warned-over-waste -materials-rules.

67. Bo Yu et al., 2020. "Effects on soil microbial community after exposure to neonicotinoid insecticides thiamethoxam and dinotefuran." *Science of the Total Environment*, vol. 725. https://doi.org/10.1016/j.scitotenv .2020.138328.

68. United States Environmental Protection Agency, 2018. *Report: EPA Unable to Assess the Impact of Hundreds of Unregulated Pollutants in Land-Applied Biosolids on Human Health and the Environment.* Office of Inspector General, Report #19-P-0002, November 15. https:// www.epa.gov/sites/production/files/2018-11/documents/_epaoig _20181115-19-p-0002.pdf.

69. Jennifer Lee, 2003. "Sludge spread on fields is fodder for lawsuits." *The New York Times*, June 26. https://www.nytimes.com/2003/06/26/us /sludge-spread-on-fields-is-fodder-for-lawsuits.html.

70. Seacoast Online, 2019. "Maine dairy farmer's blood tests high for 'forever chemicals' from toxic sludge." *Bangor Daily News*, August 16. https://bangordailynews.com/2019/08/16/news/york/maine-dairy-farmers-blood-tests-high-for-forever-chemicals-from-toxic-sludge.

71. Sharon Lerner, 2019. "Toxic PFAS chemicals found in main farms fertilized with sewage sludge." *The Intercept*, June 7. https://theintercept.com/2019/06/07/pfas-chemicals-maine-sludge.

72. Amy Lowman et al., 2013. "Land application of treated sewage sludge: Community health and environmental justice." *Environmental Health Perspectives*, vol. 121:5, pp. 537–42. https://doi.org/10.1289/ehp.1205470.

73. Tom Perkins, 2019. "Biosolids: Mix human waste with toxic chemicals, then spread on crops." *The Guardian*, October 5. https://www.theguardian.com/environment/2019/oct/05/biosolids-toxic-chemicals-pollution.

74. Esther A. Gies et al., 2018. "Retention of microplastics in a major secondary wastewater treatment plant in Vancouver, Canada." *Marine Pollution Bulletin*, vol. 133, pp. 553–61. https://doi.org/10.1016/j.marpolbul.2018.06.006.

75. Jenna Gavigan et al., 2020. "Synthetic microfiber emissions to land rival those to waterbodies and are growing." *PLOS One*, vol. 15:9, e0237839. https://doi.org/10.1371/journal.pone.0237839.

76. Alexandra Scudo et al., October 2017. *Intentionally Added Microplastics in Products: Final Report*. Amec Foster Wheeler Environment & Infrastructure UK Limited. https://ec.europa.eu/environment/chemicals/reach/pdf/39168%20Intentionally%20added%20microplastics%20-%20Final%20report%2020171020.pdf.

77. Jessica Stubenrauch and Felix Ekardt, 2020. "Plastic pollution in soils: Governance approaches to foster soil health and closed nutrient cycles." *Environments*, vol. 7: 5:38. https://doi.org/10.3390/environments7050038.

78. Marcela Calabi-Floody et al., 2018. "Smart Fertilizers as a Strategy for Sustainable Agriculture," in Donald L. Sparks (ed.), *Advances in Agronomy*, vol. 147, Academic Press, pp. 119–57. https://doi.org/10.1016/bs.agron.2017.10.003.

79. Muhammad Yasin Naz and Shaharin Anwar Sulaiman, 2016. "Slow release coating remedy for nitrogen loss from conventional urea: A review." *Journal of Controlled Release*, vol. 225, pp. 109–20. https://doi.org/10.1016/j.jconrel.2016.01.037.

80. Fabio Corradini et al., 2019. "Evidence of microplastic accumulation

in agricultural soils from sewage sludge disposal." *Science of the Total Environment*, vol. 671, pp. 411–20. https://doi.org/10.1016/j.scitotenv .2019.03.368.

81. Jill Crossman et al., 2020. "Transfer and transport of microplastics from biosolids to agricultural soils and the wider environment." *Science of the Total Environment*, vol. 724. https://doi.org/10.1016/j.scitotenv .2020.138334.

82. Dunmei Lin et al., 2020. "Microplastics negatively affect soil fauna but stimulate microbial activity: Insights from a field-based microplastic addition experiment." *Proceedings of the Royal Society*, B Biological Sciences, vol. 287:1934. https://doi.org/10.1098/rspb.2020.1268.

83. Yang Song et al., 2019. "Uptake and adverse effects of polyethylene terephthalate microplastics fibers on terrestrial snails (Achatina fulica) after soil exposure." *Environmental Pollution*, vol. 250, pp. 447–55. https://doi.org/10.1016/j.envpol.2019.04.066.

84. Dong Zhu et al., 2018. "Exposure of soil collembolans to microplastics perturbs their gut microbiota and alters their isotopic composition." *Soil Biology and Biochemistry*, vol. 116, pp. 302–10. https://doi.org /10.1016/j.soilbio.2017.10.027.

85. Dunmei Lin et al., Ibid.

86. Esperanza Huerta Lwanga et al., 2016. "Microplastics in the terrestrial ecosystem: Implications for *Lumbricus terrestris* (Oligochaeta, Lumbricidae)." *Environmental Science & Technology*, vol. 50:5. https://doi .org/10.1021/acs.est.5b05478.

87. Elma Lahive et al., 2019. "Microplastic particles reduce reproduction in the terrestrial worm *Enchytraeus crypticus* in a soil exposure." *Environmental Pollution*, vol. 255, pt. 2, pp. 113–74. https://doi.org/10.1016 /j.envpol.2019.113174.

88. Anderson Abel de Souza Machado et al., 2017. "Microplastics as an emerging threat to terrestrial ecosystems." *Global Change Biology*, vol. 24:4, pp. 1405–16. https://doi.org/10.1111/gcb.14020.

89. Xiao-Dong Sun et al., 2020. "Differentially charged nanoplastics demonstrate distinct accumulation in *Arabidopsis thaliana*." *Nature Nanotechnology*, vol. 15, pp. 755–60. https://doi.org/10.1038/s41565 -020-0707-4.

90. Simin Li et al., 2020. "Influence of long-term biosolid applications on communities of soil fauna and their metal accumulation: A field study." *Environmental Pollution*, vol. 260. https://doi.org/10.1016/j.envpol .2020.114017.

91. Global Witness, 2021. "Last line of defence: The industries causing the

climate crisis and attacks against land and environmental defenders," September 13. https://www.globalwitness.org/en/campaigns/environmental -activists/last-line-defence.

92. World Wildlife Fund, 2015. "Average EU citizen consumes 61 kg of soy per year, most from soy embedded in meat, dairy, eggs and fish." https:// wwf.panda.org/wwf_offices/brazil/?247051/WWF-Average -EU-citizen-consumes-61-kg-of-soy-per-year--most-from-soy -embedded-in-meat-dairy-eggs-and-fish.

93. Hannah Ritchie. Soy. Our World in Data. https://ourworldindata.org /soy.

94. Ibid.

95. Walter Fraanje and Tara Garnett, 2020. *Soy: Food, Feed, and Land Use Change.* Food Climate Research Network, University of Oxford (TABLE debates). https://tabledebates.org/building-blocks/soy-food -feed-and-land-use-change.

96. Jonny Hughes and Neil Burgess, 2019. "Rare wildlife in Brazil's savannah is under threat—we are all responsible." United Nations Environment Programme World Conservation Monitoring Centre (UNEP-WCMC), October 29. https://medium.com/@unepwcmc/rare-wildlife-in-brazils -savannah-is-under-threat-we-are-all-responsible-c17b21e3c0fa.

97. Raoni Rajão et al., 2020. "The rotten apples of Brazil's agribusiness." *Science*, vol. 369:6501, pp. 246–8. https://science.sciencemag.org/con tent/369/6501/246.

98. Gabriel S. Hofmann et al., 2021. "The Brazilian Cerrado is becoming hotter and drier." *Global Change Biology*, vol. 27:17, pp. 4060–73. https://doi.org/10.1111/gcb.15712.

99. André Vasconcelos et al., 2020. *Illegal Deforestation and Brazilian Soy Exports: The case of Mato Grosso.* Trase, Issue Brief 4, June 2020. http://resources.trase.earth/documents/issuebriefs/TraseIssueBrief4 _EN.pdf.

100. Nestor Ignacio Gasparri and Yann le Polain de Waroux, 2015. "The coupling of South American soybean and cattle production frontiers: New challenges for conservation policy and land change science." *Conservation Letters*, vol. 8:4, pp. 290–8. https://doi.org/10.1111/conl .12121.

101. Michael DiBartolomeis, 2019. "An assessment of acute insecticide toxicity loading (AITL) of chemical pesticides used on agricultural land in the United States." *PLOS One*, vol. 14:8, e0220029. https://doi.org /10.1371/journal.pone.0220029.

102. Thomas James Wood and Dave Goulson, 2017. "The environmental

risks of neonicotinoid pesticides: A review of the evidence post-2013." *Environmental Science and Pollution Research International*, vol. 24:21, pp. 17285–325. https://doi.org/10.1007/s11356-017-9240-x.

103. Margaret R. Douglas, Jason R. Rohr, and John F. Tooker, 2015. "Editor's Choice: Neonicotinoid insecticide travels through a soil food chain, disrupting biological control of non-target pests and decreasing soya bean yield." *Journal of Applied Ecology*, vol. 52:1, pp. 250–60. https://doi.org/10.1111/1365-2664.12372.

104. Cláudia de Lima e Silva et al., 2017. "Comparative toxicity of imidacloprid and thiacloprid to different species of soil invertebrates." *Ecotoxicology*, vol. 26, pp. 555–64. https://doi.org/10.1007/s10646-017-1790-7.

105. Samuel J. Macaulay et al., 2021. "Imidacloprid dominates the combined toxicities of neonicotinoid mixtures to stream mayfly nymphs." *Science of the Total Environment*, vol. 761, article 143263. https://doi.org/10.1016/j.scitotenv.2020.143263.

106. Verena C. Schreiner et al., 2021. "Paradise lost? Pesticide pollution in a European region with considerable amount of traditional agriculture." *Water Research*, vol. 188, article 116528. https://doi.org/10.1016/j.watres.2020.116528.

107. Pedro Cardoso et al., 2020. "Scientists' warning to humanity on insect extinctions." *Biological Conservation*, vol. 242, article 108426. https://doi.org/10.1016/j.biocon.2020.108426.

108. Caspar A. Hallmann et al., 2017. "More than 75 percent decline over 27 years in total flying insect biomass in protected areas." *PLOS One*, vol. 12:10, e0185809. https://doi.org/10.1371/journal.pone.0185809.

109. Anders Pape Møller, 2019. "Parallel declines in abundance of insects and insectivorous birds in Denmark over 22 years." *Ecology and Evolution*, vol. 9:11, pp. 6581–7. https://doi.org/10.1002/ece3.5236.

110. Caspar A. Hallmann et al., 2019. "Declining abundance of beetles, moths and caddisflies in the Netherlands." *Insect Conservation and Diversity*, vol. 13:2, pp. 127–39. https://doi.org/10.1111/icad.12377.

111. UN General Assembly, 2017. *Report of the Special Rapporteur on the Right to Food*. United Nations General Assembly Human Rights Council, Thirty-fourth session, February 27–March 24, Agenda item 3, A/HRC/34/48. https://www.pan-uk.org/site/wp-content/uploads/United-Nations-Report-of-the-Special-Rapporteur-on-the-right-to-food.pdf.

112. David L. Wagner, 2020. "Insect declines in the Anthropocene." *Annual Review of Entomology*, vol. 65, pp. 457–80. https://doi.org/10.1146/annurev-ento-011019-025151.

113. Yijia Li, Ruiqing Miao, and Madhu Khanna, 2020. "Neonicotinoids and

decline in bird biodiversity in the United States." *Nature Sustainability*, vol. 3, pp. 1027–35. https://doi.org/10.1038/s41893-020-0582-x.

114. Masumi Yamamuro et al., 2019. "Neonicotinoids disrupt aquatic food webs and decrease fishery yields." *Science*, vol. 366:6465, pp. 620–3. https://doi.org/10.1126/science.aax3442.

115. David Tilman et al., 2001. "Forecasting agriculturally driven global environmental change." *Science*, vol. 292:5515, pp. 281–4. https://doi.org/10.1126/science.1057544.

116. Mengqiu Wang et al., 2019. "The great Atlantic *Sargassum* belt." *Science*, vol. 365:6448, pp. 83–7. https://doi.org/10.1126/science.aaw7912.

117. Thais Sousa, 2020. "Fertilizers keep up with agroindustry, see higher demand." *Brazil-Arab News Agency (ANBA)*, April 28. https://anba.com.br/en/fertilizers-keep-up-with-agroindustry-see-higher-demand.

118. Virginia Institute of Marine Science. "Dead zones: Lack of oxygen a key stressor on marine ecosystems." https://www.vims.edu/research/topics/dead_zones/index.php.

119. Max Roser and Hannah Ritchie. "Phosphate fertilizer production, 1961 to 2014." Our World in Data. https://ourworldindata.org/fertilizers#phosphate-fertilizer-production.

120. Max Roser and Hannah Ritchie. "Nitrogen fertilizer production, 1961 to 2014." Our World in Data. https://ourworldindata.org/fertilizers#nitrogen-fertilizer-production.

121. Kenneth G. Cassman, Achim Dobermann, and Daniel T. Walters, 2002. "Agroecosystems, Nitrogen-use efficiency, and nitrogen management." *AMBIO: A Journal of the Human Environment*, vol. 31:2, pp. 132–40. https://doi.org/10.1579/0044-7447-31.2.132.

122. Peter Omara et al., 2019. "World cereal nitrogen use efficiency trends: Review and current knowledge." *Agrosystems, Geosciences and Environment*, vol. 2:1, pp. 1–8. https://doi.org/10.2134/age2018.10.0045.

123. Hannah Ritchie, 2021. "Excess fertilizer use: Which countries cause environmental damage by overapplying fertilizers?." Our World in Data, September 7. https://ourworldindata.org/excess-fertilizer.

124. J. J. Schröder et al., 2011. "Improved phosphorus use efficiency in agriculture: A key requirement for its sustainable use." *Chemosphere*, vol. 84:6, pp. 822–31. https://doi.org/10.1016/j.chemosphere.2011.01.065.

125. Cassman, Dobermann, and Walters, Ibid.

126. Gulshan Mahajan and Jagadish Timsina, 2011. "Effect of nitrogen rates and weed control methods on weeds abundance and yield of direct-seeded rice." *Archives of Agronomy and Soil Science*, vol. 57:3, pp. 239–50. https://doi.org/10.1080/03650340903369384.

127. Rui Catarino, Sabrina Gaba, and Vincent Bretagnolle, 2019. "Experimental and empirical evidence shows that reducing weed control in winter cereal fields is a viable strategy for farmers." *Scientific Reports*, vol. 9, article 9004. https://doi.org/10.1038/s41598-019-45315-8.

128. David Kleijn et al., 2019. "Ecological intensification: Bridging the gap between science and practice." *Trends in Ecology & Evolution*, vol. 34:2, pp. 154–66. https://doi.org/10.1016/j.tree.2018.11.002.

129. David Wuepper, Nikolaus Roleff, and Robert Finger, 2021. "Does it matter who advises farmers? Pest management choices with public and private extension." *Food Policy*, vol. 99, article 101995. https://doi.org/10.1016/j.foodpol.2020.101995.

130. Council of the European Union, 2018. *Directive (EU) 2018/2001 of the European Parliament and of the Council of 11 December 2018 on the Promotion of the Use of Energy from Renewable Sources (Text with EEA Relevance)*. Document 32018L2001. http://data.europa.eu/eli/dir/2018/2001/oj.

131. Ecovision Commercial. *The Renewable Partner for Poultry Farmers, Renewable Heating for Poultry Farms*. Ecovision Systems Ltd. https://www.ecovisionsystems.co.uk/poultry.

132. Chloe Ryan, 2018. "How to make money from renewable energy on poultry farms." *Poultry News*. http://www.poultrynews.co.uk/business-politics/business/feature-how-to-make-money-from-renewable-energy-on-poultry-farms.html.

133. Dogwood Alliance, 2017. "Destroying southern forests for international export, EU demand is stripping our forests." https://www.dogwoodalliance.org/wp-content/uploads/2017/08/Acres-of-Pellets-Fact-Sheet.pdf.

134. Hazel Sheffield, 2021. "Carbon-neutrality is a fairy tale: How the race for renewables is burning Europe's forests." *The Guardian*, January 14. https://www.theguardian.com/world/2021/jan/14/carbon-neutrality-is-a-fairy-tale-how-the-race-for-renewables-is-burning-europes-forests.

135. Aljazeera, 2020. "Romania: Rape of the forest," November 26. https://www.aljazeera.com/program/people-power/2020/11/26/rape-of-the-forest.

136. Emerging Europe, 2019. "Environmental groups warn Poland not to restart logging in Białowieża Forest," January 29. https://emerging-europe.com/news/environmental-groups-sound-alarm-over-new-polish-plans-to-log-in-bialowieza-forest/#:~:text=Environmental%20groups%20warn%20Poland%20not%20to%20restart%20logging%20in%20Bia%C5%82owie%C5%BCa%20Forest,-January%2029%2C%202019&text=A%20coalition%20of%20environmental%20

groups,Bia%C5%82owie%C5%BCa%20Forest%20in%20eastern%20 Poland.

137. Guido Ceccherini et al., 2020. "Abrupt increase in harvested forest area over Europe after 2015." *Nature*, vol. 583, pp. 72–7. https://doi.org /10.1038/s41586-020-2438-y.

138. Duncan Brack, 2017. "Woody biomass for power and heat: Impacts on the global climate." Chatham House, February 23. https://www .chathamhouse.org/2017/02/woody-biomass-power-and-heat.

139. Andrea Camia et al., 2020. *The Use of Woody Biomass for Energy Production in the EU*. European Commission's Joint Research Centre (JRC), EUR 30548 EN. https://publications.jrc.ec.europa.eu/repository/bit stream/JRC122719/jrc-forest-bioenergy-study-2021-final_online.pdf.

140. Thierry Courvoisier, President EASAC, 2018. "Letter to Jean-Claude Juncker, President of the European Commission," January 8. European Academies Science Advisory Council (EASAC). https://easac.eu/filead min/user_upload/180108_Letter_to_President_Juncker.pdf.

141. Tim Benton, Juliet Vickery, and Jeremy Wilson, 2003. "Farmland biodiversity: Is habitat heterogeneity the key?" *Trends in Ecology & Evolution*, vol. 18:4, pp. 182–8. https://doi.org/10.1016/S0169-5347(03)00011-9.

142. Doreen Gabriel et al., 2010. "Scale matters: The impact of organic farming on biodiversity at different spatial scales." *Ecology Letters*, vol. 13:7, pp. 858–69. https://doi.org/10.1111/j.1461-0248.2010.01481.x.

143. Pesticide Action Network UK. *Is Organic Better?* http://www.pan-uk .org/organic.

144. Andrew L. Neal et al., 2020. "Soil as an extended composite phenotype of the microbial metagenome." *Scientific Reports*, vol. 10, article 10649. https://doi.org/10.1038/s41598-020-67631-0.

145. Laurence G. Smith et al., 2019. "The greenhouse gas impacts of converting food production in England and Wales to organic methods." *Nature Communications*, vol. 10, article 4641. https://doi.org/10.1038 /s41467-019-12622-7.

146. Tomek de Ponti, Bert Rijk, and Martin van Ittersum, 2012. "The crop yield gap between organic and conventional agriculture." *Agricultural Systems*, vol. 108, pp. 1–9. https://doi.org/10.1016/j.agsy.2011.12.004.

147. Verena Seufert, Navin Ramankutty, and Jonathan A. Foley, 2012. "Comparing the yields of organic and conventional agriculture." *Nature*, vol. 485, pp. 229–32. https://doi.org/10.1038/nature11069.

148. Pietro Barbieri et al., 2021. "Global option space for organic agriculture is delimited by nitrogen availability." *Nature Food*, vol. 2, pp. 363–72. https://doi.org/10.1038/s43016-021-00276-y.

149. Michael Clark and David Tilman, 2017. "Comparative analysis of environmental impacts of agricultural production systems, agricultural input efficiency, and food choice." *Environmental Research Letters*, vol. 12:6. https://doi.org/10.1088/1748-9326/aa6cd5.

150. Maximilian Pieper, Amelie Michalke, and Tobias Gaugler, 2020. "Calculation of external climate costs for food highlights inadequate pricing of animal products." *Nature Communications*, vol. 11:6117. https://doi.org/10.1038/s41467-020-19474-6.

151. Hanna Tuomisto et al., 2012. "Does organic farming reduce environmental impacts? A meta-analysis of European research." *Journal of Environmental Management*, vol. 112, pp. 309–20. https://doi.org/10.1016/j.jenvman.2012.08.018.

152. Laura Cattell Noll et al., 2020. "The nitrogen footprint of organic food in the United States." *Environmental Research Letters*, vol. 15:4. https://doi.org/10.1088/1748-9326/ab7029.

153. Annisa Chand, 2020. "Organic beef lets the system down." *Nature Food*, vol. 1:253. https://doi.org/10.1038/s43016-020-0086-x.

154. H. L. Tuomisto et al., 2012. "Does organic farming reduce environmental impacts?—A meta-analysis of European research." *Journal of Environmental Management*, vol. 112, pp. 309–20. https://doi.org/10.1016/j.jenvman.2012.08.018.

155. Michael Clark and David Tilman, 2017. "Comparative analysis of environmental impacts of agricultural production systems, agricultural input efficiency, and food choice." *Environmental Research Letters*, vol. 12:6. https://doi.org/10.1088/1748-9326/aa6cd5.

156. Jon Ungoed-Thomas, 2020. "Free-range egg farms choking life out of the Wye." *The Sunday Times*, June 21. https://www.thetimes.co.uk/article/free-range-egg-farms-choking-life-out-of-the-wye-rt3c763qc.

157. Statistics for Wales, December 18, 2019. *Farm Incomes in Wales, April 2018 to March 2019*. Welsh Government, SFR 123/2019. https://gov.wales/sites/default/files/statistics-and-research/2019-12/farm-incomes-april-2018-march-2019-209.pdf.

158. Wikimedia Commons, March 29, 2009. *The Desert of Wales as Seen from the Summit of Drygarn Fawr*. https://commons.wikimedia.org/wiki/File:Desert_of_wales_from_Drygarn_Fawr.JPG.

159. F. M. Chambers et al., 2007. "Recent vegetation history of Drygarn Fawr (Elenydd SSSI), Cambrian Mountains, Wales: Implications for conservation management of degraded blanket mires." *Biodiversity and Conservation*, vol. 16, pp. 2821–46. https://doi.org/10.1007/s10531-007-9169-3.

160. Christopher J. Ellis, 2016. "Oceanic and temperate rainforest climates and their epiphyte indicators in Britain." *Ecological Indicators*, vol. 70, pp. 125–33. https://doi.org/10.1016/j.ecolind.2016.06.002.

161. Philip Shaw and D. B. A. Thompson, 2006. *The Nature of the Cairngorms: Diversity in a Changing Environment.* Stationery Office Books (TSO), Edinburgh.

162. Wesley Stephenson, 2020. "Gardens help towns and cities beat countryside for tree cover." *BBC News, Science & Environment*, October 18. https://www.bbc.co.uk/news/science-environment-54311593.

163. George Monbiot, 2017. "Explanation of the figures in grim reaping," January 11. https://www.monbiot.com/2017/01/11/explanation-of-the-figures-in-grim-reaping.

164. Savills, January 17, 2019. "Current agricultural land use in the UK." https://www.savills.co.uk/research_articles/229130/274017-0.

165. UK Department for Environment, Food and Rural Affairs (DEFRA), October 8, 2020. *Farming Statistics—Provisional Arable Crop Areas, Yields and Livestock Populations at 1 June 2020 United Kingdom.* https://assets.publishing.service.gov.uk/government/uploads/system/uploads/attachment_data/file/931104/structure-jun2020prov-UK-08oct20i.pdf.

166. Hannah Postles, 2017. "New land cover atlas reveals just six per cent of UK is built on," November 8. The University of Sheffield. https://www.sheffield.ac.uk/news/nr/land-cover-atlas-uk-1.744440.

167. UK Department for Environment, Food and Rural Affairs (DEFRA), October 8, 2020., Ibid.

168. Monbiot, "Explanation of the figures," Ibid.

169. Mark Easton, 2012. "The great myth of urban Britain," *BBC News*, June 28. https://www.bbc.co.uk/news/uk-18623096.

170. UK Department for Environment, Food and Rural Affairs (DEFRA), October 11, 2018. *Farming Statistics—Provisional Crop Areas, Yields and Livestock Populations at June 2018—United Kingdom.* https://assets.publishing.service.gov.uk/government/uploads/system/uploads/attachment_data/file/747210/structure-jun2018prov-UK-11oct18.pdf.

171. Hannah Ritchie and Max Roser, 2019. "Land use." Our World in Data, September. https://ourworldindata.org/land-use.

172. Navin Ramankutty et al., 2008. "Farming the planet: 1. geographic distribution of global agricultural lands in the year 2000." *Global Biogeochemical Cycles*, vol. 22:1. https://doi.org/10.1029/2007GB002952.

173. UN Environment Programme World Conservation Monitoring Centre (UNEP-WCMC), 2014. *Mapping the World's Special Places.* https://

www.unep-wcmc.org/featured-projects/mapping-the-worlds-special -places.

174. Tara Garnett et al., 2017. "Grazed and confused? Ruminating on cattle, grazing systems, methane, nitrous oxide, the soil carbon sequestration question—and what it all means for greenhouse gas emissions." Food Climate Research Network (FCRN). https://www.oxfordmartin.ox.ac .uk/downloads/reports/fcrn_gnc_report.pdf.

175. Our World in Data, 2018. "Land use per 100 grams of protein." https:// ourworldindata.org/grapher/land-use-protein-poore.

176. Marian Swain et al., 2018. "Reducing the environmental impact of global diets." *Science of the Total Environment*, vols. 610–11, pp. 1207–9. https://doi.org/10.1016/j.scitotenv.2017.08.125.

177. Michael Clark and David Tilman, 2017. "Comparative analysis of environmental impacts of agricultural production systems, agricultural input efficiency, and food choice." *Environmental Research Letters*, vol. 12:6. https://doi.org/10.1088/1748-9326/aa6cd5.

178. Durk Nijdam, Trudy Rood, and Henk Westhoek, 2012. "The price of protein: Review of land use and carbon footprints from life cycle assessments of animal food products and their substitutes." *Food Policy*, vol. 37:6, pp. 760–70. https://doi.org/10.1016/j.foodpol.2012 .08.002.

179. Hannah Ritchie, 2017. "How much of the world's land would we need in order to feed the global population with the average diet of a given country?" Our World in Data, October 3. https://ourworldindata.org /agricultural-land-by-global-diets.

180. J. Poore and T. Nemecek, 2018. "Reducing food's environmental impacts through producers and consumers." *Science*, vol. 360:6392, pp. 987–92. https://doi.org/10.1126/science.aaq0216.

181. Hannah Ritchie, 2021. "If the world adopted a plant-based diet we would reduce global agricultural land use from 4 to 1 billion hectares." Our World in Data, March 4. https://ourworldindata.org/land-use-diets.

182. R. Conant (compiled by), 2010. *Challenges and Opportunities for Carbon Sequestration in Grassland System—A Technical Report on Grassland Management and Climate Change Mitigation.* Food and Agriculture Organization of the United Nations (FAO). http://www .fao.org/3/a-i1399e.pdf.

183. United Nations. "Chapters of the thematic assessment of land degradation and restoration." Plenary of the Intergovernmental Science-Policy Platform on Biodiversity and Ecosystem Services (IPBES), Sixth session,

Medellin, Colombia, March 18–24, 2018, Agenda item 7. IPBES/6/INF/1/Rev.1. https://ipbes.net/sites/default/files/ipbes_6_inf_1_rev.1_2.pdf.

184. Kris Zouhar, 2003. "*Species*: Bromus tectorum." U.S. Department of Agriculture, Forest Service, Fire Effects Information System (FEIS). https://www.fs.fed.us/database/feis/plants/graminoid/brotec/all.html.

185. Alessandro Filazzola et al., 2020. "The effects of livestock grazing on biodiversity are multi-trophic: a meta-analysis." *Ecology Letters*, vol. 23:8, pp. 1298–1309. https://doi.org/10.1111/ele.13527.

186. Thomas L. Fleischner, 1994. "Ecological costs of livestock grazing in western North America." *Conservation Biology*, vol. 8:3, pp. 629–44. https://doi.org/10.1046/j.1523-1739.1994.08030629.x.

187. Knepp Wildland: Rewilding in West Sussex. https://knepp.co.uk/home

188. George Monbiot, 2019. "The Knepp Estate's beef production statistics," November 27. https://twitter.com/GeorgeMonbiot/status/119997382111 44543233.

189. George Monbiot, 2018. "Comment on Isabella Tree's article, 'If you want to save the world, veganism isn't the answer.'" *The Guardian*, August 25. https://www.theguardian.com/commentisfree/2018/aug/25/veganism-intensively-farmed-meat-dairy-soya-maize#comment-119748600.

190. Brian Machovina et al., 2015. "Biodiversity conservation: The key is reducing meat consumption." *Science of the Total Environment*, vol. 536, pp. 419–31. https://doi.org/10.1016/j.scitotenv.2015.07.022.

191. Christopher Ketcham, 2016. "The Rogue Agency, a USDA program that tortures dogs and kills endangered species." *Harper's Magazine*, March. https://harpers.org/archive/2016/03/the-rogue-agency.

192. Kristin Ruether, 2018. *Wildlife Services Kills Wolves in Sawtooth National Recreation Area to Prop Up Harmful Sheep Grazing*. Western Watersheds Project, July 17. https://www.westernwatersheds.org/wildlife-services-kills-wolves-in-sawtooth-national-recreation-area-to-prop-up-harmful-sheep-grazing.

193. Ketcham, Ibid.

194. Mariel Padilla, 2019. "Trump administration reauthorizes use of 'cyanide bombs' to kill wild animals." *The New York Times*, August 10. https://www.nytimes.com/2019/08/10/us/cyanide-bombs-animals-trump-administration.html.

195. Jimmy Tobias, 2020. "The secretive government agency planting 'cyanide bombs' across the US." *The Guardian*, June 26. https://www.theguardian.com/environment/2020/jun/26/cyanide-bombs-wildfire-services-idaho.

196. George Monbiot, 2003. *No Man's Land: An Investigative Journey through Kenya and Tanzania.* Green Books, Dartington Hall.

197. United Nations. "Chapters of the thematic assessment," Ibid.

198. Florence Pendrill et al., 2019. "Deforestation displaced: Trade in forest-risk commodities and the prospects for a global forest transition." *Environmental Research Letters*, vol. 14:5. https://doi.org/10.1088/1748-9326/ab0d41.

199. Shannon Sterling and Agnès Ducharne, 2008. "Comprehensive data set of global land cover change for land surface model applications." *Global Biogeochemical Cycles*, vol. 22:3. https://doi.org/10.1029/2007GB002959.

200. Pendrill et al., Ibid.

201. Machovina et al., Ibid.

202. William F. Laurance, Jeffrey Sayer, and Kenneth G. Cassman, 2014. "Agricultural expansion and its impacts on tropical nature." *Trends in Ecology & Evolution*, vol. 29:2, pp. 107–16. https://doi.org/10.1016/j.tree.2013.12.001.

203. Mustafa Zia et al., 2019. "Brazil once again becomes the world's largest beef exporter." U.S. Department of Agriculture, Economic Research Service, July 1. https://www.ers.usda.gov/amber-waves/2019/july/brazil-once-again-becomes-the-world-s-largest-beef-exporter.

204. David Pitt, 2020. "US lifts Brazilian beef import ban amid quality concerns." *AP News*, February 21. https://apnews.com/article/us-news-united-states-iowa-global-trade-food-safety-56c199093f898a69dc5aef2392906002.

205. Earthsight, 2021. "Amazon slaughterhouses eye greater share of American pie as Brazil beef sales surge," September 23. https://www.earthsight.org.uk/news/analysis-amazon-slaughterhouses-eye-greater-share-of-american-pie-as-brazil-beef-sales-surge?s=09.

206. André Campos et al., 2020. "Revealed: New evidence links Brazil meat giant JBS to Amazon deforestation." *The Guardian*, July 27. https://www.theguardian.com/environment/2020/jul/27/revealed-new-evidence-links-brazil-meat-giant-jbs-to-amazon-deforestation.

207. Matthew N. Hayek and Rachael D. Garrett, 2018. "Nationwide shift to grass-fed beef requires larger cattle population." *Environmental Research Letters*, vol. 13:8. https://doi.org/10.1088/1748-9326/aad401.

208. Our World in Data, 2018. "Greenhouse gas emissions per 100 grams of protein." https://ourworldindata.org/grapher/ghg-per-protein-poore.

209. J. Poore and T. Nemecek, 2018. "Reducing food's environmental

impacts through producers and consumers." *Science*, vol. 360:6392, pp. 987–92. https://doi.org/10.1126/science.aaq0216.

210. William J. Ripple et al., 2013. "Ruminants, climate change and climate policy." *Nature Climate Change*, vol. 4, pp. 2–5. https://doi.org/10.1038/nclimate2081.

211. Nijdam, Rood, and Westhoek, Ibid.

212. Michael Clark and David Tilman, 2017. "Comparative analysis of environmental impacts of agricultural production systems, agricultural input efficiency, and food choice." *Environmental Research Letters*, vol. 12:6. https://doi.org/10.1088/1748-9326/aa6cd5.

213. Hannah Ritchie, 2020. "Less meat is nearly always better than sustainable meat, to reduce your carbon footprint." Our World in Data, February 4. https://ourworldindata.org/less-meat-or-sustainable-meat.

214. Peter Scarborough et al., 2014. "Dietary greenhouse gas emissions of meat-eaters, fish-eaters, vegetarians and vegans in the UK." *Climatic Change*, vol. 125, pp. 179–92. https://doi.org/10.1007/s10584-014-1169-1.

215. M. Crippa et al., 2021. "Food systems are responsible for a third of global anthropogenic GHG emissions." *Nature Food*, vol. 2, pp. 198–209. https://doi.org/10.1038/s43016-021-00225-9.

216. Hannah Ritchie, 2021. "Emissions from food alone could use up all of our budget for 1.5°C or 2°C—but we have a range of opportunities to avoid this." Our World in Data, June 10. https://ourworldindata.org/food-emissions-carbon-budget.

217. Timothy D. Searchinger et al., 2018. "Assessing the efficiency of changes in land use for mitigating climate change." *Nature*, vol. 564, pp. 249–53. https://doi.org/10.1038/s41586-018-0757-z.

218. Matthew N. Hayek et al., 2021. "The carbon opportunity cost of animal-sourced food production on land." *Nature Sustainability*, vol. 4, pp. 21–4. https://doi.org/10.1038/s41893-020-00603-4.

219. P. J. Gerber et al., 2013. *Tackling Climate Change through Livestock: A Global Assessment of Emissions and Mitigation Opportunities*. Food and Agriculture Organization of the United Nations, Rome. http://www.fao.org/3/i3437e/i3437e.pdf.

220. Florence Pendrill et al., 2019. "Deforestation displaced: Trade in forest-risk commodities and the prospects for a global forest transition." *Environmental Research Letters*, vol. 14:5. https://doi.org/10.1088/1748-9326/ab0d41.

221. Donald Broom, 2021. "A method for assessing sustainability, with beef production as an example." *Biological Reviews*, vol. 96:5, pp. 836–1853. https://doi.org/10.1111/brv.12726.

222. Ilissa Ocko et al., 2021. "Acting rapidly to deploy readily available methane mitigation measures by sector can immediately slow global warming." *Environmental Research Letters*, vol. 16:5, article 054042. https://doi.org/10.1088/1748-9326/abf9c8.

223. Myles R. Allen et al., 2018. "A solution to the misrepresentations of CO_2-equivalent emissions of short-lived climate pollutants under ambitious mitigation." *Nature Partner Journal (NPJ) Climate and Atmospheric Science*, vol. 1, article 16. https://doi.org/10.1038/s41612 -018-0026-8.

224. Farmwel. *Ruminant Methane & GWP**. https://www.farmwel.org.uk /ruminant-methane.

225. Myles Allen et al., July 2018. *Climate Metrics for Ruminant Livestock*. Oxford Martin Programme on Climate Pollutants. https://www .oxfordmartin.ox.ac.uk/downloads/reports/Climate-metrics-for -ruminant-livestock.pdf.

226. *Nature* Editorial, 2021. "Control methane to slow global warming— fast." *Nature*, vol. 596, August 25, p. 461. doi: https://doi.org/10.1038 /d41586-021-02287-y.

227. Pete Smith and Andrew Balmford, 2020. "Climate change: 'No get out of jail free card.'" *Veterinary Record*, vol. 186:2, p. 71. https://doi.org /10.1136/vr.m190.

228. Hannah Ritchie, 2020. "You want to reduce the carbon footprint of your food? Focus on what you eat, not whether your food is local." Our World in Data, January 24. https://ourworldindata.org/food-choice -vs-eating-local.

229. Hannah Ritchie, Ibid.

230. Christopher L. Weber and H. Scott Matthews, 2008. "Food-miles and the relative climate impacts of food choices in the United States." *Environmental Science and Technology*, vol. 42:10, pp. 3508–13. https:// doi.org/10.1021/es702969f.

231. Our World in Data, 2018. "Share of global food miles by transport method." https://ourworldindata.org/grapher/share-food-miles-by-method.

232. Llorenç Milà i Canals et al., 2007. "Comparing domestic versus imported apples: A focus on energy use." *Environmental Science and Pollution Research—International*, vol. 14, pp. 338–44. https://doi.org/10.1065 /espr2007.04.412.

233. Almudena Hospido et al., 2009. "The role of seasonality in lettuce con- sumption: A case study of environmental and social aspects." *The International Journal of Life Cycle Assessment*, vol. 14, pp. 38–391. https://doi.org/10.1007/s11367-009-0091-7.

234. David Coley, Mark Howard, and Michael Winter, 2009. "Local food, food miles and carbon emissions: A comparison of farm shop and mass distribution approaches." *Food Policy*, vol. 34:2, pp. 150–5. https://doi .org/10.1016/j.foodpol.2008.11.001.

235. Allan Savory, Joseph Geni (trans.), 2013. "How to fight desertification and reverse climate change." TED: Ideas Worth Spreading. https:// www.ted.com/talks/allan_savory_how_to_fight_desertification_and _reverse_climate_change/transcript?language=en#t-1119536.

236. Allan Savory, 2013. "How to green the world's deserts and reverse climate change." YouTube (TED), March 4. https://www.youtube.com /watch?v=vpTHi7O66pI.

237. Kiss the Ground. "Awakening people to the possibilities of regeneration." https://kisstheground.com.

238. Rattan Lal, 2018. "Digging deeper: A holistic perspective of factors affecting soil organic carbon sequestration in agroecosystems." *Global Change Biology*, vol. 24:8, pp. 3285–3301. https://doi.org/10.1111 /gcb.14054.

239. Jonathan Sanderman, Tomislav Hengl, and Gregory J. Fiske, 2017. "Soil carbon debt of 12,000 years of human land use." *Proceedings of the National Academy of Sciences*, vol. 114:36, pp. 9575–80. https://doi .org/10.1073/pnas.1706103114.

240. Lal, Ibid.

241. Rattan Lal, 2010. "Managing soils and ecosystems for mitigating anthropogenic carbon emissions and advancing global food security." *BioScience*, vol. 60:9, pp. 708–21. https://doi.org/10.1525/bio.2010.60.9.8.

242. Rolf Sommer and Deborah Bossio, 2014. "Dynamics and climate change mitigation potential of soil organic carbon sequestration." *Journal of Environmental Management*, vol. 144, pp. 83–7. https://doi.org /10.1016/j.jenvman.2014.05.017.

243. Garnett et al., Ibid.

244. Gabriel Popkin, 2021. "A soil-science revolution upends plans to fight climate change." Quanta. https://www.quantamagazine.org/a-soil-science -revolution-upends-plans-to-fight-climate-change-20210727.

245. Johannes Lehmann and Markus Kleber, 2015. "The contentious nature of soil organic matter." *Nature*, vol. 528: 60–8. https://doi.org/10.1038 /nature16069.

246. Jennifer Soong et al., 2021. "Five years of whole-soil warming led to loss of subsoil carbon stocks and increased CO_2 efflux." *Science Advances*, vol. 7:21. https://www.science.org/doi/10.1126/sciadv.abd1343.

247. Andrew Nottingham et al., 2020. "Soil carbon loss by experimental

warming in a tropical forest." *Nature*, vol. 584, 234–7. https://doi.org/10.1038/s41586-020-2566-4.

248. George Monbiot, 2014. *Eat Meat and Save the World?*, August 4. https://www.monbiot.com/2014/08/04/eat-meat-and-save-the-world.

249. W. R. Teague et al., 2011. "Grazing management impacts on vegetation, soil biota and soil chemical, physical and hydrological properties in tall grass prairie." *Agriculture, Ecosystems & Environment*, vol. 141:3–4, pp. 310–22. https://doi.org/10.1016/j.agee.2011.03.009.

250. Oliver Jakoby et al., 2014. "How do individual farmers' objectives influence the evaluation of rangeland management strategies under a variable climate?" *Journal of Applied Ecology*, vol. 51:2, pp. 483–93. https://doi.org/10.1111/1365-2664.12216.

251. Christopher L. Crawford et al., 2019. "Behavioral and ecological implications of bunched, rotational cattle grazing in East African savanna ecosystem." *Rangeland Ecology & Management*, vol. 72:1, pp. 204–9. https://doi.org/10.1016/j.rama.2018.07.016.

252. Matt Barnes and Jim Howell, 2013. "Multiple-paddock grazing distributes utilization across heterogeneous mountain landscapes: A case study of strategic grazing management." *Rangelands*, vol. 35:5, pp. 52–61. https://doi.org/10.2111/RANGELANDS-D-13-00019.1.

253. David D. Briske et al., 2014. "Commentary: A critical assessment of the policy endorsement for holistic management." *Agricultural Systems*, vol. 125, pp. 50–3. https://doi.org/10.1016/j.agsy.2013.12.001.

254. Maria Nordborg and Elin Röös, 2016. *Holistic Management—A critical Review of Allan Savory's Grazing Method.* Swedish University of Agricultural Sciences (SLU), Centre for Organic Food & Farming (EPOK) & Chalmers. https://publications.lib.chalmers.se/records/fulltext/244566/local_244566.pdf.

255. Heidi-Jayne Hawkins, 2017. "A global assessment of Holistic Planned Grazing™ compared with season-long, continuous grazing: Meta-analysis findings." *African Journal of Range & Forage Science*, vol. 34:2, pp. 65–75. https://doi.org/10.2989/10220119.2017.1358213.

256. Jayne Belnap et al., 2001. *Biological Soil Crusts: Ecology and Management.* U.S. Department of the Interior, Technical Reference 1730–2, BLM/ID/ST-01/001+1730. https://www.ntc.blm.gov/krc/uploads/231/CrustManual.pdf.

257. Merlin Sheldrake, 2020. *Entangled Life: How Fungi Make Our Worlds, Change Our Minds and Shape Our Futures.* Bodley Head, London, p. 95.

258. John Carter et al., 2014. "Holistic management: Misinformation on the

science of grazed ecosystems." *International Journal of Biodiversity*, vol. 2014, article 163431. https://doi.org/10.1155/2014/163431.

259. Liming Lai and Sandeep Kumar, 2020. "A global meta-analysis of livestock grazing impacts on soil properties." *PLOS One*, vol. 15:8, e0236638. https://doi.org/10.1371/journal.pone.0236638.

260. D. Cluzeau et al., 1992. "Effects of intensive cattle trampling on soil-plant-earthworms system in two grassland types." *Soil Biology and Biochemistry*, vol. 24:12, pp. 1661–5. https://doi.org/10.1016/0038-0717(92)90166-U.

261. S. D. Warren et al., 1986. "Soil response to trampling under intensive rotation grazing." *Soil Science Society of America Journal*, vol. 50, pp. 1336–41. https://doi.org/10.2136/sssaj1986.03615995005000050050x.

262. Carter et al., Ibid.

263. Norman Ambos, George Robertson, and Jason Douglas, 2000. "Dutchwoman Butte: A relict grassland in central Arizona." *Rangelands*, vol. 22:2, pp. 3–8. http://dx.doi.org/10.2458/azu_rangelands_v22i2_ambos.

264. D. P. Fernandez, J. C. Neff, and R. L. Reynolds, 2008. "Biogeochemical and ecological impacts of livestock grazing in semi-arid southeastern Utah, USA." *Journal of Arid Environments*, vol. 72:5, pp. 777–91. https://doi.org/10.1016/j.jaridenv.2007.10.009.

265. Liping Qiu et al., 2013. "Ecosystem carbon and nitrogen accumulation after grazing exclusion in semiarid grassland." *PLOS One*, vol. 8:1, e55433. https://doi.org/10.1371/journal.pone.0055433.

266. W. W. Brady et al., 1989. "Response of a semidesert grassland to 16 years of rest from grazing." *Journal of Range Management*, vol. 42:4, pp. 284–8. https://journals.uair.arizona.edu/index.php/jrm/article/view File/8383/7995.

267. aspidoscelis, March 15, 2013 at 7:32 p.m. *Comment on thread—TED Talk: Spreading bullshit about the desert.* FreeThoughtBlogs.com. Comment 25 on thread. https://freethoughtblogs.com/pharyngula/2013/03 /15/ted-talk-spreading-bullshit-about-the-desert/#comment-580202.

268. David D. Briske et al., 2013. "The savory method can not green deserts or reverse climate change." *Rangelands*, vol. 35. pp. 72–4. https://doi .org/10.2111/rangelands-d-13-00044.1.

269. Nicholas Carter and Dr. Tushar Mehta, 2021. *Another Industry Attempt to Greenwash Beef.* Plant Based Data, January. https://plant baseddata.medium.com/the-failed-attempt-to-greenwash-beef -7dfca9d74333.

270. Jessica Scott-Reid, 2020. *How Grass-Fed Beef is Duping Consumers, Again.* Sentient Media, October 27. https://sentientmedia.org/how-grass -fed-beef-is-duping-consumers-again.

271. James Temple, 2020. "Why we can't count on carbon-sucking farms to slow climate change." *MIT Technology Review*, June 3. https://www.technologyreview.com/2020/06/03/1002484/why-we-cant-count-on-carbon-sucking-farms-to-slow-climate-change.

272. Shan Goodwin, 2021. "Microsoft buys carbon credits from NSW cattle operation." *Farm Weekly*, January 29. https://www.farmweekly.com.au/story/7105542/microsoft-buys-carbon-credits-from-nsw-cattle-operation/?cs=5151.

273. Robert Paarlberg, 2021. "President Biden, please don't get into carbon farming." *Wired*, January 22. https://www.wired.com/story/president-biden-please-dont-get-into-carbon-farming.

274. Gustaf Hugelius et al., 2020. "Large stocks of peatland carbon and nitrogen are vulnerable to permafrost thaw." *Proceedings of the National Academy of Sciences*, vol. 117:34, pp. 20438–46. https://doi.org/10.1073/pnas.1916387117.

275. Myron King et al., 2018. "Northward shift of the agricultural climate zone under 21st-century global climate change." *Scientific Reports*, vol. 8, article 7904. https://doi.org/10.1038/s41598-018-26321-8.

276. Lee Hannah et al., 2020. "The environmental consequences of climate-driven agricultural frontiers." *PLOS One*, vol. 15:7, e0236028. https://doi.org/10.1371/journal.pone.0236028.

277. Emily Chung, 2020. "Canada could be a huge climate change winner when it comes to farmland." Canadian Broadcasting Corporation (CBC), February 12. https://www.cbc.ca/news/technology/climate-change-farming-1.5461275.

278. "Land grant: Govt drafts bill for '1-hectare per Russian' in Far East." *RT (Rossiya Segodnya—Russia Today)*, November 17, 2015. https://www.rt.com/russia/322404-government-drafts-bill-on-free.

279. "Chinese firm to rent Russian land in Siberia for crops." *BBC News*, June 19, 2015. https://www.bbc.co.uk/news/world-asia-33196396.

280. Government of Northwest Territories Department of Industry, Tourism and Investment, 2017. *The Business of Food: A Food Production Plan, 2017–2022*. Northwest Territories Agriculture Strategy Tabled document 314–18(2). https://www.iti.gov.nt.ca/sites/iti/files/agriculture_strategy.pdf.

281. White Rock Consulting & Communications, 2017. *Agriculture Industry Supported by Increased Access to Crown Land*, February 17. http://www.whiterocknl.com/agriculture-industry-supported-by-increased-access-to-crown-land.

282. Tim G. Benton et al., 2021. "Food system impacts on biodiversity loss: Three levers for food system transformation in support of nature." Chatham House, the Royal Institute of International Affairs, Research Paper, Energy, Environment and Resources Programme. https://www.chathamhouse.org/sites/default/files/2021-02/2021-02-03-food-system-biodiversity-loss-benton-et-al_0.pdf.

283. Tim Newbold et al., 2015. "Global effects of land use on local terrestrial biodiversity." *Nature*, vol. 520, pp. 45–50. https://doi.org/10.1038/nature14324.

284. Florian Zabel et al., 2019. "Global impacts of future cropland expansion and intensification on agricultural markets and biodiversity." *Nature Communications*, vol. 10:1. https://doi.org/10.1038/s41467-019-10775-z.

285. Laura Kehoe et al., 2017. "Biodiversity at risk under future cropland expansion and intensification." *Nature Ecology & Evolution*, vol. 1, pp. 1129–35. https://doi.org/10.1038/s41559-017-0234-3.

286. Zabel et al., Ibid.

287. Walter Willett et al., 2019. "Food in the Anthropocene: The EAT—Lancet Commission on healthy diets from sustainable food systems." *The Lancet Commissions*, vol. 393:10170, pp. 447–92. https://doi.org/10.1016/S0140-6736(18)31788-4.

288. Machovina et al., Ibid.

289. Bruce M. Campbell et al., 2017. "Agriculture production as a major driver of the Earth system exceeding planetary boundaries." *Ecology and Society* vol. 22:4. https://doi.org/10.5751/ES-09595-220408.

290. Rosamunde Almond, Monique Grooten, and Tanya Petersen, 2020. *Living Planet Report 2020—Bending the Curve of Biodiversity Loss.* World Wildlife Fund. https://www.wwf.org.uk/sites/default/files/2020-09/LPR20_Full_report.pdf.

291. Hannah Ritchie and Max Roser, 2020. "Environmental impacts of food and agriculture." Our World in Data. https://ourworldindata.org/environmental-impacts-of-food#environmental-impacts-of-food-and-agriculture.

292. Yinon M. Bar-On, Rob Phillips, and Ron Milo, 2018. "The biomass distribution on Earth." *Proceedings of the National Academy of Sciences*, vol. 115:25, pp. 6506–11. https://doi.org/10.1073/pnas.1711842115

293. Bar-On, Phillips, and Milo, Ibid.

294. Andrew Balmford et al., 2018. "The environmental costs and benefits of high-yield farming." *Nature Sustainability*, vol. 1, pp. 477–85. https://doi.org/10.1038/s41893-018-0138-5.

295. David P. Edwards et al., 2015. "Land-sparing agriculture best protects avian phylogenetic diversity." *Current Biology*, vol. 25:18, pp. 2384–91. https://doi.org/10.1016/j.cub.2015.07.063.

296. Ben Phalan et al., 2011. "Reconciling food production and biodiversity conservation: Land sharing and land sparing compared." *Science*, vol. 333:6047, pp. 1289–91. https://doi.org/10.1126/science.1208742.

297. M. Pfeifer et al., 2017. "Creation of forest edges has a global impact on forest vertebrates." *Nature*, vol. 551, pp. 187–91. https://doi.org/10.1038/nature24457.

298. Zabel et al., Ibid.

299. Laura Kehoe et al., 2017. "Biodiversity at risk under future cropland expansion and intensification." *Nature Ecology & Evolution*, vol. 1, pp. 1129–35. https://doi.org/10.1038/s41559-017-0234-3.

300. John M. Halley et al., 2016. "Dynamics of extinction debt across five taxonomic groups." *Nature Communications*, vol. 7, article 12283. https://doi.org/10.1038/ncomms12283.

301. Ingo Grass et al., 2019. "Land-sharing/-sparing connectivity landscapes for ecosystem services and biodiversity conservation." *People and Nature*, vol. 1:2, pp. 262–72. https://doi.org/10.1002/pan3.21.

302. Laurance, Sayer, and Cassman, Ibid.

303. Ivette Perfecto and John Vandermeer, 2010. "The agroecological matrix as alternative to the land-sparing/agriculture intensification model." *Proceedings of the National Academy of Sciences*, vol. 107:13, pp. 5786–91. https://doi.org/10.1073/pnas.0905455107.

304. Grass et al., Ibid.

305. Lucas A. Garibaldi et al., 2013. "Wild pollinators enhance fruit set of crops regardless of honey bee abundance." *Science*, vol. 339:6127, pp. 1608–11. https://doi.org/10.1126/science.1230200.

306. Gail MacInnis and Jessica R. K. Forrest, 2019. "Pollination by wild bees yields larger strawberries than pollination by honey bees." *Journal of Applied Ecology*, vol. 56:4, pp. 824–32. https://doi.org/10.1111/1365-2664.13344.

307. Denis Vasiliev and Sarah Greenwood, 2020. "Pollinator biodiversity and crop pollination in temperate ecosystems, implications for national pollinator conservation strategies: Mini review." *Science of the Total Environment*, vol. 744. https://doi.org/10.1016/j.scitotenv.2020.140880.

308. Maxime Eeraerts et al., 2019. "Pollination efficiency and foraging behaviour of honey bees and non-*Apis* bees to sweet cherry." *Agricul-

tural and Forest Entomology, vol. 22:1, pp. 75–82. https://doi.org/10.1111/afe.12363.

309. Cedric Alaux, Yves Le Conte, and Axel Decourtye, 2019. "Pitting wild bees against managed honey bees in their native range, a losing strategy for the conservation of honey bee biodiversity." *Frontiers in Ecology and Evolution*, vol. 7. https://doi.org/10.3389/fevo.2019.00060.

310. Alfredo Valido, María C. Rodríguez-Rodríguez, and Pedro Jordano, 2019. "Honeybees disrupt the structure and functionality of plant-pollinator networks." *Scientific Reports*, vol. 9, article 4711. https://doi.org/10.1038/s41598-019-41271-5.

311. Océane Bartholomée et al., 2020. "Pollinator presence in orchards depends on landscape-scale habitats more than in-field flower resources." *Agriculture, Ecosystems & Environment*, vol. 293. https://doi.org/10.1016/j.agee.2019.106806.

312. Adara Pardo and Paulo A. V. Borges, 2020. "Worldwide importance of insect pollination in apple orchards: A review." *Agriculture, Ecosystems & Environment*, vol. 93. https://doi.org/10.1016/j.agee.2020.106839.

313. M. G. Ceddia et al., 2013. "Sustainable agricultural intensification or Jevons paradox? The role of public governance in tropical South America." *Global Environmental Change*, vol. 23:5, pp. 1052–63. https://doi.org/10.1016/j.gloenvcha.2013.07.005.

314. Laura Vang Rasmussen et al., 2018. "Social-ecological outcomes of agricultural intensification." *Nature Sustainability*, vol. 1, pp. 275–82. https://doi.org/10.1038/s41893-018-0070-8.

315. Benton et al., Ibid.

316. Nigel Dudley and Sasha Alexander, 2017. "Agriculture and biodiversity: A review." *Biodiversity*, vol. 18:2–3, pp. 45–9. https://doi.org/10.1080/14888386.2017.1351892.

317. Laurance, Sayer, and Cassman, Ibid.

318. Bruce M. Campbell et al., 2017. "Agriculture production as a major driver of the Earth system exceeding planetary boundaries." *Ecology and Society*, vol. 22:4, article 8. https://doi.org/10.5751/ES-09595-220408.

CHAPTER 4

1. Victoria County History, June 2019. *VCH Oxfordshire Texts in Progress, Whitchurch*. Introduction: Landscape, Settlement, and Buildings. https://www.history.ac.uk/sites/default/files/file-uploads/2019-06/whitchurch_intro_web_june_2019.pdf.

2. Michael Redley, 2016. *The Real Mr. Toad: Merchant Venturer and Radical in the Age of Gold*. Self-published, available from the Bell Bookshop in Henley, Garlands in Pangbourne, and the Hardwick Estate Office. https://hardwickestate.wordpress.com/history.

3. Hardwick Estate. *About Sir Julian*. https://hardwickestate.wordpress.com/sir-julian.

4. Agforward (Organic Research Centre). *Silvoarable Agroforestry in the UK*. https://www.agforward.eu/index.php/en/silvoarable-agroforestry-in-the-uk.html.

5. Louis Sutter et al., 2017. "Enhancing plant diversity in agricultural landscapes promotes both rare bees and dominant crop-pollinating bees through complementary increase in key floral resources." *Journal of Applied Ecology*, vol. 54:6, pp. 1856–64. https://doi.org/10.1111/1365-2664.12907.

6. Matthias Albrecht et al., 2020. "The effectiveness of flower strips and hedgerows on pest control, pollination services and crop yield: A quantitative synthesis." *Ecology Letters*, vol. 23:10, pp. 1488–98. https://doi.org/10.1111/ele.13576.

7. Jeroen Scheper et al., 2013. "Environmental factors driving the effectiveness of European agri-environmental measures in mitigating pollinator loss—a meta-analysis." *Ecology Letters*, vol. 16:7, pp. 912–20. https://doi.org/10.1111/ele.12128.

8. Ingo Grass et al., 2019. "Land-sharing/-sparing connectivity landscapes for ecosystem services and biodiversity conservation." *People and Nature*, vol. 1:2, pp. 262–72. https://doi.org/10.1002/pan3.21.

9. Eusun Han et al., 2021. "Can precrops uplift subsoil nutrients to topsoil?" *Plant and Soil*, vol. 463, pp. 329–45. https://doi.org/10.1007/s11104-021-04910-3.

10. David Weisberger, Virginia Nichols, and Matt Liebman, 2019. "Does diversifying crop rotations suppress weeds? A meta-analysis." *PLOS One*, vol. 14:7, e0219847. https://doi.org/10.1371/journal.pone.0219847.

11. Chloe MacLaren et al., 2020. "An ecological future for weed science to sustain crop production and the environment: A review." *Agronomy for Sustainable Development*, issue 40, article 24. https://doi.org/10.1007/s13593-020-00631-6.

12. Sally Westaway, 2020. "Ramial woodchip in agricultural production," WOOFS Technical Guide 2. Organic Research Centre, November 2020. https://www.organicresearchcentre.com/wp-content/uploads/2020/12/WOOFS_TG2_Final.pdf.

13. S. Jeffery and F. G. A. Verheijen, 2020. "A new soil health policy

paradigm: Pay for practice not performance!" *Environmental Science & Policy*, vol. 112, pp. 371–3. https://doi.org/10.1016/j.envsci.2020 .07.006.

14. Y. Kuzyakov, J. K. Friedel, and K. Stahr, 2000. "Review of mechanisms and quantification of priming effects." *Soil Biology and Biochemistry*, vol. 32:11–12, pp. 1485–98. https://doi.org/10.1016/S0038-0717(00) 00084-5.

15. Céline Caron, Gilles Lemieux, and Lionel Lachance. *Regenerating Soils with Ramial Chipped Wood*. The Dirt Doctor, Howard Garrett. https:// www.dirtdoctor.com/organic-research-page/Regenerating-Soils-with -Ramial-Chipped-Wood_vq4462.htm.

16. Westaway, Ibid.

17. Michael Clark and David Tilman, 2017. "Comparative analysis of environmental impacts of agricultural production systems, agricultural input efficiency, and food choice." *Environmental Research Letters*, vol. 12:6. https://doi.org/10.1088/1748-9326/aa6cd5.

18. Soil Association, 2021. *Soil Association Standards Farming and Growing—Version 18.6: Updated on 12th February 2021*. https://www .soilassociation.org/media/15931/farming-and-growing-standards.pdf.

19. Department for Environment, Food & Rural Affairs, 2019. *Guidance— Broiler (meat) Chickens: Welfare Recommendations*. https://www.gov .uk/government/publications/poultry-on-farm-welfare/broiler-meat -chickens-welfare-recommendations.

20. Soil Association, Ibid.

21. Kenneth G. Cassman, Achim Dobermann, and Daniel T. Walters, 2002. "Agroecosystems, nitrogen-use efficiency, and nitrogen management." *AMBIO: A Journal of the Human Environment*, vol. 31:2, pp. 132–40. https://doi.org/10.1579/0044-7447-31.2.132.

22. X. P. Pang and J. Letey, 2000. "Organic farming challenge of timing nitrogen availability to crop nitrogen requirements." *Soil Science Society of America Journal*, vol. 64:1, pp. 247–53. https://doi.org/10.2136 /sssaj2000.641247x.

23. Wendy J. Binder and Blaire Van Valkenburgh, 2010. "A comparison of tooth wear and breakage in Rancho La Brea sabertooth cats and dire wolves across time." *Journal of Vertebrate Paleontology*, vol. 30:1, pp. 255–61. https://doi.org/10.1080/02724630903413016.

24. George Monbiot, 2014. "Is this all humans are? Diminutive monsters of death and destruction?" *The Guardian*, March 24. https://www.the guardian.com/commentisfree/2014/mar/24/humans-diminutive -monster-destruction.

25. Gareth Grundy 2011. "Building a boat in the back yard—in pictures." *The Guardian*, July 17. https://www.theguardian.com/lifeandstyle/gallery /2011/jul/17/homemade-boat-in-pictures.

26. Cornelia Rumpel et al., 2020. "The 4p1000 initiative: Opportunities, limitations and challenges for implementing soil organic carbon sequestration as a sustainable development strategy." *Ambio*, vol. 49, pp. 350–60. https://doi.org/10.1007/s13280-019-01165-2.

27. Pete Smith et al., 2019. "How to measure, report and verify soil carbon change to realize the potential of soil carbon sequestration for atmospheric greenhouse gas removal." *Global Change Biology*, vol. 26:1, pp. 219–41. https://doi.org/10.1111/gcb.14815.

28. Sigbert Huber et al., 2008. *Environmental Assessment of Soil for Monitoring Volume I: Indicators & Criteria*. JRC Institute for Environment and Sustainability. EUR 23490 EN/1, Luxembourg, OPOCE, JRC47184. https://doi.org/10.2788/93515.

29. Nicolas P. A. Saby et al., 2008. "Will European soil-monitoring networks be able to detect changes in topsoil organic carbon content?" *Global Change Biology*, vol. 14:10, pp. 2432–42. https://doi.org/10.1111 /j.1365-2486.2008.01658.x.

30. Pete Smith, 2004. "How long before a change in soil organic carbon can be detected?" *Global Change Biology*, vol. 10:11, pp. 1878–83. https://doi.org/10.1111/j.1365-2486.2004.00854.x.

31. W. Amelung et al., 2020. "Towards a global-scale soil climate mitigation strategy." *Nature Communications*, vol. 11, article 5427. https:// doi.org/10.1038/s41467-020-18887-7.

32. Ibid.

33. Jens Leifeld et al., 2013. "Organic farming gives no climate change benefit through soil carbon sequestration." *Proceedings of the National Academy of Sciences*, vol. 110:11, article E984. https://doi.org/10.1073 /pnas.1220724110.

34. Jan Willem van Groenigen et al., 2017. "Sequestering soil organic carbon: A nitrogen dilemma." *Environmental Science & Technology*, vol. 51:9, pp. 4738–9. https://doi.org/10.1021/acs.est.7b01427.

35. Cassman, Dobermann, and Walters, Ibid.

36. Emanuele Lugato, Adrian Leip, and Arwyn Jones, 2018. "Mitigation potential of soil carbon management overestimated by neglecting N2O emissions." *Nature Climate Change*, vol. 8, pp. 219–23. https://doi.org /10.1038/s41558-018-0087-z.

37. Elvir Tenic, Rishikesh Ghogare, and Amit Dhingra, 2020. "Biochar—A

panacea for agriculture or just carbon?" *Horticulturae*, vol. 6:3, p. 37. https://doi.org/10.3390/horticulturae6030037.

38. SoilFixer Products. *Biochar Fine Granules (0-2-mm)*. SoilFixer. https://www.soilfixer.co.uk/Biochar-fine-granules.

39. Forage. *Learn to Make Charcoal*. Forage Open-Source Charcoal. http://foragejournalism.org/make-charcoal/#production-1.

40. Dominic Woolf, 2010. "Sustainable biochar to mitigate global climate change." *Nature Communications*, vol. 1, article 56. https://doi.org/10.1038/ncomms1053.

41. Kyle S. Hemes et al., 2019. "Assessing the carbon and climate benefit of restoring degraded agricultural peat soils to managed wetlands." *Agricultural and Forest Meteorology*, vol. 268, pp. 202–14. https://doi.org/10.1016/j.agrformet.2019.01.017.

42. Lucas E. Nave et al., 2018. "Reforestation can sequester two petagrams of carbon in US topsoils in a century." *Proceedings of the National Academy of Sciences*, vol. 115:11, pp. 2776–81. https://doi.org/10.1073/pnas.1719685115.

43. La Via Campesina, International Peasants' Movement. https://viacampesina.org/en.

44. Les Levidow et al., 2019. *Transitions towards a European Bioeconomy: Life Sciences versus Agroecology Trajectories. Ecology, Capitalism and the New Agricultural Economy: The Second Great Transformation*. London: Routledge, pp. 181–203. http://oro.open.ac.uk/58109/20/LL_Transitions%20towards%20a%20European%20bioeconomy_2019.pdf.

45. Michel P. Pimbert and Nina Isabella Moeller, 2018. "Absent agroecology aid: On UK agricultural development assistance since 2010." *Sustainability*, 10:2, p. 505. https://doi.org/10.3390/su10020505.

46. H. K. Gibb and J. M. Salmon, 2015. "Mapping the world's degraded lands." *Applied Geography*, vol. 57, pp. 12–21. https://doi.org/10.1016/j.apgeog.2014.11.024.

47. Carlos Guerra et al., 2020. "Blind spots in global soil biodiversity and ecosystem function research." *Nature Communications*, vol. 11, p. 3870. https://doi.org/10.1038/s41467-020-17688-2.

48. Carlos Guerra et al., 2021. "Tracking, targeting, and conserving soil biodiversity." *Science*, vol. 371:6526, pp. 239–41. https://science.org/doi/10.1126/science.abd7926.

49. Marcia DeLonge, Albie Miles, and Liz Carlisle, 2016. "Investing in the transition to sustainable agriculture." *Environmental Science and*

Policy, vol. 55, pt. 1, pp. 266–73. https://doi.org/10.1016/j.envsci.2015
.09.013.

50. Jennifer Clapp, 2021. "The problem with growing corporate concentration and power in the global food system." *Nature Food*, vol. 2, pp. 404–8. https://doi.org/10.1038/s43016-021-00297-7.

51. The Vegan-Organic Network, 2007. *The Stockfree-Organic Standards.* https://veganorganic.net/wp-content/uploads/Organic-Stockfree-Veganic-Standards-September-2018.pdf.

52. Ruth Kelly et al., 2012. *The Hunger Grains.* Oxfam Briefing Paper 161. https://oxfamilibrary.openrepository.com/bitstream/handle/10546/242997/bp161-the-hunger-grains-170912-en.pdf;jsessionid=59A3BFB0B7A7B7E0A65838095E08972C?sequence=1.

53. Chris Malins, 2020. "Biofuel to the fire—The impact of continued expansion of palm and soy oil demand through biofuel policy." *Rainforest Foundation Norway.* https://d5i6is0eze552.cloudfront.net/documents/RF_report_biofuel_0320_eng_SP_update.pdf.

54. Krystof Obidzinski et al., 2012. "Environmental and social impacts of oil palm plantations and their implications for biofuel production in Indonesia." *Ecology and Society*, vol. 17:1. https://www.jstor.org/stable/26269006.

55. Janis Brizga, Klaus Hubacek, and Kuishuang Feng, 2020. "The unintended side effects of bioplastics: Carbon, land, and water footprints." *One Earth*, vol. 3:4, pp. 515–16. https://doi.org/10.1016/j.oneear.2020.06.016.

56. Tolhurst Organic, 2018. *Tolly's Rambles: Plastic Confessional.* Tolhurst Organic Partnership, May 3. https://www.tolhurstorganic.co.uk/2018/05/tollys-rambles-plastic-confessional.

CHAPTER 5

1. "Rivercide: The world's first live investigative documentary." https://www.youtube.com/watch?v=5ID0VAUNANA.

2. Hannah Ritchie, July 12, 2021. "Three billion people cannot afford a healthy diet." https://ourworldindata.org/diet-affordability#a-healthy-diet-three-billion-people-cannot-afford-one.

3. South Oxfordshire Food & Education Academy. SOFEA. https://www.sofea.uk.com.

4. FareShare. https://fareshare.org.uk.

5. FareShare, pers comm.

6. Jo Dyson, head of food at FareShare, pers comm.

7. Feedback Global, 2018. *Farmers Talk Food Waste. Supermarkets' Role in Crop Waste on UK Farms.* https://feedbackglobal.org/wp-content/uploads/2018/08/Farm_waste_report_.pdf.

8. Charlie Spring, 2020. "How foodbanks went global." *New Internationalist*, November 10. https://newint.org/immersive/2020/11/10/how-foodbanks-went-global.

9. Jenny Gustavsson et al., 2011. *Global Food Losses and Food Waste: Extent, Causes and Prevention.* Food and Agriculture Organization of the United Nations (FAO). http://www.fao.org/3/i2697e/i2697e.pdf.

10. Brian Lipinski et al., 2017. *SDG Target 12.3 on Food Loss and Waste: 2017 Progress Report.* World Resources Institute—Champions 12.3. https://champions123.org/sites/default/files/2020-09/champions-12-3-2017-progress-report.pdf.

11. Waste & Resources Action Programme, 2008. *The Food We Waste.* https://wrap.s3.amazonaws.com/the-food-we-waste-executive-summary.pdf.

12. Food and Agriculture Organization of the United Nations (FAO). *Technical Platform on the Measurement and Reduction of Food Loss and Waste.* http://www.fao.org/platform-food-loss-waste/en.

13. Melanie Saltzman, Christopher Livesay, and Mark Bittman, 2019. "Is France's groundbreaking food-waste law working?" *PBS Newshour*, Pulitzer Center, September 1. https://pulitzercenter.org/stories/frances-groundbreaking-food-waste-law-working.

14. Yanne Goossens, Alina Wegner, and Thomas Schmidt, 2019. *Sustainability Assessment of Food Waste Prevention Measures: Review of Existing Evaluation Practices.* Frontiers in Sustainable Food Systems, vol. 3. https://doi.org/10.3389/fsufs.2019.00090.

15. UN Food and Agriculture Organization, 2015. *Food Wastage Footprint & Climate Change.* https://www.fao.org/3/bb144e/bb144e.pdf.

16. Purabi R. Ghosh et al., 2015. "An Overview of Food Loss and Waste: Why Does It Matter?" *COSMOS*, vol. 11:1, pp. 89–103. https://doi.org/10.1142/S0219607715500068.

17. William F. Laurance, Miriam Goosem, and Susan G. W. Laurance, 2009. "Impacts of roads and linear clearings on tropical forests." *Trends in Ecology & Evolution*, vol. 24:12, pp. 659–69. https://doi.org/10.1016/j.tree.2009.06.009.

18. William F. Laurance, Jeffrey Sayer, and Kenneth G. Cassman, 2014. "Agricultural expansion and its impacts on tropical nature." *Trends in Ecology & Evolution*, vol. 29:2, pp. 107–16. https://doi.org/10.1016/j.tree.2013.12.001.

19. Ibid.

20. William F. Laurance et al., 2006. "Impacts of roads and hunting on Central African rainforest mammals." *Conservation Biology*, vol. 20:4, pp. 1251–61. https://doi.org/10.1111/j.1523-1739.2006.00420.x.

21. Walter Willett et al., 2019. "Food in the Anthropocene: the EAT–Lancet Commission on healthy diets from sustainable food systems." *The Lancet*, vol. 393:10170, pp. 447–92. https://doi.org/10.1016/S0140-6736(18)31788-4.

22. Anjum Klair, 2020. "The five-week wait for first payment of universal credit is unnecessary and unacceptable." Trades Union Congress (TUC), November 10. https://www.tuc.org.uk/blogs/five-week-wait-first-payment-universal-credit-unnecessary-and-unacceptable.

23. Rachel Loopstra and Doireann Lalor, 2017. "Financial insecurity, food insecurity, and disability: The profile of people receiving emergency food assistance from The Trussell Trust Foodbank Network in Britain." The Trussell Trust. https://www.trusselltrust.org/wp-content/uploads/sites/2/2017/07/OU_Report_final_01_08_online2.pdf.

24. Alice Goisis, Amanda Sacker, and Yvonne Kelly, 2015. "Why are poorer children at higher risk of obesity and overweight? A UK cohort study." *European Journal of Public Health*, vol. 26:1, pp. 7–13. https://doi.org/10.1093/eurpub/ckv219.

25. Helen P. Booth, Judith Charlton, and Martin C. Gulliford, 2017. "Socioeconomic inequality in morbid obesity with body mass index more than 40kg/m2 in the United States and England." *SSM—Population Health*, vol. 3, pp. 172–8. https://doi.org/10.1016/j.ssmph.2016.12.012.

26. Felicity Lawrence, 2013. *Not on the Label—What Really Goes into the Food on Your Plate*. Penguin Books, London. https://www.penguin.co.uk/books/54815/not-on-the-label/9780241967829.html.

27. Food and Agriculture Organization of the United Nations (FAO), the International Fund for Agricultural Development (IFAD), the United Nations Children's Fund (UNICEF), the World Food Programme (WFP), and the World Health Organization (WHO), 2020. *The State of Food Security and Nutrition in the World 2020: Transforming Food Systems for Affordable Healthy Diets*. Rome, FAO. https://doi.org/10.4060/ca9692en.

28. Rocco Barazzoni and Gianluca Gortan Cappellari, 2020. "Double burden of malnutrition in persons with obesity." *Reviews in Endocrine and Metabolic Disorders*, vol. 21, pp. 307–13. https://doi.org/10.1007/s11154-020-09578-1.

29. Pilyoung Kim et al., 2017. "How socioeconomic disadvantages get

under the skin and into the brain to influence health development across the lifespan," in N. Halfon, C. Forrest, R. Lerner and E. Faustman (eds.), *Handbook of Life Course Health Development*. Springer, Cham. https://doi.org/10.1007/978-3-319-47143-3_19.

30. The Equality Trust. "Obesity—Obesity is less common in more equal societies." https://www.equalitytrust.org.uk/obesity.

31. Anthony Rodgers et al., 2018. "Prevalence trends tell us what did not precipitate the US obesity epidemic." *The Lancet Public Health*, vol. 3:4, E162–3. https://doi.org/10.1016/S2468-2667(18)30021-5.

32. Carl Baker, 2021. *Research Briefing—Obesity Statistics*. House of Commons Library, January 11. https://commonslibrary.parliament.uk/research-briefings/sn03336/#fullreport.

33. Ferris Jabr, 2016. "How sugar and fat trick the brain into wanting more food." *Scientific American*, January 1. https://www.scientificamerican.com/article/how-sugar-and-fat-trick-the-brain-into-wanting-more-food.

34. Kevin Hall et al., 2019. "Ultra-processed diets cause excess calorie intake and weight gain: An inpatient randomized controlled trial of ad libitum food intake." *Cell Metabolism*, vol. 30, pp. 67–77. https://doi.org/10.1016/j.cmet.2019.05.008.

35. Sarah Boseley, 2018. "Food firms could face litigation over neuro-marketing to hijack brains." *The Guardian*, May 25. https://www.theguardian.com/society/2018/may/25/food-firms-may-face-litigation-over-neuromarketing-to-hijack-brains.

36. Anahad O'Connor, 2015. "Coca Cola funds scientists who shift blame for obesity away from bad diets." *The New York Times*, August 9. https://well.blogs.nytimes.com/2015/08/09/coca-cola-funds-scientists-who-shift-blame-for-obesity-away-from-bad-diets.

37. Cristin E. Kearns, Laura A. Schmidt, and Stanton A. Glantz, 2016. "Sugar industry and coronary heart disease research—A historical analysis of internal industry documents." *JAMA Internal Medicine*, vol. 176:11, pp. 1680–5. https://doi.org/10.1001/jamainternmed.2016.5394.

38. WhyHunger, 2020. *We Can't "Foodbank" Our Way Out of Hunger: From Charity to a Social Justice Funding Model*. Sustainable Agriculture and Food Systems Funders (SAFSF), October 22. https://www.agandfoodfunders.org/event/we-cant-foodbank-our-way-out-of-hunger.

39. The Trussell Trust, 2021. "Trussell Trust data briefing on end-of-year statistics relating to use of food banks: April 2020—March 2021," April 22. https://www.trusselltrust.org/wp-content/uploads/sites/2/2021/04/Trusell-Trust-End-of-Year-stats-data-briefing_2020_21.pdf.

40. Spring, Ibid.

41. UK Food Standards Agency, 2020. *Covid19 Research Tracker—Wave Four*, August 12. https://data.food.gov.uk/catalog/datasets/da60fd93-be85 -4a6b-8fb6-63eddf32eeab.

42. Oxford Real Farming Conference, 2020. *#ORFC20 A Food Strategy for the UK: Local Food Systems and Access to Healthy, Affordable Food*, January 18. https://www.youtube.com/watch?v=jA8kGRLAWu0.

43. Alastair Smith, 2021. "Why global food prices are higher today than for most of modern history," September 27. https://theconversation .com/why-global-food-prices-are-higher-today-than-for-most -of-modern-history-168210.

44. Willett et al., Ibid.

45. George Monbiot, 2018. "The UK government wants to put a price on nature—but that will destroy it." *The Guardian*, May 15. https://www .theguardian.com/commentisfree/2018/may/15/price-natural-world -destruction-natural-capital.

46. Nyéléni Forum 2007. *Declaration of the Forum for Food Sovereignty, Nyéléni, Sélingué, Mali*, February 27. https://nyeleni.org/en/declaration -of-nyeleni.

47. Ibid.

48. Oxford Real Farming Conference, 2020. "ORFC 2020 Patrick Holden in conversation with George Monbiot," January 16. https://www.you tube.com/watch?v=fB2F5GsOUCU.

49. David P. Edwards et al., 2015. "Land-sparing agriculture best protects avian phylogenetic diversity." *Current Biology*, vol. 25:18, pp. 2384– 91. https://doi.org/10.1016/j.cub.2015.07.063.

50. Andrew Balmford et al., 2018. "The environmental costs and benefits of high-yield farming." *Nature Sustainability*, vol. 1, pp. 477–85. https://doi.org/10.1038/s41893-018-0138-5.

51. Pat Mooney et al., 2021. *A Long Food Movement: Transforming Food Systems by 2045.* The International Panel of Experts on Sustainable Food Systems (IPES-Food) and ETC Group. http://www.ipes-food .org/_img/upload/files/LongFoodMovementEN.pdf.

52. La Via Campesina, 2021. "Global Solidarity Actions demand equal ownership of land, recognition of women's work and a world free of violence," March 25. https://viacampesina.org/en/08-march-global -solidarity-actions-demand-equal-ownership-of-land -recognition-of-womens-work-and-a-world-free-of-violence.

53. La Via Campesina, 2021. "The path of peasant and popular feminism in La Via Campesina." https://viacampesina.org/en/publication-the-path-of -peasant-and-popular-feminism-in-la-via-campesina/#:~:text=La%20

Via%20Campesina%2C%20presents%20the,tool%20against%20 oppression%20and%20violence.

54. La Via Campesina, 2021. "Food sovereignty, a manifesto for the future of our planet," October 13. https://viacampesina.org/en/food-sovereignty -a-manifesto-for-the-future-of-our-planet-la-via-campesina.

55. Pekka Kinnunen et al., 2020. "Local food crop production can fulfil demand for less than one-third of the population." *Nature Food*, vol. 1, pp. 229–37. https://doi.org/10.1038/s43016-020-0060-7.

56. Richard Reynolds. *Guerrilla Gardening blog*. http://www.guerrillagar dening.org.

57. Mooney et al., Ibid.

58. George Monbiot, 2010. "Towering lunacy," August 16. https://www .monbiot.com/2010/08/16/towering-lunacy.

59. Chris Beytes, 2017. *FarmedHere Shuts Down*. Grower Talks, March 29. https://www.growertalks.com/Article/?articleid=22890.

60. Urvaksh Karkaria, 2016. "Bloom to bust: The birth and death of Atlanta startup PodPonics." *Atlanta Business Chronicle*, June 20. https://www.bizjournals.com/atlanta/print-edition/2016/06/17/bloom -to-bust-the-birth-and-death-of-an-atlanta.html.

61. Jennifer Marston, 2019. "What Plantagon's bankruptcy could tell us about the future of large-scale vertical farming." *The Spoon*, March 1. https://thespoon.tech/what-plantagons-bankruptcy-could-tell-us -about-the-future-of-large-scale-vertical-farming.

62. Edward Game et al., 2014. "Conservation in a wicked complex world: Challenges and solutions." *Conservation Letters*, vol. 7, pp. 271–7. https://doi.org/10.1111/conl.12050.

CHAPTER 6

1. Joseph M. Awika, 2011. *Major Cereal Grains Production and Use around the World*. Advances in Cereal Science: Implications to Food Processing and Health Promotion, ACS Symposium Series, vol. 1089, pp. 1–13. https://doi.org/10.1021/bk-2011-1089.ch001.

2. Green Alliance, 2021. *Net Zero Policy Tracker, April 2021 Update*. https://green-alliance.org.uk/resources/Net_zero_policy_tracker_April _2021.pdf.

3. Niels Corfield, 2020. *Wet on Top Dry Underneath*. https://nielscorfield .com/soil-health/wet-on-top-dry-underneath.

4. David C. Coleman, 2017. *Fundamentals of Soil Ecology*. Academic

Press, Cambridge, MA. https://www.elsevier.com/search-results?query=9780128052518.

5. Jan Willem van Groenigen et al., 2014. "Earthworms increase plant production: A meta-analysis." *Scientific Reports*, vol. 4, article 6365. https://doi.org/10.1038/srep06365.

6. Lorna J. Cole, Jenni Stockan, and Rachel Helliwell, 2020. "Managing riparian buffer strips to optimise ecosystem services: A review." *Agriculture, Ecosystems & Environment*, vol. 296. https://doi.org/10.1016/j.agee.2020.106891.

7. Humberto Blanco-Canqui and Rattan Lal, 2010. "Buffer strips," in *Principles of Soil Conservation and Management*, Springer, Dordrecht, pp. 223–57. https://doi.org/10.1007/978-1-4020-8709-7_9.

8. Eduardo González et al., 2017. "Integrative conservation of riparian zones." *Biological Conservation*, vol. 211, part B, pp. 20–9. https://doi.org/10.1016/j.biocon.2016.10.035.

9. María Sol Balbuena et al., 2015. "Effects of sublethal doses of glyphosate on honeybee navigation." *Journal of Experimental Biology*, vol. 218:17, pp. 2799–805. https://doi.org/10.1242/jeb.117291.

10. Pingli Dai et al., 2018. "The herbicide glyphosate negatively affects midgut bacterial communities and survival of honeybee during larvae reared in vitro." *Journal of Agricultural and Food Chemistry*, vol. 66:29, pp. 7786–93. https://doi.org/10.1021/acs.jafc.8b02212.

11. Abbas Güngördü, Miraç Uçkun, and Ertan Yoloğlu, 2016. "Integrated assessment of biochemical markers in premetamorphic tadpoles of three amphibian species exposed to glyphosate- and methidathion-based pesticides in single and combination forms." *Chemosphere*, vol. 144, pp. 2024–35. https://doi.org/10.1016/j.chemosphere.2015.10.125.

12. Sonia Soloneski, Celeste Ruiz de Arcaute, and Marcelo L. Larramendy, 2016. "Genotoxic effect of a binary mixture of dicamba- and glyphosate-based commercial herbicide formulations on *Rhinella arenarum* (Hensel, 1867) (*Anura, Bufonidae*) late-stage larvae." *Environmental Science and Pollution Research*, vol. 23, pp. 17811–21. https://doi.org/10.1007/s11356-016-6992-7.

13. Rafael Zanelli Rissoli et al., 2016. "Effects of glyphosate and the glyphosate-based herbicides Roundup Original® and Roundup Transorb® on respiratory morphophysiology of bullfrog tadpoles." *Chemosphere*, vol. 156, pp. 37–44. https://doi.org/10.1016/j.chemosphere.2016.04.083.

14. Ming-Hui Li et al., 2017. "Metabolic profiling of goldfish (*Carassius auratis*) after long-term glyphosate-based herbicide exposure." *Aquatic*

Toxicology, vol. 188, pp. 159–69. https://doi.org/10.1016/j.aquatox.2017.05.004.

15. Sofia Guilherme et al., 2009. "Tissue specific DNA damage in the European eel (*Anguilla anguilla*) following a short-term exposure to a glyphosate-based herbicide." *Toxicology Letters*, vol. 189, supplement, p. S212. https://doi.org/10.1016/j.toxlet.2009.06.550.

16. Jimena L. Frontera et al., 2011. "Effects of glyphosate and polyoxyethylenamine on growth and energetic reserves in the freshwater crayfish *Cherax quadricarinatus (Decapoda, Parastacidae)*." *Archives of Environmental Contamination and Toxicology*, vol. 61, pp. 590–8. https://doi.org/10.1007/s00244-011-9661-3.

17. Valerio Matozzo et al., 2018. "Ecotoxicological risk assessment for the herbicide glyphosate to non-target aquatic species: A case study with the mussel *Mytilus galloprovincialis*." *Environmental Pollution*, vol. 233, pp. 623–32. https://doi.org/10.1016/j.envpol.2017.10.100.

18. Cong Wang et al., 2016. "Differential growth responses of marine phytoplankton to herbicide glyphosate." *PLOS One*, vol. 11:3. https://doi.org/10.1371/journal.pone.0151633.

19. Louis Carles et al., 2019. "Meta-analysis of glyphosate contamination in surface waters and dissipation by biofilms." *Environment International*, vol. 124, pp. 284–93. https://doi.org/10.1016/j.envint.2018.12.064.

20. Philip Mercurio et al., 2014. "Glyphosate persistence in seawater." *Marine Pollution Bulletin*, vol. 85:2, pp. 385–90. https://doi.org/10.1016/j.marpolbul.2014.01.021.

21. Valerio Matozzo, Jacopo Fabrello, and Maria Gabriella Marin, 2020. "The effects of glyphosate and its commercial formulations to marine invertebrates: A review." *Journal of Marine Science and Engineering*, vol. 8:6, p. 399. https://doi.org/10.3390/jmse8060399.

22. Todd Funke et al., 2006. "Molecular basis for the herbicide resistance of Roundup Ready crops." *Proceedings of the National Academy of Sciences*, vol. 103:35, pp. 13010–15. https://doi.org/10.1073/pnas.0603638103.

23. A. H. C. Van Bruggen et al., 2018. "Environmental and health effects of the herbicide glyphosate." *Science of the Total Environment*, vols. 616–17, pp. 255–68. https://doi.org/10.1016/j.scitotenv.2017.10.309.

24. Laura Arango et al., 2014. "Effects of glyphosate on the bacterial community associated with roots of transgenic Roundup Ready® soybean." *European Journal of Soil Biology*, vol. 63, pp. 41–8. https://doi.org/10.1016/j.ejsobi.2014.05.005.

25. M. Druille et al., 2015. "Glyphosate vulnerability explains changes in

root-symbionts propagules viability in pampean grasslands." *Agriculture, Ecosystems & Environment*, vol. 202, pp. 48–55. https://doi.org/10.1016/j.agee.2014.12.017.

26. María C. Zabaloy et al., 2015. "Soil ecotoxicity assessment of glyphosate use under field conditions: Microbial activity and community structure of Eubacteria and ammonia-oxidising bacteria." *Pest Management Science*, vol. 72:4, pp. 684–91. https://doi.org/10.1002/ps.4037.

27. Tsuioshi Yamada et al., 2009. "Glyphosate interactions with physiology, nutrition, and diseases of plants: Threat to agricultural sustainability?," *European Journal of Agronomy*, vol. 31:3, pp. 111–13. http://www.lcb.esalq.usp.br/publications/articles/2009/2009ejav31n3p111-113.pdf.

28. Elena Okada, José Luis Costa, and Francisco Bedmar, 2019. "Glyphosate dissipation in different soils under no-till and conventional tillage." *Pedosphere*, vol. 29:6, pp. 773–83. https://doi.org/10.1016/S1002-0160(17)60430-2.

29. Yong-Guan Zhu et al., 2019. "Soil biota, antimicrobial resistance and planetary health." *Environment International*, vol. 131. https://doi.org/10.1016/j.envint.2019.105059.

30. Paulo Durão, Roberto Balbontín, and Isabel Gordo, 2018. "Evolutionary mechanisms shaping the maintenance of antibiotic resistance." *Trends in Microbiology*, vol. 26:8, pp. 677–91. https://doi.org/10.1016/j.tim.2018.01.005.

31. Hanpeng Liao et al., 2021. "Herbicide selection promotes antibiotic resistance in soil microbiomes." *Molecular Biology and Evolution*, vol. 38:6, pp. 2337–50. https://doi.org/10.1093/molbev/msab029.

32. Anne Mendler et al., 2020. "Mucosal-associated invariant T-Cell (MAIT) activation is altered by chlorpyrifos- and glyphosate-treated commensal gut bacteria." *Journal of Immunotoxicology*, vol. 17:1, pp. 10–20. https://doi.org/10.1080/1547691X.2019.1706672.

33. Sebastian T. Soukup et al., 2020. "Glyphosate and AMPA levels in human urine samples and their correlation with food consumption: Results of the cross-sectional KarMeN study in Germany." *Archives of Toxicology*, vol. 94, pp. 1575–84. https://doi.org/10.1007/s00204-020-02704-7.

34. Braeden Van Deynze, Scott Swinton, and David Hennessy, 2018. "Are glyphosate-resistant weeds a threat to conservation agriculture? Evidence from tillage practices in soybean." Conference Paper/Presentation, Agricultural and Applied Economics Association, no. 274360. https://doi.org/10.22004/ag.econ.274360.

35. Mailin Gaupp-Berghausen et al., 2015. "Glyphosate-based herbicides

reduce the activity and reproduction of earthworms and lead to increased soil nutrient concentrations." *Nature Scientific Reports*, vol. 5, article 12886. https://doi.org/10.1038/srep12886.

36. María Jesús I. Briones and Olaf Schmidt, 2017. "Conventional tillage decreases the abundance and biomass of earthworms and alters their community structure in a global meta-analysis." *Global Change Biology*, vol. 23:10, pp. 4396–419. https://doi.org/10.1111/gcb.13744.

37. Niki Grigoropoulou, Kevin R. Butt, and Christopher N. Lowe, 2008. "Effects of adult *Lumbricus terrestris* on cocoons and hatchlings in Evans' boxes." *Pedobiologia*, vol. 51:5–6, pp. 343–9. https://doi.org/10.1016/j.pedobi.2007.07.001.

38. Maria A. Tsiafouli et al., 2015. "Intensive agriculture reduces soil biodiversity across Europe." *Global Change Biology*, vol. 21:2, pp. 973–985. https://doi.org/10.1111/gcb.12752.

39. Stacy M. Zuber and María B. Villamil, 2016. "Meta-analysis approach to assess effect of tillage on microbial biomass and enzyme activities." *Soil Biology and Biochemistry*, vol. 97, pp. 176–87. https://doi.org/10.1016/j.soilbio.2016.03.011.

40. UK Government Natural Capital Committee, April 2017. *How to Do It: A Natural Capital Workbook, Version 1*. Department for Environment, Food & Rural Affairs. https://assets.publishing.service.gov.uk/government/uploads/system/uploads/attachment_data/file/957503/ncc-natural-capital-workbook.pdf.

41. Lucinda Dann, 2018. "Video: Crimper roller trial aims to kill cover crops without herbicides." *Farmers Weekly*, June 28. https://www.fwi.co.uk/arable/land-preparation/cover-crops/video-crimper-roller-trial-aims-to-kill-cover-crops-without-herbicides.

42. Damian Carrington, 2021. "Killer farm robot dispatches weeds with electric bolts." *The Guardian*, April 29. https://www.theguardian.com/environment/2021/apr/29/killer-farm-robot-dispatches-weeds-with-electric-bolts.

43. FarmWise, 2021. "Titan FT35 pay-per-acre model." https://farmwise.io/services.

44. Cameron M. Pittelkow et al., 2015. "When does no-till yield more? A global meta-analysis." *Field Crops Research*, vol. 183, pp. 156–68. https://doi.org/10.1016/j.fcr.2015.07.020.

45. Ademir Calegari et al., 2014. "Conservation agriculture in Brazil," in *Conservation Agriculture: Global Prospects and Challenges*, ch. 3, pp. 54–87. https://doi.org/10.1079/9781780642598.0054.

46. Peter R. Hobbs, Ken Sayre, and Raj Gupta, 2007. "The role of conservation agriculture in sustainable agriculture." *Philosophical Transactions*

of the Royal Society B: Biological Sciences, vol. 363:1491. https://doi.org/10.1098/rstb.2007.2169.

47. Peter R. Hobbs and Raj K. Gupta, 2003. "Resource-conserving technologies for wheat in the rice-wheat system," in Improving the Productivity and Sustainability of Rice-Wheat Systems: Issues and Impacts. https://doi.org/10.2134/asaspecpub65.c7.

48. Krutika Pathi and Arvind Chhabra, 2020. "Stubble burning: Why it continues to smother north India." BBC News, November 30. https://www.bbc.co.uk/news/world-asia-india-54930380.

49. Cameron M. Pittelkow et al., 2015. "Productivity limits and potentials of the principles of conservation agriculture." Nature, vol. 517, pp. 365–8. https://doi.org/10.1038/nature13809.

50. N. Verhulst et al., 2010. "Conservation agriculture, improving soil quality for sustainable production systems?," in Food Security and Soil Quality. https://www.taylorfrancis.com/chapters/edit/10.1201/EBK 1439800577-7/conservation-agriculture-improving-soil-quality -sustainable-production-systems-verhulst-govaerts-verachtert -castellanos-navarrete-mezzalama-wall-chocobar-deckers-sayre.

51. Humberto Blanco-Canqui and Sabrina J. Ruis, 2018. "No-tillage and soil physical environment." Geoderma, vol. 326, pp. 164–200. https://doi.org/10.1016/j.geoderma.2018.03.011.

52. Zhongkui Luo, Enli Wang, and Osbert J. Sun, 2010. "Can no-tillage stimulate carbon sequestration in agricultural soils? A meta-analysis of paired experiments." Agriculture, Ecosystems & Environment, vol. 139:1–2, pp. 224–31. https://doi.org/10.1016/j.agee.2010.08.006.

53. Stefani Daryanto, Lixin Wang, and Pierre-André Jacinthe, 2017. "Impacts of no-tillage management on nitrate loss from corn, soybean and wheat cultivation: A meta-analysis." Scientific Reports, vol. 7, article 12117. https://doi.org/10.1038/s41598-017-12383-7.

54. Kenneth G. Cassman, Achim Dobermann, and Daniel T. Walters, 2002. "Agroecosystems, nitrogen-use efficiency, and nitrogen management." Ambio—A Journal of Environment and Society, vol. 31:2, pp. 132–40. https://doi.org/10.1579/0044-7447-31.2.132.

55. Daniel Elias, Lixin Wang, and Pierre-Andre Jacinthe, 2018. "A meta-analysis of pesticide loss in runoff under conventional tillage and no-till management." Environmental Monitoring and Assessment, vol. 190, article 79. https://doi.org/10.1007/s10661-017-6441-1.

56. Rolf Derpsch et al., 2010. "Current status of Adoption of no-till farming in the world and some of its main benefits." International Journal

of Agricultural and Biological Engineering, vol. 3:1. https://ijabe.org/index.php/ijabe/article/viewFile/223/113.

57. Hobbs, Sayre, and Gupta, Ibid.

58. Olaf Erenstein et al., 2012. "Conservation agriculture in maize- and wheat-based systems in the (sub)tropics: Lessons from adaptation initiatives in South Asia, Mexico, and Southern Africa." *Journal of Sustainable Agriculture*, vol. 36:2, pp. 180–206. https://doi.org/10.1080/10440046.2011.620230.

59. Hobbs, Sayre, and Gupta, Ibid.

60. George Rapsomanikis, 2015. "The economic lives of smallholder farmers: An analysis based on household data from nine countries." Food and Agriculture Organization of the United Nations, Rome. http://www.fao.org/3/i5251e/i5251e.pdf.

61. Emma Hamer, 2016. "Can we continue to grow oilseed rape in the UK?" National Farmers Union. https://www.nfuonline.com/sectors/crops/crops-news/blog-can-we-continue-to-grow-oilseed-rape-in-the.

62. Timothy M. Bowles, 2020. "Long-term evidence shows that crop-rotation diversification increases agricultural resilience to adverse growing conditions in North America." *One Earth*, vol. 2:3, pp. 284–93. https://doi.org/10.1016/j.oneear.2020.02.007.

63. Paula Sanginés de Cárcer et al., 2019. "Long-term effects of crop succession, soil tillage and climate on wheat yield and soil properties." *Soil and Tillage Research*, vol. 190, pp. 209–19. https://doi.org/10.1016/j.still.2019.01.012.

64. Jeremy Cherfas, 2020. "The worst thing since sliced bread: The Chorleywood bread process." Dublin Gastronomy Symposium, Disruptive Technology, Food and Disruption. https://doi.org/10.21427/99cm-eb95.

65. Felicity Lawrence, 2013. *Not on the Label: What Really Goes into the Food on Your Plate*. Penguin, London. https://www.penguin.co.uk/books/548/54815/not-on-the-label/9780241967829.html.

66. Cherfas, Ibid.

67. Pat M. Burton et al., 2011. "Glycemic impact and health: New horizons in white bread formulations." *Critical Reviews in Food Science and Nutrition*, vol. 51:10, pp. 965–82. https://doi.org/10.1080/10408398.2010.491584.

68. Andrew S. Ross, 2019. *A Shifting Climate for Grains and Flour*. Cereals & Grains Association. https://www.cerealsgrains.org/publications/cfw/2019/September-October/Pages/CFW-64-5-0050.aspx.

69. Steve Gliessman, 2011. "Transforming food systems to sustainability

with agroecology." *Journal of Sustainable Agriculture*, vol. 35:8, pp. 823–5. https://doi.org/10.1080/10440046.2011.611585.

70. Colin Ray Anderson, 2019. "From transition to domains of transformation: Getting to sustainable and just food systems through agroecology." *Sustainability*, vol. 11:19, pp. 5272. https://doi.org/10.3390/su11195272.

71. Mateo Mier y Terán Giménez Cacho et al., 2018. "Bringing agroecology to scale: Key drivers and emblematic cases." *Agroecology and Sustainable Food Systems*, vol. 42:6, pp. 637–65. https://doi.org/10.1080/21683565.2018.1443313.

72. Laura Pereira, Rachel Wynberg, and Yuna Reis, 2018. "Agroecology: The future of sustainable farming?" *Environment: Science and Policy for Sustainable Development*, vol. 60:4, pp. 4–17. https://doi.org/10.1080/00139157.2018.1472507.

73. Oliver Rackham, 2020. *The History of the Countryside*. Weidenfeld & Nicolson, London. https://www.weidenfeldandnicolson.co.uk/titles/oliver-rackham/the-history-of-the-countryside/9781474614023.

74. FarmED. *Farm & Food Education*. https://www.farm-ed.co.uk.

75. Li Guo Qiang et al., 2009. "Dynamic analysis on response of dry matter accumulation and partitioning to nitrogen fertilizer in wheat cultivars with different plant types." *Acta Agronomica Sinica*, vol. 35:12, pp. 2258–65. https://doi.org/10.3724/SP.J.1006.2009.02258.

76. Tejendra Chapagain, Laura Super, and Andrew Riseman, 2014. "Root architecture variation in wheat and barley cultivars." *Journal of Experimental Agriculture International*, vol. 4:7, pp. 849–56. https://doi.org/10.9734/AJEA/2014/9462.

77. S. Jeffery and F. G. A. Verheijen, 2020. "A new soil health policy paradigm: Pay for practice not performance!" *Environmental Science & Policy*, vol. 112, pp. 371–3. https://doi.org/10.1016/j.envsci.2020.07.006.

78. I. K. S. Andrew, J. Storkey, and D. L Sparkes, 2015. "A review of the potential for competitive cereal cultivars as a tool in integrated weed management." *Weed Research*, vol. 55:3, pp. 239–48. https://doi.org/10.1111/wre.12137.

79. Jo Smith et al., 2017. *Lessons Learnt: Silvoarable Agroforestry in the UK (Part 1)*. AgForward, Agroforestry for Europe. https://www.agforward.eu/documents/LessonsLearnt/WP4_UK_Silvoarable_1_lessons_learnt.pdf.

80. Peter R. Shewry, Till K. Pellny, and Alison Lovegrove, 2016. "Is modern wheat bad for health?" *Nature Plants*, vol. 2, article 16097. https://doi.org/10.1038/nplants.2016.97.

81. Peter Shewry et al., 2020. "Do modern types of wheat have lower quality for human health?" *Nutrition Bulletin*, vol. 45:4, pp. 362–73. https://doi.org/10.1111/nbu.12461.

82. Peter Shewry et al., 2017. "Defining genetic and chemical diversity in wheat grain by 1H-NMR spectroscopy of polar metabolites." *Molecular Nutrition & Food Research*, vol. 61, article 1600807. https://doi.org/10.1002/mnfr.201600807.

83. J. I. Macdiarmid and S. Whybrow, 2019. "Nutrition from a climate change perspective. Conference on 'Getting energy balance right,' Symposium 5: Sustainability of food production and dietary recommendations." *Proceedings of the Nutrition Society*, vol. 78:3, pp. 380–7. https://doi.org/10.1017/S0029665118002896.

84. Andrew S. Ross, 2019. *A Shifting Climate for Grains and Flour*. Cereals & Grains Association. https://www.cerealsgrains.org/publications/cfw/2019/September-October/Pages/CFW-64-5-0050.aspx.

85. Burton et al., Ibid.

86. Hannah Ritchie, 2021. "If the world adopted a plant-based diet we would reduce global agricultural land use from 4 to 1 billion hectares." Our World in Data, viewed 2022. Land use per kilogram of food product. https://ourworldindata.org/grapher/land-use-per-kg-poore.

87. Mier y Terán Giménez Cacho et al., Ibid.

88. Meagan E. Schipanski et al., 2016. "Realizing resilient food systems." *BioScience*, vol. 66:7, pp. 600–10. https://doi.org/10.1093/biosci/biw052.

89. Steven Lawry et al., 2017. "The impact of land property rights interventions on investment and agricultural productivity in developing countries: A systematic review." *Journal of Development Effectiveness*, vol. 9:1, pp. 61–81. https://doi.org/10.1080/19439342.2016.1160947.

90. Ward Anseeuw and Giulia Maria Baldinelli, 2020. *Uneven Ground: Land Inequality at the Heart of Unequal Societies*. International Land Coalition and Oxfam. https://www.oxfam.org/en/research/uneven-ground-land-inequality-heart-unequal-societies.

91. Mauricio Betancourt, 2020. "The effect of Cuban agroecology in mitigating the metabolic rift: A quantitative approach to Latin American food production." *Global Environmental Change*, vol. 63, article 102075. https://doi.org/10.1016/j.gloenvcha.2020.102075.

92. Miguel A. Altieri, Fernando R. Funes-Monzote, and Paulo Petersen, 2012. "Agroecologically efficient agricultural systems for smallholder farmers: Contributions to food sovereignty." *Agronomy for Sustainable*

Development, vol. 32, pp. 1–13. https://doi.org/10.1007/s13593-011 -0065-6.

93. Marc-Olivier Martin-Guay et al., 2018. "The new Green Revolution: Sustainable intensification of agriculture by intercropping." *Science of the Total Environment*, vol. 615, pp. 767–72. https://doi.org/10.1016 /j.scitotenv.2017.10.024.

94. Brian Machovina, Kenneth J. Feeley, and William J. Ripple, 2015. "Biodiversity conservation: The key is reducing meat consumption." *Science of the Total Environment*, vol. 536, pp. 419–31. https://doi.org/10 .1016/j.scitotenv.2015.07.022.

95. Christian Dupraz et al., 2011. "To mix or not to mix: Evidences for the unexpected high productivity of new complex agrivoltaic and agroforestry systems." 5th World Congress of Conservation Agriculture incorporating 3rd Farming Systems Design Conference. https://vtech works.lib.vt.edu/bitstream/handle/10919/70121/5015_WCCA_FSD _2011.pdf?sequence=1&is#page=219.

96. Matthew Heron Wilson and Sarah Taylor Lovell, 2016. "Agroforestry— The next step in sustainable and resilient agriculture." *Sustainability*, vol. 8:6, p. 574. https://doi.org/10.3390/su8060574.

97. Nicholas Carter. *The Secret to Farming for the Climate*. A Well-Fed World.   https://awellfedworld.org/issues/climate-issues/farming-for -climate.

98. Michael Langemeier and Elizabeth Yeager, 2016. *International Benchmarks for Wheat Production*. Purdue University Center for Commercial Agriculture. https://ag.purdue.edu/commercialag/home/wp-content /uploads/2016/09/201609_Langemeier_InternationalBenchmarksfor WheatProduction2016.pdf.

99. Monique Kleinhuizen, 2017. "Kernz-huh? Could the perennial promise of Kernza benefit food, beer, and the world?" *The Growler*, February 27. https://images.app.goo.gl/T9kjbo8v4NdGFSXy5.

100. The Land Institute. https://landinstitute.org.

101. Yanming Zhang et al., 2011. "Potential of perennial crop on environmental sustainability of agriculture." *Procedia Environmental Sciences*, vol. 10, pt. B, pp. 1141–7. https://doi.org/10.1016/j.proenv .2011.09.182.

102. Frank Rasche et al., 2017. "A preview of perennial grain agriculture: Knowledge gain from biotic interactions in natural and agricultural ecosystems." *Ecosphere*, vol. 8:12. https://doi.org/10.1002 /ecs2.2048.

103. Brandon Schlautman, 2018. "Perennial grain legume domestication phase I: Criteria for candidate species selection." *Sustainability*, vol. 10:3, p. 730. https://doi.org/10.3390/su10030730.

104. Timothy E. Crews and Douglas J. Cattani, 2018. "Strategies, advances, and challenges in breeding perennial grain crops." *Sustainability*, vol. 10:7, article 2192. https://doi.org/10.3390/su10072192.

105. The Land Institute. *Perennial Crops: New Hardware for Agriculture.* https://landinstitute.org/our-work/perennial-crops.

106. Schlautman, Ibid.

107. David L. Van Tassel et al., 2017. "Accelerating *Silphium* domestication: An opportunity to develop new crop ideotypes and breeding strategies informed by multiple disciplines." *Crop Science*, vol. 57:3, pp. 1274–84. https://doi.org/10.2135/cropsci2016.10.0834.

108. David Van Tassel and Lee DeHaan, 2013. "Wild plants to the rescue." *American Scientist*, vol. 101:3, p. 218. https://www.americanscientist.org/article/wild-plants-to-the-rescue.

109. The Land Institute. *Kernza® Grain.* https://landinstitute.org/our-work/perennial-crops/kernza.

110. Frank Rasche et al., 2017. "A preview of perennial grain agriculture: knowledge gain from biotic interactions in natural and agricultural ecosystems." *Ecosphere*, vol. 8:12. https://doi.org/10.1002/ecs2.2048.

111. Alicia Ledo et al., 2020. "Changes in soil organic carbon under perennial crops." *Global Change Biology*, vol. 26:7, pp. 4158–68. https://doi.org/10.1111/gcb.15120.

112. Timothy E. Crews and Brian E. Rumsey, 2017. "What agriculture can learn from native ecosystems in building soil organic matter: A review." *Sustainability*, vol. 9:4, p. 578. https://doi.org/10.3390/su9040578.

113. S. W. Culman et al., 2010. "Long-term impacts of high-input annual cropping and unfertilized perennial grass production on soil properties and belowground food webs in Kansas, USA, agriculture." *Ecosystems & Environment*, vol. 137:1–2, pp. 13–24. https://doi.org/10.1016/j.agee.2009.11.008.

114. Jerry D. Glover et al., 2010. "Harvested perennial grasslands provide ecological benchmarks for agricultural sustainability." *Agriculture, Ecosystems & Environment*, vol. 137:1–2, pp. 3–12. https://doi.org/10.1016/j.agee.2009.11.001.

115. C. Emmerling, 2014. "Impact of land-use change towards perennial energy crops on earthworm population." *Applied Soil Ecology*, vol. 84, pp. 12–15. https://doi.org/10.1016/j.apsoil.2014.06.006.

116. Frank Rasche et al., 2017. "A preview of perennial grain agriculture: Knowledge gain from biotic interactions in natural and agricultural ecosystems." *Ecosphere*, vol. 8:12. https://doi.org/10.1002/ecs2.2048.

117. Ibid.

118. Virginia Nichols et al., 2015. "Weed dynamics and conservation agriculture principles: A review." *Field Crops Research*, vol. 183, pp. 56–68. https://doi.org/10.1016/j.fcr.2015.07.012.

119. Iris Lewandowski et al., 2003. "The development and current status of perennial rhizomatous grasses as energy crops in the US and Europe." *Biomass and Bioenergy*, vol. 25:4, pp. 335–61. https://doi.org/10.1016/S0961-9534(03)00030-8.

120. Marisa Lanker, Michael Bell, and Valentin D. Picasso, 2019. "Farmer perspectives and experiences introducing the novel perennial grain Kernza intermediate wheatgrass in the US Midwest." *Renewable Agriculture and Food Systems*, vol. 35:6, pp. 653–62. https://doi.org/10.1017/S1742170519000310.

121. Richard G. Smith, 2015. "A succession-energy framework for reducing non-target impacts of annual crop production." *Agricultural Systems*, vol. 133, pp. 14–21. https://doi.org/10.1016/j.agsy.2014.10.006.

122. Caterina Batello et al., 2014. *Perennial Crops for Food Security. Proceedings of the FAO Expert Workshop.* Food and Agriculture Organization of the United Nations (FAO). http://www.fao.org/3/i3495e/i3495e.pdf.

123. C. M. Cox, K. A. Garrett, and W. W. Bockus, 2005. "Meeting the challenge of disease management in perennial grain cropping systems." *Renewable Agriculture and Food Systems*, vol. 20:1, *Special Issue: Perennial Grain Crops: An Agricultural Revolution*, pp. 15–24. https://doi.org/10.1079/RAF200495.

124. Rasche et al., Ibid.

125. Batello et al., Ibid.

126. Van Tassel and DeHaan, Ibid.

127. J. D. Glover et al., 2010. "Increased food and ecosystem security via perennial grains." *Science*, vol. 328:5986, pp. 1638–9. https://doi.org/10.1126/science.1188761.

128. Huang Guangfu et al., 2018. "Performance, economics and potential impact of perennial rice PR23 relative to annual rice cultivars at multiple locations in Yunnan Province of China." *Sustainability* (Switzerland), vol. 10:4. https://doi.org/10.3390/su10041086.

129. The Land Institute. *Perennial Rice.* https://landinstitute.org/our-work/perennial-crops/perennial-rice.

130. Ibid.

131. Claudia Ciotir et al., 2019. "Building a botanical foundation for perennial agriculture: Global inventory of wild, perennial herbaceous *Fabaceae* species." *Plants, People, Planet*, vol. 1:4, pp. 375–86. https://doi.org/10.1002/ppp3.37.

132. The Land Institute. *Perennial Wheat.* https://landinstitute.org/our-work/perennial-crops/perennial-wheat.

133. Schlautman, Ibid.

134. Matthias Albrecht, 2020. "The effectiveness of flower strips and hedgerows on pest control, pollination services and crop yield: A quantitative synthesis." *Ecology Letters*, vol. 23:10, pp. 1488–98. https://doi.org/10.1111/ele.13576.

135. Dominik Ganser, Eva Knop, and Matthias Albrecht, 2019. "Sown wildflower strips as overwintering habitat for arthropods: Effective measure or ecological trap?" *Agriculture, Ecosystems & Environment*, vol. 275, pp. 123–31. https://doi.org/10.1016/j.agee.2019.02.010.

136. David Kleijn et al., 2019. "Ecological intensification: Bridging the gap between science and practice." *Trends in Ecology & Evolution*, vol. 34:2, pp. 154–66. https://doi.org/10.1016/j.tree.2018.11.002.

137. Richard F. Pywell et al., 2015. "Wildlife-friendly farming increases crop yield: Evidence for ecological intensification." *Proceedings of the Royal Society B: Biological Sciences*, vol. 282:1816. https://doi.org/10.1098/rspb.2015.1740.

138. K. L. Collins et al., 2002. "Influence of beetle banks on cereal aphid predation in winter wheat." *Agriculture, Ecosystems & Environment*, vol. 93:1–3, pp. 337–50. https://doi.org/10.1016/S0167-8809(01)00340-1.

139. Albrecht, Ibid.

140. B. A. Woodcock et al., 2010. "Impact of habitat type and landscape structure on biomass, species richness and functional diversity of ground beetles." *Agriculture, Ecosystems & Environment*, vol. 139:1–2, pp. 181–6. https://doi.org/10.1016/j.agee.2010.07.018.

141. Martin H. Schmidt et al., 2007. "Contrasting responses of arable spiders to the landscape matrix at different spatial scales." *Journal of Biogeography*, vol. 35:1, pp. 157–66. https://doi.org/10.1111/j.1365-2699.2007.01774.x.

142. Rui Catarino et al., 2019. "Bee pollination outperforms pesticides for oilseed crop production and profitability." *Proceedings of the Royal Society B: Biological Sciences*, vol. 286:1912. https://doi.org/10.1098/rspb.2019.1550.

143. Chunlong Shi et al., 2011. "Prospect of perennial wheat in agro-ecological system." *Procedia Environmental Sciences*, vol. 11, pt. C, pp. 1574–9. https://doi.org/10.1016/j.proenv.2011.12.237.

144. Crews and Rumsey, Ibid.

145. Lanker, Bell, and Picasso, Ibid.

CHAPTER 7

1. Channel 4, 2020. *Apocalypse Cow: How Meat Killed the Planet*, January 8. https://www.channel4.com/programmes/apocalypse-cow-how-meat-killed-the-planet.

2. John Litchfield, 1967. "Nutrition in life support systems for space exploration," in International Congress of Nutrition, *Problems of World Nutrition*, vol. 4, pp. 1068–74.

3. John Foster and John Litchfield, 1967. "Engineering requirements for culturing of *Hydrogenomonas* bacteria." SAE Technical Paper 70854. https://doi.org/10.4271/670854.

4. Jani Sillman et al., 2019. "Bacterial protein for food and feed generated via renewable energy and direct air capture of CO_2: Can it reduce land and water use?" *Global Food Security*, vol. 22, pp. 25–32. https://doi.org/10.1016/j.gfs.2019.09.007.

5. Tomas Linder, 2019. "Making the case for edible microorganisms as an integral part of a more sustainable and resilient food production system." *Food Security*, vol. 11, pp. 265–78. https://doi.org/10.1007/s12571-019-00912-3.

6. Tomas Linder, 2019. "Edible Microorganisms—An Overlooked Technology Option to Counteract Agricultural Expansion." *Frontiers in Sustainable Food Systems*, vol. 3, pp. 32. https://doi.org/10.3389/fsufs.2019.00032.

7. Solar Foods, 2019. *Food Out of Thin Air*. https://solarfoods.fi.

8. Bart Pander et al., 2020. "Hydrogen oxidising bacteria for production of single-cell protein and other food and feed ingredients." *Engineering Biology*, vol. 4:2, pp. 21–4. https://doi.org/10.1049/enb.2020.0005.

9. Tom Linder, 2019. "Making food without photosynthesis." Biology Fortified Inc., August 18. https://biofortified.org/2019/08/food-without-photosynthesis.

10. Bernardo B. N. Strassburg, 2020. "Global priority areas for ecosystem restoration." *Nature*, vol. 586, pp. 724–9. https://doi.org/10.1038/s41586-020-2784-9.

11. Akanksha Mishra et al., 2020. "Power-to-protein: Carbon fixation

with renewable electric power to feed the world." *Joule*, vol. 4:6, pp. 1142–7. https://doi.org/10.1016/j.joule.2020.04.008.

12. Xiaona Hu et al., 2020. "Microbial protein out of thin air: Fixation of nitrogen gas by an autotrophic hydrogen-oxidizing bacterial enrichment." *Environmental Science & Technology*, vol. 54:6, pp. 3609–17. https://doi.org/10.1021/acs.est.9b06755.

13. Sillman et al., Ibid.

14. Jacob Knutson, 2020. "Global need for copper is pitting clean energy against the wilderness." *Axios*, September 5. https://www.axios.com/copper-renewable-energy-mining-environment-8f2bf6b4-8557-4937-8020-7b6f39dc3fc2.html.

15. Laura Millan Lombrana, 2019. *Saving the Planet with Electric Cars Means Strangling this Desert.* Bloomberg Green, June 11. https://www.bloomberg.com/news/features/2019-06-11/saving-the-planet-with-electric-cars-means-strangling-this-desert.

16. P. J. Gerber et al., 2013. "Tackling climate change through livestock: A global assessment of emissions and mitigation opportunities." Food and Agriculture Organization of the United Nations, Rome. http://www.fao.org/3/i3437e/i3437e.pdf.

17. Antti Nyyssölä et al., 2021. "Production of endotoxin-free microbial biomass for food applications by gas fermentation of gram-positive H_2-oxidizing bacteria." *American Chemical Society (ACS) Food Science & Technology*, vol. 1:3, pp. 470–9. https://doi.org/10.1021/acsfoodscitech.0c00129.

18. Marja Nappa et al., 2020. "Solar-powered carbon fixation for food and feed production using microorganisms—A comparative techno-economic analysis." *American Chemical Society (ACS) Omega*, vol. 5:51, pp. 33242–52. https://doi.org/10.1021/acsomega.0c04926.

19. Claudia Hitaj, 2017. *Energy Consumption and Production in Agriculture.* U.S. Department of Agriculture (USDA), Economic Research Service, February 6. https://www.ers.usda.gov/amber-waves_2017_januaryfebruary_energy-consumption-and-production-in-agriculture.pdf.

20. Deepak Yadav and Rangan Banerjee, 2020. "Net energy and carbon footprint analysis of solar hydrogen production from the high-temperature electrolysis process." *Applied Energy*, vol. 262, article 114503. https://doi.org/10.1016/j.apenergy.2020.114503.

21. Farid Safari and Ibrahim Dincer, 2020. "A review and comparative evaluation of thermochemical water splitting cycles for hydrogen production." *Energy Conversion and Management*, vol. 205, article 112182. https://doi.org/10.1016/j.enconman.2019.112182.

22. Seyed Ehsan Hosseini and Mazlan Abdul Wahid, 2019. "Hydrogen from solar energy, a clean energy carrier from a sustainable source of energy." *International Journal of Energy Research*, vol. 44:6, pp. 4110–31.

23. S. Shahab Naghavi, Jiangang He, and C. Wolverton, 2020. "CeTi2O6— A promising oxide for solar thermochemical hydrogen production." *American Chemical Society (ACS) Applied Materials & Interfaces*, vol. 12:19, pp. 21521–7. https://doi.org/10.1021/acsami.0c01083.

24. International Atomic Energy Agency (IAEA), 2013. *Hydrogen Production Using Nuclear Energy*. IAEA Nuclear Energy Series, no. NP-T-4.2. IAEA Vienna. https://www-pub.iaea.org/MTCD/Publica tions/PDF/Pub1577_web.pdf.

25. Sillman et al., Ibid.

26. Tara Garnett et al., 2017. "Grazed and Confused? Ruminating on cattle, grazing systems, methane, nitrous oxide, the soil carbon sequestration question—and what it all means for greenhouse gas emissions." Food Climate Research Network (FCRN), University of Oxford. https:// www.oxfordmartin.ox.ac.uk/downloads/reports/fcrn_gnc_report.pdf.

27. UK Government, 2014. *Nutrient Intakes*. Office for National Statistics. https://assets.publishing.service.gov.uk/government/uploads/system /uploads/attachment_data/file/384775/familyfood-method-rni -11dec14.pdf.

28. FAOSTAT, 2021. *New Food Balances*. Food and Agriculture Organi- zation of the United Nations (FAO), April 14. http://www.fao.org /faostat/en/#data/FBS.

29. Sumedha Minocha et al., 2019. "Supply and demand of high-quality protein foods in India: Trends and opportunities." *Global Food Secu- rity*, vol. 23, pp. 139–48. https://doi.org/10.1016/j.gfs.2019.05.004.

30. Safiu Adewale Suberu et al., 2020. *Prevalence and Associated Factors for Protein Energy Malnutrition Among Children Below 5 Years Admitted at Jinja Regional Referral Hospital, Uganda*. Research Square, version 1. https://doi.org/10.21203/rs.3.rs-68882/v1.

31. Sebastian Hermann, Asami Miketa, and Nicolas Fichaux, 2014. "Esti- mating the renewable energy potential in Africa: A GIS-based approach." International Renewable Energy Agency (IRENA) Secre- tariat working paper, Abu Dhabi. https://www.irena.org/-/media/Files /IRENA/Agency/Publication/2014/IRENA_Africa_Resource_Poten tial_Aug2014.pdf.

32. Michael J. Puma et al., 2015. "Assessing the evolving fragility of the global food system." *Environmental Research Letters*, vol. 10:2. https:// doi.org/10.1088/1748-9326/10/2/024007.

33. Dirk Helbing, 2013. "Globally networked risks and how to respond." *Nature*, vol. 497, pp. 51–9. https://doi.org/10.1038/nature12047.

34. David C. Denkenbergera and Joshua M. Pearce, 2015. "Feeding everyone: Solving the food crisis in event of global catastrophes that kill crops or obscure the sun." *Futures*, vol. 72, pp. 57–68. https://doi.org/10.1016/j.futures.2014.11.008.

35. David C. Denkenberger and Joshua M. Pearce, 2016. "Cost-effectiveness of interventions for alternate food to address agricultural catastrophes globally." *International Journal of Disaster Risk Science*, vol. 7, pp. 205–15. https://doi.org/10.1007/s13753-016-0097-2.

36. Pasi Vainikka, 2019. *Food Out of Thin Air*. Solar Foods Presentation. https://cdn2.hubspot.net/hubfs/4422035/Solar-Foods-presentation-03-2019.pdf.

37. Solar Foods. *Questions and Answers*. https://solarfoods.fi/pressroom/material-bank/qa.

38. A. Parodi et al., 2018. "The potential of future foods for sustainable and healthy diets." *Nature Sustainability*, vol. 1, pp. 782–9. https://doi.org/10.1038/s41893-018-0189-7.

39. T. G. Volova and V. A. Barashkov, 2010. "Characteristics of proteins synthesized by hydrogen-oxidizing microorganisms." *Applied Biochemistry and Microbiology*, vol. 46, pp. 574–9. https://doi.org/10.1134/S0003683810060037.

40. Anneli Ritala et al., 2017. "Single cell protein—State-of-the-art, industrial landscape and patents 2001–2016." *Frontiers in Microbiology*, vol. 9, p. 2009. https://doi.org/10.3389/fmicb.2017.02009.

41. A. T. Nasseri et al., 2011. "Single cell protein: Production and process." *American Journal of Food Technology*, vol. 6:2, pp. 103–16. https://doi.org/10.3923/ajft.2011.103.116.

42. Isabella Pali-Schöll et al., 2019. "Allergenic and novel food proteins: State of the art and challenges in the allergenicity assessment." *Trends in Food Science & Technology*, vol. 84, pp. 45–8. https://doi.org/10.1016/j.tifs.2018.03.007.

43. Kasia J. Lipska, Joseph S. Ross, and Holly K. Van Houten, 2014. "Use and out-of-pocket costs of insulin for type 2 diabetes mellitus from 2000 through 2010." *Journal of the American Medical Association (JAMA)*, vol. 311:22, pp. 2331–3. https://doi.org/10.1001/jama.2014.6316.

44. Jon Entine and XiaoZhi Lim, 2018. "Cheese: The GMO food die-hard GMO opponents love, but don't want to label." *Genetic Literacy Project*, November 2. https://geneticliteracyproject.org/2018/11/02

/cheese-gmo-food-die-hard-gmo-opponents-love-and-oppose
-a-label-for.

45. *Soil Association Standards Food and Drink. Version 18.6: updated on
12th February 2021.* Soil Association. https://www.soilassociation.org
/media/15883/food-and-drink-standards.pdf.

46. Jeanne Yacoubou, 2012. *Microbial Rennets and Fermentation Produced Chymosin (FPC): How Vegetarian Are They?* The Vegetarian
Resource Group (VRG), August 21. https://www.vrg.org/blog/2012
/08/21/microbial-rennets-and-fermentation-produced-chymosin-fpc
-how-vegetarian-are-they.

47. Roman Pawlak et al., 2013. "How prevalent is vitamin B12 deficiency
among vegetarians?" *Nutrition Reviews*, vol. 71:2, pp. 110–17. https://
doi.org/10.1111/nure.12001.

48. Global Agriculture. *Health: Food or Cause of Illness?* https://www
.globalagriculture.org/report-topics/health.html.

49. Raychel E. Santo et al., 2020. "Considering plant-based meat substitutes and cell-based meats: A public health and food systems
perspective." *Frontiers in Sustainable Food Systems*, vol. 4, p. 134.
https://doi.org/10.3389/fsufs.2020.00134.

50. Ujué Fresán et al., 2019. "Water footprint of meat analogs: Selected
indicators according to life cycle assessment." *Water*, vol. 11:4, p. 728.
https://doi.org/10.3390/w11040728.

51. Xueqin Zhu and Ekko C. van Ierland, 2004. "Protein chains and environmental pressures: A comparison of pork and novel protein foods."
Environmental Sciences, vol. 1:3, pp. 254–76. https://doi.org/10.1080
/15693430412331291652.

52. Benjamin M. Bohrer, 2019. "An investigation of the formulation and
nutritional composition of modern meat analogue products." *Food Science and Human Wellness*, vol. 8:4, pp. 320–9. https://doi.org/10.1016
/j.fshw.2019.11.006.

53. Felicity Curtain and Sara Grafenauer, 2019. "Plant-based meat substitutes in the flexitarian age: An audit of products on supermarket
shelves." *Nutrients*, vol. 11:11, p. 2603. https://doi.org/10.3390
/nu11112603.

54. Lieven Thorrez and Herman Vandenburgh, 2019. "Challenges in the
quest for 'clean meat.'" *Nature Biotechnology*, vol. 37, pp. 215–16.
https://doi.org/10.1038/s41587-019-0043-0.

55. Sghaier Chriki and Jean-François Hocquette, 2020. "The myth of cultured meat: A review." *Frontiers in Nutrition*, vol. 7, p. 7. https://doi
.org/10.3389/fnut.2020.00007.

56. Neil Stephens, 2018. "Bringing cultured meat to market: Technical, socio-political, and regulatory challenges in cellular agriculture." *Trends in Food Science & Technology*, vol. 78, pp. 155–66. https://doi .org/10.1016/j.tifs.2018.04.010.

57. Ilse Fraeye et al., 2020. "Sensorial and nutritional aspects of cultured meat in comparison to traditional meat: Much to be inferred." *Frontiers in Nutrition*, vol. 7, p. 35. https://doi.org/10.3389/fnut.2020.00035

58. Joe Fassler, 2021. "Lab-grown meat is supposed to be inevitable: The science tells a different story." *The Counter*, September 22. https://the counter.org/lab-grown-cultivated-meat-cost-at-scale.

59. Liz Specht, 2020. "An analysis of culture medium costs and production volumes for cultivated meat." The Good Food Institute. https://gfi.org /wp-content/uploads/2021/01/clean-meat-production-vol.-and -medium-cost.pdf.

60. Carsten Gerhardt et al., 2019. *How Will Cultured Meat and Meat Alternatives Disrupt the Agricultural and Food Industry?* AT Kearney Business Consulting. https://www.kearney.com/documents/20152 /2795757/How+Will+Cultured+Meat+and+Meat+Alternatives+Dis rupt+the+Agricultural+and+Food+Industry.pdf/06ec385b-63a1-71d2 -c081-51c07ab88ad1?t=1559860712714.

61. Impossible Foods (IF™). *What is Soy Leghemoglobin, or Heme?* https://faq.impossiblefoods.com/hc/en-us/articles/360019100553 -What-is-soy-leghemoglobin-or-heme-.

62. Zoe Williams, 2021. "3D-printed steak, anyone? I taste test this 'gamechang ing' meat mimic." *The Guardian*, November 16. https://www.theguardian .com/food/2021/nov/16/3d-printed-steak-taste-test-meat-mimic.

63. Hoxton Farms. https://hoxtonfarms.com.

64. Bee Wilson, 2019. *The Way We Eat Now: Strategies for Eating in a World of Change.* 4th Estate, London.

65. Faunalytics, 2018. *Global Chicken Slaughter Statistics and Charts.* https://faunalytics.org/global-chicken-slaughter-statistics-and -charts.

66. Gerhardt et al., Ibid.

67. Don Close, 2014. *Ground Beef Nation: The Effect of Changing Con sumer Tastes and Preferences on the U.S. Cattle Industry.* Rabobank AgFocus. https://www.beefcentral.com/wp-content/uploads/2014/06 /Ground-Beef-Nation.pdf.

68. Catherine Tubb and Tony Seba, 2019. *Rethinking Food and Agricul ture 2020–2030: The Second Domestication of Plants and Animals, the Disruption of the Cow, and the Collapse of Industrial Livestock Farming.*

Rethink Food & Agriculture, A RethinkX Sector Disruption Report. https://www.rethinkx.com/food-and-agriculture#food-and-agriculture -download.

69. Tubb and Seba, Ibid.

70. Kat Smith. *Dairy-Identical Vegan Cheese Is Coming to Save the Cows.* Livekindly. https://www.livekindly.co/dairy-identical-vegan-cheese-is -coming-to-save-cows.

71. Elaine Watson, 2020. "Perfect Day secures no objections letter from FDA for non-animal whey protein." FoodNavigator-USA, April 14. https://www.foodnavigator-usa.com/Article/2020/04/14/Perfect-Day -secures-no-objections-letter-from-FDA-for-non-animal-whey-protein.

72. Simon Sharpe and Timothy Lenton, 2021. "Upward-scaling tipping cascades to meet climate goals: Plausible grounds for hope." *Climate Policy*, vol. 21:4, pp. 421–33. https://doi.org/10.1080/14693062.2020 .1870097.

73. Timothy Lenton et al., 2021. "Operationalising positive tipping points towards global sustainability." Working paper, in press. https://www .exeter.ac.uk/media/universityofexeter/globalsystemsinstitute/docu ments/Lenton_et_al_-_Operationalising_positive_tipping_points.pdf.

74. Akanksha Mishra et al., 2020. "Power-to-protein: Carbon fixation with renewable electric power to feed the world." *Joule*, vol. 4:6, pp. 1142–7. https://doi.org/10.1016/j.joule.2020.04.008.

75. Vesa Ruuskanen et al., 2021. "Neo-Carbon Food concept: A pilot-scale hybrid biological-inorganic system with direct air capture of carbon dioxide." *Journal of Cleaner Production*, vol. 278, article 123423. https://doi.org/10.1016/j.jclepro.2020.123423.

76. Tomas Linder, 2019. "Making the case for edible microorganisms as an integral part of a more sustainable and resilient food production sys- tem." *Food Security*, vol. 11, pp. 265–78. https://doi.org/10.1007/s12571 -019-00912-3.

77. Luis P. da Silva and Vanessa A. Mata, 2019. "Stop harvesting olives at night—it kills millions of songbirds." *Nature* (Correspondence), vol. 569, May 7, p. 192. https://doi.org/10.1038/d41586-019-01456-4.

78. Erik Meijaard et al., 2020. *Coconut Oil, Conservation and the Consci- entious Consumer.* Social Science Research Network (SSRN) Current Biology. http://dx.doi.org/10.2139/ssrn.3575129.

79. Benjamin S. Halpern, 2021. "The long and narrow path for novel cell-based seafood to reduce fishing pressure for marine ecosystem recov- ery." *Fish and Fisheries*, vol. 22:3, pp. 652–64. https://doi.org/10.1111 /faf.12541.

80. Inez Blackburn. *Speed to Market—Capitalizing on Demand.* U Connect 08. http://www.markettechniques.com/assets/pdf/Speed2Market.pdf.

81. Sonia Oreffice, 2015. "The contraceptive pill was a revolution for women and men." *The Conversation*, February 6. https://theconversa tion.com/the-contraceptive-pill-was-a-revolution-for-women -and-men-37193.

82. Damon Centola et al., 2018. "Experimental evidence for tipping points in social convention." *Science*, vol. 360:6393, pp. 1116–19. https:// www.science.org/doi/abs/10.1126/science.aas8827.

83. Wei Lan and Chunlei Yang, 2019. "Ruminal methane production: Associated microorganisms and the potential of applying hydrogen-utilizing bacteria for mitigation." *Science of the Total Environment*, vol. 654, pp. 1270–83. https://doi.org/10.1016/j.scitotenv.2018.11.180.

84. George Monbiot, 2021. "Capitalism is killing the planet—it's time to stop buying into our own destruction." *The Guardian*, October 30. https:// www.theguardian.com/environment/2021/oct/30/capitalism-is-killing -the-planet-its-time-to-stop-buying-into-our-own-destruction.

85. Michael Pollan, 2006. "Six rules for eating wisely." *Time Magazine*, June 4. https://michaelpollan.com/articles-archive/six-rules-for-eating -wisely.

86. Anna Jones, Saturday Kitchen. "Green peppercorn and lemongrass coconut broth." *BBC Food.* https://www.bbc.co.uk/food/recipes/pep percorn_coconut_broth_84709.

87. Fumio Watanabe et al., 2014. "Vitamin B12-containing plant food sources for vegetarians." *Nutrients*, vol. 6:5, pp. 1861–73. https://doi .org/10.3390/nu6051861.

88. Lori Marino, 2017. "Thinking chickens: A review of cognition, emotion, and behavior in the domestic chicken." *Animal Cognition*, vol. 20, pp. 127–47. https://doi.org/10.1007/s10071-016-1064-4.

89. FAOSTAT, 2021. *Livestock Primary.* Food and Agriculture Organization of the United Nations (FAO), February 19. http://www.fao.org /faostat/en/#data/QL.

90. Niccolò Machiavelli, 1532. *The Prince.* Early Modern Texts. https:// www.earlymoderntexts.com/assets/pdfs/machiavelli1532.pdf.

91. State of Arkansas House Bill, 2019. "An act to require truth in labeling of agricultural 10 products that are edible by humans; and for other purposes." 92nd General Assembly, House Bill 1407. https://www .arkleg.state.ar.us/Acts/Document?type=pdf&act=501&ddBiennium Session=2019%2F2019R.

92. Louisiana Senate Bill, 2019. "Provides for truth in labeling require-

ments of agricultural products." LA SB152. https://legiscan.com/LA/text/SB152/2019.

93. State of Montana House Bill, 2019. *Real Meat Act.* MT HB327. https://legiscan.com/MT/text/HB327/2019.

94. European Parliament, 2018/19. *Regulation (EU) No 1308/2013. Annex VII—part I a (new).* https://www.europarl.europa.eu/doceo/document/A-8-2019-0198_EN.pdf#page=169.

95. Press Release No 63/17—Court of Justice of the European Union, Luxembourg, June 14, 2017. "Purely plant-based products cannot, in principle, be marketed with designations such as 'milk,' 'cream,' 'butter,' 'cheese' or 'yoghurt,' which are reserved by EU law for animal products." Judgment in Case C-422/16, Verband Sozialer Wettbewerb eV v TofuTown.com GmbH. https://curia.europa.eu/jcms/upload/docs/application/pdf/2017-06/cp170063en.pdf.

96. Caroline Sommerfelt, 2020. "EU parliament rejects 'veggie burger ban' but supports 'dairy ban' against vegan producers." Natural Products Global, October 26. https://www.naturalproductsglobal.com/europe/eu-parliament-rejects-veggie-burger-ban-but-supports-dairy-ban-against-vegan-producers.

97. Nicolas Treich, 2021. "Cultured meat: Promises and challenges." *Environmental and Resource Economics*, vol. 79, pp. 33–61. https://doi.org/10.1007/s10640-021-00551-3.

98. George Lakoff, 2004. *Don't Think of an Elephant: Know Your Values and Frame the Debate.* Chelsea Green Publishing, Vermont. https://georgelakoff.com/books/dont_think_of_an_elephant_know_your_values_and_frame_the_debatethe_essential_guide_for_progressives-119190455949080.

99. Joanna Blythman, 2020. "They are expensive and tasteless, don't fall for the fake-meat veggie burger fad." *The Herald*, September 5. https://www.heraldscotland.com/news/18687082.opinion-joanna-blythman-expensive-tasteless-dont-fall-fake-meat-veggie-burger-fad.

100. International Panel of Experts on Sustainable Food Systems. *A Long Food Movement: Transforming Food Systems by 2045.* http://www.ipes-food.org/pages/LongFoodMovement.

101. Food and Agriculture Organization of the United Nations (FAO), the International Fund for Agricultural Development (IFAD), the United Nations Children's Fund (UNICEF), the World Food Programme (WFP) and the World Health Organization (WHO), 2020. *The State of Food Security and Nutrition in the World 2020. Transforming Food*

Systems for Affordable Healthy Diets. FAO, Rome. https://doi.org/10.4060/ca9692en.

102. Euromonitor, Lifestyles Survey, 2020-2021. Proprietary data. https://www.euromonitor.com/voice-of-the-consumer-lifestyles-survey-2021-key-insights/report.

103. Alisa Jordan, 2018. "Apartments don't come with kitchens in Germany." *Alisa Jordan Writes,* August 6. https://alisajordanwrites.com/2018/08/06/apartments-dont-come-with-kitchens-in-germany.

104. United States Census Bureau, 2018. *Selected Housing Characteristics.* American Community Survey, Table ID: DP04. https://data.census.gov/cedsci/table?d=ACS%205-Year%20Estimates%20Data%20Profiles&tid=ACSDP5Y2018.DP04&vintage=2018.

105. Corey Mintz, 2019. "Do homes without kitchens mark the end of human civilization?" *The Ontario Educational Communications Authority (TVO),* August 13. https://www.tvo.org/article/do-homes-without-kitchens-mark-the-end-of-human-civilization.

106. *In Our Time,* September 21, 2017. *Kant's Categorical Imperative.* BBC Radio 4. https://www.bbc.co.uk/programmes/b0952zl3.

107. Robert Paarlberg, 2020. "Foodies and factory farmers have formed an unholy alliance." *Wired,* November 8. https://www.wired.com/story/foodies-and-factory-farmers-have-formed-an-unholy-alliance.

108. Jan Dutkiewicz and Gabriel N. Rosenberg, 2020. "Burgers won't save the planet—but fast food might." *Wired,* July 8. https://www.wired.com/story/opinion-burgers-wont-save-the-planet-but-fast-food-might.

109. Treich, Ibid.

110. Philip Howard et al., 2021. "'Protein' industry convergence and its implications for resilient and equitable food systems." *Frontiers in Sustainable Food Systems,* vol. 5, article 684181. https://doi.org/10.3389/fsufs.2021.684181.

111. Garrett M. Broad, 2019. "Plant-based and cell-based animal product alternatives: An assessment and agenda for food tech justice." *Geoforum,* vol. 107, pp. 223–6. https://doi.org/10.1016/j.geoforum.2019.06.014.

112. Malte Rödl, 2019. "Why the meat industry could win big from the switch to veggie lifestyles." *The Conversation,* March 5. https://theconversation.com/why-the-meat-industry-could-win-big-from-the-switch-to-veggie-lifestyles-112714.

113. Audrey Enjoli, 2021. "Meat giant Tyson Foods introduces its first plant-based burger." *Live Kindly,* May 5. https://www.livekindly.co/tyson-foods-first-plant-based-burger.

114. Petra Moser, 2016. "Patents and innovation in economic history." *Annual Review of Economics*, vol. 8:1, pp. 241–58. https://doi.org/10.1146/annurev-economics-080315-015136.

115. John H. Barton and Peter Berger, 2001. "Patenting agriculture." *Issues in Science and Technology* (Arizona State University), vol. XVII:4. https://issues.org/barton.

116. Science|Business Viewpoint Debate, Richard L. Hudson, 2016. "The great IP debate: Do patents do more harm than good?" *Science|Business*, July 28. https://sciencebusiness.net/news/79887/The-Great-IP-Debate%3A-Do-patents-do-more-harm-than-good%3F.

117. Fiona Mischel, 2021. "Who owns CRISPR in 2021? It's even more complicated than you think." *SynBioBeta*, April 27. https://synbiobeta.com/read/who-owns-crispr-in-2021-its-even-more-complicated-than-you-think.

118. Broad, Ibid.

119. LUT University News. "Food from air with a new process—power-to-x solution and pilot equipment by LUT and VTT." LUT University, Finland and VTT Technical Research Centre of Finland. https://www.vttresearch.com/en/news-and-ideas/food-air-new-process-power-x-solution-and-pilot-equipment-lut-and-vtt.

120. Sally Ho, 2021. "Solar Foods bags €10m from Finnish climate fund to commercialise air protein by 2023." *Green Queen*, April 13. https://www.greenqueen.com.hk/solar-foods-bags-e10m-from-finnish-climate-fund-to-commercialise-air-protein-by-2023.

121. Jan Dutkiewicz, 2019. "Socialize lab meat." *Jacobin*, August 11. https://jacobinmag.com/2019/08/lab-meat-socialism-green-new-deal.

122. Livia Gershon, 2020. "Why does meatpacking have such bad working conditions?" *JSTOR Daily*, May 8. https://daily.jstor.org/why-does-meatpacking-have-such-bad-working-conditions.

123. Business & Human Rights Resource Centre News, 2020. "Europe: Poor working & housing conditions at meat packing plants responsible for COVID-19 outbreak among workforce, report alleges." Business & Human Rights Resource Centre, July 8. https://www.business-humanrights.org/en/latest-news/europe-poor-working-housing-conditions-at-meat-packing-plants-responsible-for-covid-19-outbreak-among-workforce-report-alleges.

124. Tubb and Seba, Ibid.

125. George Monbiot, 2014. *Feral: Rewilding the Land, Sea and Human Life*. Penguin, London. https://www.penguin.co.uk/books/180586/feral/9780141975580.html.

126. Felisa Smith et al., 2016. "Megafauna in the Earth system." *Ecography*, vol. 39, pp. 99–108. https://dx.doi.org/10.1111/ecog.02156.

127. Jacquelyn Gill, 2014. "Ecological impacts of the late Quaternary mega-herbivore extinctions." *New Phytologist*, vol. 201, pp. 1163–9. https://doi.org/10.1111/nph.12576.

128. Christopher Doughty et al., 2016. "Global nutrient transport in a world of giants." *Proceedings of the National Academy of Sciences*, vol. 113:4, pp. 868–73. https://doi.org/10.1073/pnas.1502549112.

129. Elisabeth Bakker and Jens-Christian Svenning, 2018. "Trophic rewilding: Impact on ecosystems under global change." *Philosophical Transactions of the Royal Society B: Biological Sciences*, vol. 373, p. 1761. https://doi.org/10.1098/rstb.2017.0432.

130. Rebecca Wrigley, 2021. *Rewilding and the Rural Economy: How Nature-Based Economies Can Help Boost and Sustain Local Communities*. Rewilding Britain. https://s3.eu-west-2.amazonaws.com/assets .rewildingbritain.org.uk/documents/nature-based-economies -rewilding-britain.pdf.

131. Susan Cook-Patton et al., 2020. "Mapping carbon accumulation potential from global natural forest regrowth." *Nature*, vol. 585, pp. 545–50. https://doi.org/10.1038/s41586-020-2686-x.

132. Eric Dinerstein et al., 2020. "A 'global safety net' to reverse biodiversity loss and stabilize Earth's climate." *Science Advances*, vol. 6, p. 36. https://doi.org/10.1126/sciadv.abb2824.

133. Simon Lewis et al., 2019. "Restoring natural forests is the best way to remove atmospheric carbon." *Nature*, vol. 568, pp. 25–8. doi: https://doi.org/10.1038/d41586-019-01026-8.

134. Lucas Nave et al., 2018. "Reforestation can sequester two petagrams of carbon in US topsoils in a century." *Proceedings of the National Academy of Sciences*, vol. 115:11, pp. 2776–81. https://doi.org/10.1073/pnas .1719685115.

135. *IPBES-IPCC co-sponsored Workshop on Biodiversity and Climate Change—Scientific Outcome*. https://www.ipbes.net/sites/default/files /2021-06/2021_IPCC-IPBES_scientific_outcome_20210612.pdf.

CHAPTER 8

1. Amanda Mummert et al., 2011. "Stature and robusticity during the agricultural transition: Evidence from the bioarchaeological record." *Economics & Human Biology*, vol. 9:3, pp. 284–301. https://doi.org /10.1016/j.ehb.2011.03.004.

2. Stephanie Marciniak et al., 2021. "An integrative skeletal and paleogenomic analysis of prehistoric stature variation suggests relatively reduced health for early European farmers." *bioRxiv*—the preprint server for *Biology*, March 31. https://doi.org/10.1101/2021.03.31.437881.

3. Iain Mathieson et al., 2015. "Genome-wide patterns of selection in 230 ancient Eurasians." *Nature*, vol. 528, pp. 499–503. https://doi.org/10.1038/nature16152.

4. Mark Dyble et al., 2019. "Engagement in agricultural work is associated with reduced leisure time among Agta hunter-gatherers." *Nature Human Behaviour*, vol. 3, pp. 792–6. https://doi.org/10.1038/s41562-019-0614-6.

5. Peter Gray, 2009. "Play as a foundation for hunter-gatherer social existence." *American Journal of Play*, vol. 1:4, pp. 476–522. https://www.nifplay.org/play-as-a-foundation-for-huntergatherer-social-existence.

6. James Suzman, 2019. *Affluence Without Abundance—What We Can Learn from the World's Most Successful Civilisation*. Bloomsbury Publishing, London. https://www.bloomsbury.com/uk/affluence-without-abundance-9781526609311.

7. Yadvinder Malhi, 2014. "The metabolism of a human-dominated planet," in Ian Goldin (ed.), *Is the Planet Full?* Oxford University Press.

8. Ron Pinhasi, Vered Eshed, and Noreen von Cramon-Taubadel, 2015. "Incongruity between affinity patterns based on mandibular and lower dental dimensions following the transition to agriculture in the Near East, Anatolia and Europe." *PLOS One*, vol. 10:2. https://doi.org/10.1371/journal.pone.0117301.

9. See e.g. Theocritus, 3rd century BC. *Idyll XXIX, The Aeolic Love Poems, The First Love Poem*. https://www.theoi.com/Text/TheocritusIdylls5.html.

10. Genesis 4:2.

11. Numbers 27:17.

12. Psalm 95:7.

13. Nahum 3:1.

14. Virgil, 37 BC. *The Eclogues, Eclogue IV, Pollio*. http://classics.mit.edu/Virgil/eclogue.4.iv.html.

15. John 10:11.

16. John 1:29.

17. John 21:17.

18. Edmund Spenser, 1579. *The Shepheardes Calender I: Januarye*. http://spenserians.cath.vt.edu/TextRecord.php?action=GET&textsid=15.

19. Keith Thomas, 1983. *Man and the Natural World: Changing Attitudes in England 1500–1800*. Pantheon Books, New York, p. 46.

20. John Milton, 1637. *Lycidas*. https://www.poetryfoundation.org/poems /44733/lycidas.

21. Percy Bysshe Shelley, 1821. *Adonais: An Elegy on the Death of John Keats*. https://www.poetryfoundation.org/poems/45112/adonais-an-elegy -on-the-death-of-john-keats.

22. George Crabbe, 1783. *The Village: Book I*. https://www.poetryfounda tion.org/poems/44041/the-village-book-i.

23. Jeremy Lent, 2017. *The Patterning Instinct: A Cultural History of Humanity's Search for Meaning*. Prometheus, New York.

24. George Monbiot, 2015. "It's time to wean ourselves off the fairytale version of farming." *The Guardian*, May 29. https://www.theguardian .com/environment/georgemonbiot/2015/may/29/its-time-to-wean -ourselves-off-the-fairytale-version-of-farming.

25. Zach Hrynowski, 2019. "What percentage of Americans are vegetar- ian?" Gallup, September 27. https://news.gallup.com/poll/267074 /percentage-americans-vegetarian.aspx.

26. Teresa Steckler, 2018. "Survey: Nearly one-third of Americans support ban on slaughterhouses." Illinois Extension, University of Illinois, Feb- ruary 11. https://extension.illinois.edu/blogs/cattle-blog/2018-02-11 -survey-nearly-one-third-americans-support-ban-slaughterhouses.

27. W. H. Auden, June 3, 1965. *Et in Arcadia Ego*. https://www.nybooks .com/articles/1965/06/03/et-in-arcadia-ego.

28. The Farm Business Survey in Wales, 2019. *Wales Farm Income Book- let 2018/19 Results*. Institute of Biological, Environmental and Rural Sciences, Aberystwyth University. https://www.aber.ac.uk/en/media /departmental/ibers/farmbusinesssurvey/FBS_Booklet_2019 _Web.pdf.

29. Charlie Reeve, 2021. "Farm business incomes increasingly reliant on direct payments." Agriculture and Horticulture Development Board (AHDB), January 7. https://ahdb.org.uk/news/farm-business-incomes -increasingly-reliant-on-direct-payments.

30. Department for Environment, Food & Rural Affairs and Department (DEFRA), 2020. *Farm Business Income by type of farm, England, 2019/20*, December 16. https://assets.publishing.service.gov.uk/govern ment/uploads/system/uploads/attachment_data/file/944352/fbs -businessincome-statsnotice-16dec20.pdf.

31. Statistics for Wales & Welsh Government, 2019. "Farm incomes in Wales, April 2018 to March 2019." Statistical First Release, SFR 123/

2019, December 18. https://gov.wales/sites/default/files/statistics-and-research/2019-12/farm-incomes-april-2018-march-2019-209.pdf.

32. Eric Hobsbawm, 2014. *Fractured Times: Culture and Society in the Twentieth Century*. Little, Brown, London. https://www.littlebrown.co.uk/titles/eric-hobsbawm-2/fractured-times/9780349139098.

33. Christopher Ketcham, 2021. "Capitol attackers have long threatened violence in rural American west." *The Guardian*, January 9. https://www.theguardian.com/environment/2021/jan/09/us-capitol-attackers-violence-rural-west.

34. Aruna Viswanatha and Brett Wolf, 2012. "HSBC to pay $1.9 billion U.S. fine in money-laundering case." Reuters, December 11. https://www.reuters.com/article/us-hsbc-probe/hsbc-to-pay-1-9-billion-u-s-fine-in-money-laundering-case-idUSBRE8BA05M20121211.

35. Food and Agriculture Organization of the United Nations, United Nations Development Programme and United Nations Environment Programme, 2021. "A multi-billion-dollar opportunity—Repurposing agricultural support to transform food systems." http://www.fao.org/3/cb6562en/cb6562en.pdf.

36. Christophe Bellmann, 2019. *Subsidies and Sustainable Agriculture: Mapping the Policy Landscape*. Hoffmann Centre for Sustainable Resource Economy, Chatham House. https://www.chathamhouse.org/sites/default/files/Subsidies%20and%20Sustainable%20Ag%20-%20Mapping%20the%20Policy%20Landscape%20FINAL-compressed.pdf.

37. UN Climate Change News, 2021. "UN climate chief urges countries to deliver on USD 100 billion pledge," June 7. https://unfccc.int/news/un-climate-chief-urges-countries-to-deliver-on-usd-100-billion-pledge.

38. Timothy D. Searchinger et al., 2020. *Revising Public Agricultural Support to Mitigate Climate Change*. World Bank, Development Knowledge and Learning, Washington DC. https://openknowledge.worldbank.org/bitstream/handle/10986/33677/K880502.pdf?sequence=4&isAllowed=y.

39. Pavel Ciaian et al., 2021. "The capitalization of agricultural subsidies into land prices." *Annual Review of Resource Economics*, vol. 13, pp. 17–38. https://doi.org/10.1146/annurev-resource-102020-100625.

40. Murray W. Scown, Mark V. Brady, and Kimberly A. Nicholas, 2020. "Billions in misspent EU agricultural subsidies could support the sustainable development goals." *One Earth*, vol. 3:2, pp. 237–50. https://doi.org/10.1016/j.oneear.2020.07.011.

41. Selam Gebrekidan, Matt Apuzzo, and Benjamin Novak, 2019. "The money farmers: How oligarchs and populists milk the E.U. for mil-

lions." *The New York Times*, November 3. https://www.nytimes.com /2019/11/03/world/europe/eu-farm-subsidy-hungary.html.

42. European Court of Auditors, 2016. "Is the Commission's system for performance measurement in relation to farmers' incomes well designed and based on sound data?" Publications Office of the European Union, Luxembourg. https://doi.org/10.2865/72393.

43. Searchinger et al., Ibid.

44. Congressional Research Service (CRS), 2013. *The Pigford Cases: USDA Settlement of Discrimination Suits by Black Farmers*. Every-CRSReport, August 25, 2005–May 29, 2013, RS20430. https://www .everycrsreport.com/reports/RS20430.html.

45. Raj Patel, 2019. "A green new deal for agriculture," April 9. https:// rajpatel.org/2019/04/09/a-green-new-deal-for-agriculture.

46. Food Empowerment Project, n.d. *Slavery in the U.S.* https://foodis power.org/human-labor-slavery/slavery-in-the-us.

47. Shruti Bhogal and Shreya Sinha, 2021. "India protests: Farmers could switch to more climate-resilient crops—but they have been given no incentive." *The Conversation*, February 12. https://theconversation .com/india-protests-farmers-could-switch-to-more-climate-resilient -crops-but-they-have-been-given-no-incentive-154700.

48. Kaitlyn Spangler, Emily K. Burchfield, and Britta Schumacher, 2020. "Past and current dynamics of U.S. agricultural land use and policy." *Frontiers in Sustainable Food Systems*, vol. 4, article 98. https://doi.org /10.3389/fsufs.2020.00098.

49. Francis Annan and Wolfram Schlenker, 2015. "Federal crop insurance and the disincentive to adapt to extreme heat." *American Economic Review*, vol. 105:5, pp. 262–6. https://doi.org/10.1257/aer.p20151031.

50. UK Rural Payments Agency, 2019. *Basic Payment Scheme: Rules for 2019*. https://assets.publishing.service.gov.uk/government/uploads/system /uploads/attachment_data/file/915511/BPS_2019_scheme_rules_v3.0.pdf.

51. George Monbiot, 2017. "The hills are dead," January 4. monbiot.com /2017/01/04/the-hills-are-dead/#_ftnref8.

52. George Monbiot, 2016. "The shocking waste of cash even leavers won't condemn." *The Guardian*, June 21. https://www.theguardian.com /commentisfree/2016/jun/21/waste-cash-leavers-in-out-land-subsidie.

53. Greenpeace, 2019. *Feeding the Problem: The Dangerous Intensifica-tion of Animal Farming in Europe*. Greenpeace European Unit. https:// www.greenpeace.org/static/planet4-eu-unit-stateless/2019/02 /83254ee1-190212-feeding-the-problem-dangerous-intensification-of -animal-farming-in-europe.pdf.

54. Searchinger et al., Ibid.

55. European Court of Auditors, 2017. "Special Report n°21/2017: Greening: A more complex income support program, not yet environmentally effective." Publications Office of the European Union, Luxembourg, December 12. https://www.eca.europa.eu/en/Pages/Doc Item.aspx?did=44179.

56. Charlie Reeve, 2021. "Farm business incomes increasingly reliant on direct payments." Agriculture and Horticulture Development Board (AHDB), January 7. https://ahdb.org.uk/news/farm-business-incomes -increasingly-reliant-on-direct-payments.

57. Ibid.

58. BirdLife International, European Environmental Bureau and Greenpeace, 2021. "Does the new CAP measure up? NGOs assessment against 10 tests for a Green Deal-compatible EU Farming Policy." https://www .greenpeace.org/static/planet4-eu-unit-stateless/2021/06/874e7b56 -2021-06-29-cap-10-tests-green-deal-compatible-farm-policy.pdf.

59. Jack Peat, 2018. "The EU has archived all of the 'Euromyths' printed in UK media—and it makes for some disturbing reading." *The London Economic*, November 14. https://www.thelondoneconomic.com/news /the-eu-have-archived-all-of-the-euromyths-printed-in-uk-media -and-it-makes-for-some-disturbing-reading-108942.

60. The Newsroom, 2016. "After Brexit, farm subsidies could be even bigger: Villiers." *Belfast News Letter*, March 5. https://www.newsletter .co.uk/news/after-brexit-farm-subsidies-could-be-even-bigger-villiers -1261493.

61. John Mulgrew, 2016. "'No stability in voting to stay in EU,' Owen Paterson warns farmers." *Belfast Telegraph*, June 8. https://www.bel fasttelegraph.co.uk/business/brexit/no-stability-in-voting-to-stay -in-eu-owen-paterson-warns-farmers-34781544.html.

62. Peter Teffer, 2019. "EU promotes meat, despite climate goals." European Data Journalism Network, March 13. https://www.europeandatajour nalism.eu/eng/News/Data-news/EU-promotes-meat-despite -climate-goals.

63. Maeve Campbell, 2020. "'Become a beefatarian' says controversial EU-funded red meat campaign." Euronews, November 25. https://www .euronews.com/green/2020/11/25/become-a-beefatarian-says -controversial-eu-funded-red-meat-campaign.

64. Research Executive Agency (REA), 2017. *EU Lamb Campaign*. European Commission. https://ec.europa.eu/chafea/agri/campaigns/eu-lamb -campaign.

65. Lamb. Try it, love it, n.d. *Sheep Farming in Europe: A Positive Contribution to Biodiversity!* https://www.trylamb.co.uk/sustainability/sheep-farming-in-europe-biodiversity.

66. Ibid.

67. Paul Kingsnorth, 2011. *The Quants & The Poets.* https://www.paulkingsnorth.net/quants.

68. George Eliot, 1871. *Middlemarch.* Penguin, London, p. 272.

69. George Monbiot, 2019. "The new political story that could change everything." TEDSummit, July. https://www.ted.com/talks/george_monbiot_the_new_political_story_that_could_change_everything?language=en#t-1591.

70. Giuliana Viglione, 2021. "Climate justice: The challenge of achieving a 'just transition' in agriculture." *CarbonBrief,* October 6. https://www.carbonbrief.org/climate-justice-the-challenge-of-achieving-a-just-transition-in-agriculture.

71. P. Jepson, F. Schepers, and W. Helmer, 2018. "Governing with nature: A European perspective on putting rewilding principles into practice." *Philosophical Transactions of the Royal Society B: Biological Sciences,* vol. 373:1761. http://doi.org/10.1098/rstb.2017.0434.

72. Alastair Driver. "Rewilding boosts jobs and volunteering opportunities, study shows." Rewilding Britain. https://www.rewildingbritain.org.uk/news-and-views/press-releases-and-media-statements/rewilding-boosts-jobs-and-volunteering-opportunities-study-shows.

73. Forest Isbell et al., 2019. "Deficits of biodiversity and productivity linger a century after agricultural abandonment." *Nature Ecology & Evolution,* vol. 3, pp. 1533–8. https://doi.org/10.1038/s41559-019-1012-1.

74. José M. Rey Benayas, 2009. "Enhancement of biodiversity and ecosystem services by ecological restoration: A meta-analysis." *Science,* vol. 325:5944, pp. 1121–4. https://doi.org/10.1126/science.1172460.

CHAPTER 9

1. David Bowker, 2014. "Ice saints and the Spring Northerlies." *Weather,* vol. 69:10, pp. 272–4. https://doi.org/10.1002/wea.2271.

Index

Greenpeace, 66
Gryphaea (Jurassic oyster fossil), 5
Guardian newspaper, 61–2, 202
guerrilla gardens, 146

Haber-Bosch process, 72*
Hagberg Falling Number, 165
Hardwick Estate, 94–5, 97–112,
    116–21, 122–8, 131, 172, 177,
    184, 226
hedgehogs, 6
hedgerows, 91, 94, 98, 100, 111, 118,
    125, 173
Helsinki, 186
heptagenids, 57
herbal leys, 171, 172, 181
herbicides, 34, 35–6, 95, 171, 178,
    193; glyphosate, 159–61; machine
    alternatives to, 161–2, 181; and
    "no-till" farming, 158–62, 227;
    unnecessary, 72–3
Hesiod (poet), 212–13
High-Jones, Dr. Christine, 58*
Hobsbawm, Eric, 218
Hominidae (great apes), 12
honeydew, 10, 155
hormones, plant, 16, 19, 22
Hoxton Farms, 195
Hu, Professor Fengyi, 183
hunter-gatherer ancestors, 212–13
hunting of wild animals, 135
Hutchinsonian hypervolume, 24–5
hydrogen production, 188, 190–1

Ibsen, Henrik, *An Enemy of the
    People*, 233
Idai, Cyclone, 45
immune system, human: and bacteria
    in gut, 20, 160, 200; in infants, 17
immune systems, plant, 19–20, 22, 104
Impossible Foods, 194
India, 27–8, 33, 42, 43, 47–8, 54,
    144–5, 162, 176, 220, 222

Indonesia, 43, 53
Induced Systemic Resistance, 19*,
    19–20
Indus River farming system, 47–8
inequality: disproportionate impact of
    climate crisis, 42–5, 53–4, 191;
    equality principle of agroecology,
    177; and farm subsidies, 220; food
    as too expensive, 129–31, 141–2,
    143, 191, 206–8; food justice and
    the environment, 130, 134–5; and
    "food sovereignty" movement,
    143–5; food waste in poor
    countries, 134; lack of political
    willpower over, 141; land grabbing
    by ultra-rich, 38–9; land ownership
    and power, 145, 176–7; need for
    fairness in new food industries,
    206–8, 210; obscene
    maldistribution of wealth in UK,
    142; and protein, 44–5, 191–2, 209;
    protein consumption as poorly
    distributed, 191; redistributing
    surplus food as controversial,
    142–3; return of hunger/
    malnutrition, 39–40, 42, 44–5, 140;
    and severe soil erosion, 53–4; and
    switch to farming perennials, 182;
    in UK today, 130–4, 142;
    undercutting of farmers by richer
    nations, 33, 145
insect life: biological control of pests,
    227; collapse of, 70–1; at Hardwick
    Estate, 99–100; insect-eating
    nematodes, 19; insecticides, 18, 54,
    70–1, 91–2, 102, 156, 164; and
    perennial plants, 184–5; plant pests,
    18, 19, 22, 23, 103–5, 155–6, 164,
    184–5
insulin production, 193
intellectual property rights, 35, 36,
    208–9
International Monetary Fund, 37